Individuation, Process, and Scientific Practices

Individuation, Process, and Scientific Practices

Edited by

Otávio Bueno

Ruey-Lin Chen

Melinda Bonnie Fagan

OXFORD
UNIVERSITY PRESS

OXFORD
UNIVERSITY PRESS

Oxford University Press is a department of the University of Oxford. It furthers the University's objective of excellence in research, scholarship, and education by publishing worldwide. Oxford is a registered trade mark of Oxford University Press in the UK and certain other countries.

Published in the United States of America by Oxford University Press
198 Madison Avenue, New York, NY 10016, United States of America.

Library of Congress Cataloging-in-Publication Data
Names: Bueno, Otávio, editor. | Chen, Ruey-lin, editor. |
Fagan, Melinda Bonnie, editor.
Title: Individuation, process, and scientific practices /
edited by Otávio Bueno, Ruey-lin Chen, and Melinda Bonnie Fagan.
Description: New York, NY, United States of America : Oxford University
Press, [2018] | Includes bibliographical references and index.
Identifiers: LCCN 2018008358 (print) | LCCN 2018028845 (ebook) |
ISBN 978-0-19-063684-5 (online content) | ISBN 978-0-19-063682-1 (updf) |
ISBN 978-0-19-063683-8 (epub) | ISBN 978-0-19-063681-4 (cloth)
Subjects: LCSH: Haecceity (Philosophy) | Individuation (Philosophy) |
Philosophy and science. | Science—Philosophy.
Classification: LCC BD395.5 (ebook) | LCC BD395.5 .I53 2018 (print) | DDC 111—dc23
LC record available at https://lccn.loc.gov/2018008358

{ CONTENTS }

PART III Individuation in Philosophical Approaches to Science: Realism, Anti-Realism, Environmentalism

{ ACKNOWLEDGMENTS }

In the past several decades, philosophers of science have taken a strong interest in metaphysical problems that emerge from the sciences (including physics, biology, and chemistry, among others). Several of these problems have, in the past, particularly in the first half of the twentieth century, been relegated to the class of unsolvable (or, even worse, meaningless) issues. The current situation has changed significantly since the heyday of logical positivism. The authors in this volume are concerned with the issue of individuality in physics, biology, and other sciences, and they share the view that this issue should be explored from the perspective of an analysis of scientific practice. This perspective naturally leads them to consider the notion of individuation and to connect, in many instances, individuation with particular processes. Most contributors presented an earlier draft of their chapters at the "Taiwan Conference on Scientific Individuation in Physical, Biological, and Experimental Sciences," initiated by Alexandre Guay and Ruey-Lin Chen and held at the Department of Philosophy at National Chung Cheng University in Taiwan on December 8–9, 2014. This conference received financial aid from Taiwan's Minister of Science and Technology (MOST) and National Chung Cheng University. We hereby thank both organizations for their role in producing this volume.

After the conference in Taiwan, Otávio Bueno, Ruey-Lin Chen, and Melinda Bonnie Fagan took the responsibility of editing the present volume. Each chapter has been reviewed by two anonymous reviewers, and their comments have helped the authors revise their articles accordingly. We express our sincere gratitude to those philosophers who generously accepted our requests and participated in the review process. In addition, we thank our authors for their diligent and impressive work. Without the efforts of authors and reviewers, this volume would not have the features it currently has. We also express our gratitude to the anonymous reviewers (for the entire volume) secured by Oxford University Press for their comments and advice. Finally, our thanks go to the superb team at Oxford University Press: our publishing editor, Lucy Randall, for her encouragement from the start and amazing support throughout; assistant editor Hannah Doyle, for her enormous help; and all of the team who has helped to produce this volume.

Last, but not least, Ruey-Lin Chen would like to thank especially Otávio Bueno and Melinda Bonnie Fagan for their willingness to co-edit this volume. He learned much from the collaboration with them. Many thanks go out to Alexandre Guay who guided him to the topic of individuality and individuation. He would also like to thank Hsiang-Ke Chao and other colleagues for their private encouragement. Otávio Bueno would similarly like to thank Ruey-Lin Chen and Melinda Bonnie Fagan for their amazing work. It was truly a pleasure to collaborate with them, and a very inspiring and enlightening experience. He would also like to thank his colleagues at the University of Miami's Philosophy Department for their understanding and support, as well as Patrícia, Julia, and Olivia, for all the ways they make life such a great adventure. Melinda Bonnie Fagan thanks her co-editors, Ruey-Lin Chen and Otávio Bueno, from whom she learned a great deal throughout this, her first foray into editing a collection of papers; her colleagues at the University of Utah, for a wonderful working environment and great advice throughout this project; and Thomas Pradeu and his ImmunoConcEpT team at the University of Bordeaux, for valuable comments and feedback.

{ CONTRIBUTORS }

Otávio Bueno is Professor of Philosophy and Chair of the Philosophy Department at the University of Miami. His research concentrates in philosophy of science, philosophy of mathematics, philosophy of logic, and epistemology. He has published more than 160 papers in these areas in journals and collections such as *Noûs, Mind, British Journal for the Philosophy of Science, Philosophical Studies, Philosophy of Science, Synthese, Journal of Philosophical Logic, Studies in History and Philosophy of Science, Analysis, Studies in History of Philosophy of Modern Physics, Erkenntnis, Monist*, and *Studies in History and Philosophy of Biological and Biomedical Sciences*. He is the author or editor of several books and the editor in chief of *Synthese*.

Ruey-Lin Chen is Professor of Philosophy at National Chung Cheng University, Taiwan. His current research interest is in the philosophy of science across physical, biological, and experimental cases. He is the author of five books in history and philosophy of science and has published more than fifty articles and book chapters in Chinese. He has also published a number of journal articles (*Studies in History and Philosophy of Science, East Asian Science, Technology, and Society*) and book chapters (*Mechanism and Causality in Biology and Economics, Individuals Across the Sciences*) in English.

John Dupré is Professor of Philosophy of Science at the University of Exeter and Director of Egenis. His publications include *The Disorder of Things: Metaphysical Foundations of the Disunity of Science* (1993); *Human Nature and the Limits of Science* (2001); *Humans and Other Animals* (2002); *Darwin's Legacy: What Evolution Means Today* (2003); *Genomes and What to Make of Them* (with Barry Barnes, 2008); and *Processes of Life: Essays on the Philosophy of Biology*. He is a former president of the British Society for the Philosophy of Science and a fellow of the American Association for the Advancement of Science.

Melinda Bonnie Fagan is Associate Professor of Philosophy at the University of Utah. Her research focuses on experimental practice in biology (particularly stem cell and developmental biology), explanation, and philosophical conceptions of objectivity and evidence. She is the author of *Philosophy of Stem Cell Biology* (2013) and more than forty articles and book chapters on topics in philosophy of science and biology. She is currently working on a view of explanation focused on concepts of collaboration and mutuality.

James Griesemer is Professor and Chair of the Department of Philosophy at the University of California, Davis, and in the Technology Studies Program, the Center for Science and Innovation Studies, the Cultural Studies Graduate Group, the Population Biology Graduate Group, and the Center for Population Biology. He is a past president of the International Society for History, Philosophy, and Social Studies of Biology and is a member of the KLI in Klosterneuberg, Austria. His primary interests are philosophical, historical, and social understanding of the biological sciences, especially evolutionary biology, genetics, developmental biology, ecology, and systematics. He is writing a book, *Reproduction in the Evolutionary Process*.

Alexandre Guay is Professor of Philosophy of Natural Sciences and Analytical Philosophy at the Université catholique de Louvain (Belgium), where he is also president of the Institut Supérieur de Philosophie. Most of his research focuses on ontological puzzles in physics. He is the Editor in Chief of *Lato Sensu: revue de la Société de Philosophie des Sciences* and the President of the *Société de philosophie analytique*. He edited *Autour des Principia Mathematica de Russell et Whitehead* and co-edited *Individuals Across the Sciences* and *Science, Philosophie, Société*.

Jonathon Hricko is an assistant professor in the Education Center for Humanities and Social Sciences at National Yang-Ming University in Taiwan. His research focuses on the history and philosophy of science. In particular, he is interested in the history of chemistry in the late eighteenth and early nineteenth centuries, the scientific realism debate, and the semantics of theoretical terms. He received his PhD from Johns Hopkins University, where he wrote his dissertation on hypothetical entities in the sciences.

Marie I. Kaiser is Professor of Philosophy of Science at the University of Bielefeld in Germany. Her main research interests are the philosophy of biology, the general philosophy of science, and the metaphysics of biological practice. In particular, her work focuses on the concept of reductive explanation in biology, mechanisms, part–whole relations, causal modeling, complex systems, biological individuality, and the methodology of philosophy of science. Contact Dr. Kaiser at kaiser.m@uni-bielefeld.de.

Alan C. Love is Professor of Philosophy at the University of Minnesota and Director of the Minnesota Center for Philosophy of Science. His research focuses on conceptual issues in developmental and evolutionary biology, including conceptual change, explanatory pluralism, the structure of scientific questions and theories, reductionism, the nature of historical science, and interdisciplinary epistemology. He is the editor of *Conceptual Change in Biology: Scientific and Philosophical Perspectives on Evolution and Development* (2015) and more than forty articles and book chapters on topics in philosophy of science and biology.

Roberta L. Millstein is Professor of Philosophy and Science and Technology Studies at the University of California, Davis. Her research is in the philosophy of science, the history and philosophy of biology, and environmental ethics. Her research interests include the way that general topics in the philosophy of science, such as causation, illuminate and are illuminated by topics in evolutionary biology and ecology; her work also examines particular concepts within those sciences, such as "fitness" and "population." She is co-editor (with Hsiang-Ke Chao and Szu-Ting Chen) of *Mechanism and Causality in Biology and Economics*. She is working on two books, one on the history and philosophy of random genetic drift (with Michael R. Dietrich and Robert A. Skipper) and one that develops and defends a new interpretation of Aldo Leopold's land ethic. She is an Editor for the online, open-access journal, *Philosophy, Theory, and Practice in Biology*.

John Pemberton is an Associate at the Centre for the Philosophy of the Natural and Social Sciences at the LSE, a Research Associate of the Powers Structuralism project at Corpus Christi College Oxford, and an Associate at the Centre for Humanities Engaging Science and Society at Durham University. The central focus of his research is processes, powers, change, causation, mechanisms (nomological machines), and laws—this work straddles the boundary between philosophy of science and metaphysics. A further strand of his work is focused on the foundations of finance and economics, making use of his extensive experience in these practice areas.

Olivier Sartenaer graduated in physics and in philosophy and defended a PhD dissertation on emergence at the Higher Institute of Philosophy of the Catholic University of Louvain (Belgium) in 2013. He then pursued his research on emergence in physics and in biology as a postdoctoral research fellow at Columbia University (Fulbright/BAEF), the Catholic University of Louvain (FNRS) and, now, the University of Cologne (von Humboldt). He is the author of *Qu'est-ce que l'émergence?* (2017) and is currently working on a book project about diachronic emergence, an overlooked variety of the notion that allows for revamping the debates on emergence and reduction, both in metaphysics and in philosophy of science.

C. Kenneth Waters is Canada Research Chair in Logic and the Philosophy of Science and is Professor of Philosophy at the University of Calgary. He is the former president of the Philosophy of Science Association (2015–2016), and he has been named AAAS Fellow. His research centers on the epistemology of biological sciences. He has written on reductionism, pluralism, experimentation, conceptual and investigative practices, and causal reasoning.

Individuation, Process, and Scientific Practices

{1}

Individuation, Process, and Scientific Practices

Otávio Bueno, Ruey-Lin Chen, and Melinda Bonnie Fagan

This book is concerned with a classic philosophical question: "What things count as individuals?" Rather than addressing it from the perspective of analytic metaphysics, this volume proposes to reformulate the question and answer it from the perspective of *scientific practices*. So reformulated, the new question is: "How do scientists individuate the things they investigate and thus count them as individuals?" More precisely, our reformulated approach involves three themes:

1. *Experimental practice*: Many of this volume's contributions focus on practices of individuation in the sciences with a pronounced experimental character (e.g., stem cell biology) or consider the conception of individuality from the perspective of experimental practices more generally.
2. *Process*: Several chapters argue explicitly that individuals as such should be ontologically viewed as processes. Others understand "individuation" in the metaphysical sense as referring to a kind of process (that is, the formation, composition, emergence, and maintenance of a thing). Other chapters emphasize practical or epistemic senses of individuation—the active or cognitive processes by which scientists investigate the metaphysical individuation of their objects of study.
3. *Pluralism*: Most of the contributing authors allow for the possibility of multiple criteria of scientific individuation, rejecting the monism of traditional metaphysics. A number of authors explicitly defend pluralism about criteria of individuals or individuality *within* a science and *a fortiori across* the sciences.

The three themes together comprise a unique approach to the classic problem of individuality and exhibit the strengths of a *practice-based* philosophy of science. This volume thus examines a core philosophical question from a new angle, building on and consolidating important recent work in the philosophy of science.

To grasp the aims and motivation of this new approach, a review of the background literature and the concept of individuation is helpful. We begin by briefly surveying the approach of analytic metaphysics to questions of individuality and individuation. We then contrast this with approaches taken by recent philosophy of science, highlighting both the continuity and novelty of the approach taken in this volume.

1.1 Individuality and Individuation in Metaphysics

Questions of individuality and individuation have been a concern of metaphysics from ancient Greece to the present. Contemporary analytic metaphysicians aim to discover general criteria of individuality or general principles of individuation in terms of logic, conceptual, and semantic analysis. They construe the concepts of individuality and individuation as closely related. (Russell 1940; Sellars 1952; Strawson 1959; Loux 1978; Adams 1979; Wiggins 1980, 2001; van Cleve 1985; Gracia 1988; van Inwagen 1990; Tooley 1999; Mackie 2006; Lowe 1989, 1998, 2009). For our purpose, we begin by examining their semantic analysis of the term "individuation."

Metaphysicians distinguish metaphysical and epistemic senses of the term "individuation." E. J. Lowe (2005) provides a clear definition for each sense. "In the epistemic sense, individuation is a cognitive activity. . . . For someone to *individuate* an object, in this sense, is for that person to 'single out' that object as a distinct object of perception, thought, or linguistic reference" (Lowe 2005: 75). However, "individuation in this epistemic sense *presupposes* individuation in the metaphysical sense," which "is an ontological relationship between entities: what 'individuates' an object, in this sense, is whatever it is that makes it the single object that it is—whatever it is that makes it *one* object, distinct from others, and the very object that it is as opposed to any other thing" (Lowe 2005: 75).

What the two definitions reveal is no more than the intensions of "individuation." A semantic analysis of "individuation" from the perspective of extension is to be considered. According to metaphysician Jorge Gracia, the term "individuation" in the literature refers to two things: "(1) the process by which something becomes individual, and (2) individuality itself, that is, the character of being individual" (Gracia 1988: 4) However, Gracia thinks that "individuation" should be used only in sense (1) because it refers primarily to the process that is expressed by the verb. He interprets it by invoking universals: " 'individuation' can be taken to refer primarily to the process whereby something universal (say human being) becomes individual (for example, Leticia or Clarisa)" (Gracia 1988: 4). The extension of individuation presupposes the concept of a universal and refers to a *process* in which universals *become* particular individuals.

Empirically inclined philosophers may question whether there are such universals and, if so, whether we can have reasonable knowledge of the processes in which they become individuals. From the scientific perspective of contemporary natural science, the meanings of "process" and "becoming" in this definition are obscure and seem unrelated to real-world processes that we learn about using scientific methods. More sympathetically, one could say that, in the metaphysical definition, the term "individuation" is used in the abstractive sense of principle or theory in which the nature and the criterion of individuals are implied or presupposed, and, according to the criterion, objects are said to be individuated or to be identified as individuals. As a consequence, metaphysical studies in the principle of individuation are no more than a theoretic (in a logic-semantic sense) construction of the nature of individuality and its attendant criterion. This claim can be illustrated by a number of metaphysicians' works (see Wiggins 1980, 2001; Gracia 1988; Lowe 1989, 2005). For example, commenting on a theory of individuation, Lowe notes that "[s]ortal concepts are characteristically governed by criteria of individuation and identity—semantic principles which determine what are to count as individual instances of the sorts or kinds in question and the conditions for their identity or diversity at a time and (where this is applicable) over time" (Lowe 1989: 1). On this approach, then, the question of individuality is prior to that of individuation, and there is no distinct problem of individuation apart from the problem of individuality.

To solve that problem, metaphysicians have proposed a number of theories of individuation that suggest different kinds of individuals and different criteria for identifying what individuals are. At least six such theories can be identified: hylomorphism (matter-form) theory, bare particular theory, haecceity or primitive thisness theory, bundle (of universals, properties, or tropes) theory, space-time trajectories theory, and sortal concepts theory (Tooley 1999: ix–xi; Lowe 2005: 80–89). Each of these theories may have different versions. Their variation depends on how to interpret the metaphysical concepts such as *property, trope, universal, particular, substance, substratum, time, space, sort* or *kind* and their *relationships* to the concept of individual. These concepts set the terms for mainstream philosophical debates over the issue of individuality and individuation.

For our purpose, the approach of analytic metaphysics has two distinctive features: (1) criteria of individuality and the principles of individuation are supposed to be universally applicable to all entities in the world,[1] and (2) the concept of individuation presupposes the concept of individuality, so that the conceptual analysis of individuality and the determination of criteria for an entity being an individual suffice to solve the individuation problem. Recent work in the philosophy of science on questions of individuality challenges (1). However, there has as

[1] Also see Gracia's statement: "anyone who wishes to present a systematic philosophical discussion of individuality must also deal with another important issue, the so called 'problem of individuation.' There are two questions which are usually taken up in this context: (1) the identification of the principle or cause of individuation, and (2) the determination of whether this principle, or cause, is the same for all entities" (Gracia 1988: 16).

yet not been sustained challenge to (2). Many of the essays in this volume do so, in diverse ways.

1.2 Individuality and Individuation in the Philosophy of Science

Recently, a number of philosophers of science have taken an interest in this topic of individuality. One motivation for this is that many entities posited by contemporary sciences challenge the accounts proposed by traditional analytic metaphysicians. For example, there are many diverse cases of organismal entities (corals, slime molds, fungi, etc.), supraorganismal entities (colonies, groups, populations, species, etc.), and suborganismal entities (stem cells, genes, gene networks, genomes, etc.) all of which plausibly qualify as living individuals. No set of metaphysical criteria of individuality proposed to date seems to be universally suitable for all these living entities. In physics, there are strange quantum particles that appear, according to prominent analytic accounts, to be either individuals or nonindividuals. Those particles and other physical objects such as waves, fields, forces, and the like, together challenge feature (1): the universal applicability of traditional criteria.

Many philosophers of science think that questions of individuality can be answered only via investigating the outcomes of scientific research. Accordingly, philosophers of physics examine those metaphysical theories of individuality (i.e., criteria of individuals) by referring to classical mechanics, quantum mechanics, and the theory of relativity, which are among our best physical theories now. Naturalistic philosophers tend to exclude those transcendental theories such as hylomorphism theory, haecceity theory, and bare particular theory because their criteria cannot be tested empirically by scientific theories. Some of them adopt bundle theory as the most promising one, argue that quantum particles cannot meet the discernibility criterion, and conclude that quantum particles are nonindividuals in general (French and Krause 2006; Saunders 2006, 2016; Ladyman and Don Ross 2007; Ladyman and Bigaj 2010; Ladyman 2016; French 2014, 2016).[2] Some others argue that we cannot but take transcendental concepts of individuality (e.g., "primitive thisness") to save the individuality of quantum particles (French and Redhead 1988; Morganti 2009, 2013, 2016; Pylkkänen, Hiley, and Pättiniemi 2016). Still other philosophers suggest new conceptions such as "structural individuality" (Glick 2016) or "experimental individuality" (Chen 2016) as alternative answers to the individuality problem of some quantum objects, but it is not clear whether their answers can be applied to all entities in quantum theories. Reviewing this long-term dispute, James Ladyman comments: "On such

[2] Saunders (2006) distinguishes between absolute discernibility, relative discernibility, and weak discernibility, which in turn become a baseline for others' discussion (see Ladyman and Ross 2007; Ladyman 2016).

a naturalistic view, if quantum particles and/or space-time points are individuals, it is not in virtue of weak discernibility or any other general principle of individuation, but because the practice of science treat them as being so in effective descriptions applicable to certain regimes" (Ladyman 2016: 204). This view *envisages* the direction of our volume. The key idea is to abandon the search for universal principles or criteria of individuality for all entities and instead focus on how scientists individuate their objects of study.

In a sense, the current philosophy of physics is an extensive application of analytic metaphysics to physical theories. Philosophers of physics seldom propose new conceptions of individuality or theories of individuation and tend to follow the logic-semantic techniques of analytic metaphysics. What they have done is *test* metaphysical theories in terms of physical theories. As a consequence, no consensus on which metaphysical theory, whether bundle theory or space-time theory, can be suitable for all macroscopic objects and microentities in the quantum scale. In the domain of physics, thus, philosophers of physics have argued that the criteria of analytic metaphysics are not universally applicable. Philosophers of biology, in contrast, seldom use metaphysical criteria to treat biological objects such as organisms—one important kind of individual.

As early as the 1970s, philosophers of biology began discussing the problem of individuality by debating whether biological species are individuals (Ghiselin 1974; Hull 1978). Recently, philosophers of biology have published a great number of articles and volumes related to biological individuality (e.g., Dupré and O'Malley 2007, 2009; Folse and Roughgarden 2010; Godfrey-Smith 2009, 2013, 2016; Dupré 2010, 2012; Pradeu 2010, 2012, 2013; Clarke 2010, 2012, 2013; Calcott and Sterelny 2011; Ereshefsky and Pedroso 2013, 2016; Haber 2013, 2016; Bouchard and Huneman 2013; Paternotte 2016; Lidgard and Nyhart 2017). Much of this work focuses on organisms, the paradigmatic biological individuals, and the difficulty of formulating criteria of organismality and, accordingly, biological individuality. In this work, three influential conceptions of individuality and their attendant criteria can be distinguished (Pradeu 2012: 230–231; Wilson and Barker 2013):

(a) *Phenomenal individuality*: Biological objects count as individuals according to common sense or empirical observation. Proposed criteria for individuality in this sense are: three-dimensional spatial boundaries, enduring for some period of time, bearing properties, and participating in processes and events.

(b) *Physiological individuality*: Biological objects count as individuals according to physiological investigations. Proposed criteria for this sense of individuality are functional integration, autonomy, and immune responsiveness.

(c) *Evolutionary individuality*: Biological objects count as individuals just in case they are units of selection according to the terms of evolutionary theory.

The phenomenal conception of individuality has little current support because the criterion of common sense is vague, unreliable, and faces many counterexamples.[3] Some philosophers of biology believe that biological theories or models may offer better criteria. The question is which theory or set of theories should be used. The main alternatives are physiological theories and the theory of evolution by natural selection. However, within each group variants have been proposed. For example, some philosophers highlight organisms as units of functional integration (Sober 1991; Wilson 1999; Lewontin 2000), while others argue that immunity offers a better criterion for individuality (Pradeu 2010, 2012). Similarly, the evolutionary theory approach encompasses the replicative reproducer version (Godfrey-Smith 2009), the interactor version (Hull 1980; Ereshefsky and Pedroso 2016), and the developmental reproducer version (Griesemer 2000). Other proposals combine different conditions from different conceptions into a general criterion of individuality, for example, the functional interpretations of policing and demarcation mechanisms (see Clarke 2013).

As this brief survey indicates, philosophers of biology seldom appeal to the criteria of analytic metaphysics. Nonetheless, most current work addresses the problem of formulating criteria of individuality through examination of different biological theories. Some works explore biological individuals in the evolutionary transitions beyond organisms, such as the edited volumes, *The Major Transitions in Evolution Revisited* (Calcott and Sterelny 2011) and *From Groups to Individuals: Evolution and Emerging Individuality* (Bouchard and Huneman 2013). Other works contribute to the topic through exploring atypical cases such as bacteria, stem cells, biofilms, populations, and so forth (Dupré and O'Malley 2007, 2009; Millstein 2009; Ereshefsky and Pedroso 2013, 2016; Fagan 2016; Paternotte 2016). But, for most part, organisms remain the conceptual focus (Wilson and Barker 2013). Insofar as organisms are taken to be the paradigmatic individuals, most current philosophy of biology does not challenge the second distinctive feature of analytic metaphysics: the concept of individuality is prior to that of individuation.

This survey of recent literature on individuality and individuation reveals two general trends: first, most philosophers in metaphysics, philosophy of physics, and philosophy of biology tend to invoke a dominant theory, whether metaphysical, physical, or biological, in proposing a criterion to solve the problem of individuality, and, second, most philosophical treatments approach the issue within a single field, say, metaphysics, physics, or biology (or general common sense, with less success). A notable exception to the second trend is *Individuals across the sciences* (Guay and Pradeu 2016) edited by Alexandre Guay and Thomas Pradeu,

[3] Many organisms such as corals, fungi, slime molds, and aspens are difficult to identify. A number of philosophers of biology reject the use of the conception of phenomenal individuality as a single criterion because they think that those relevant conditions based on common sense and observation cannot be trusted (Hull 1992; Lewontin 2000; for a brief discussion, see Pradeu 2012).

which explores the issue across metaphysics, physics, and biology. A few chapters in that volume, such as Fagan (2016) and Chen (2016), are exception to this trend, exploring the issue of individuality in terms of experimental practices rather than theories.

This volume adopts and extends the strategy of crossing disciplinary borders in addressing questions of individuality. The following chapters, taken together, engage not only analytic metaphysics, physics, and biology, but also in chemistry, environmental science, engineering, and ethics. The goal of this expanded treatment is not, however, to compare different accounts of individuality from different sciences and fields of philosophy, reinforcing divisions among those fields. Rather, the idea is to extend the cross-disciplinary approach while moving away from the emphasis on theory. Many chapters in this volume take experimental practices to be primary in resolving questions of individuality. This focus on practice can be expressed as giving priority to concepts and problems of individuation within and across diverse sciences. As noted earlier, "individuation" traditionally has both metaphysical and epistemic senses, which are seen as closely connected. Our approach gives primacy to the latter, focusing on the diverse methods that scientists use in practice to identify individuals as objects of study.[4] This approach, reversing the usual philosophical priorities, offers a fresh perspective on the philosophical problems of individuation with and across diverse sciences, indicating new points of contact between scientific practice and philosophical argumentation.

Expanding on Lowe's distinction between metaphysical and epistemic senses of individuation noted earlier, on the "practice approach," it is helpful to distinguish three senses of individuation (see also Chen, Chapter 9, this volume):

1. *Practical individuation*: The practical process in which scientists manipulate, target, track, present, or produce an individual by means of scientific procedures. Most of the chapters in Part II (see later discussion) address this aspect of individuation.
2. *Epistemic individuation*: The cognitive process by which scientists identify, distinguish, and individuate a thing from its environment and other things. Most of the chapters in this volume deal with this aspect, in different ways and in reference to different scientific contexts.
3. *Metaphysical individuation*: The process through which an individual comes into being or persists until it perishes. Nearly all the chapters in this volume involve this aspect in some way or other. Most do so by focusing on how individuation in this sense is discovered or produced in the practice of particular sciences.

The metaphysical sense of individuation, as stated here, has an interesting consequence: in order to understand individuation as process, we also have to

[4] Love and Brigandt (2017) suggest a similar view and approach to ours.

understand individuals in themselves as processes. This is because the metaphysical process of individuation comprises the formation (or emergence or composition), the persistence, or the maintenance of individuals. This process as a whole is a process of an individual per se. That is, *an individual emerges from a process of individuation in the metaphysical sense. Epistemic and practical individuation, then, are processes that aim to uncover stages of that metaphysical process.* A number of chapters in this volume engage explicitly with this strong process view.

1.3 Individuation, Individuals, and Process Metaphysics

The significance of our volume, in the light of this background, is that it approaches individuation and individuals in a significantly different way from most existing literature, which focuses primarily on individuals. Our primary approach is to reverse the order of the classical philosophical question, asking "How do researchers (and, in particular, scientists) individuate things and thus count them as individuals?" The new question implies a new perspective: individuals are understood as products or outcomes of individuation as a process, rather than identified by principles or criteria of individuality applied to entities. The basic idea of the new perspective is that, in individuating, individuality is endowed to things in various specific ways, thereby making them be individuals. For example, an egg is fertilized by a sperm, becomes a zygote, develops into an organism, and lives for a period of time. The whole process can be understood as individuation, which consists of a unitary whole of sequential stages. Our knowledge of this process, and its stages, comes from scientific research—crucially, from individuation in the epistemic and practical senses. Since we understand "individuation" as process, the question arises whether individuals, as the products of individuation, themselves are also processes or not. This question involves us in the metaphysics of processes—another important theme of this volume.

Process metaphysics is a view or an approach which treats beings as dynamic and centralizes the concept of process in understanding all entities. The ultimate goal of process metaphysicians is to provide a comprehensive theory of reality.[5] However, the topic is as yet relatively unexplored so that there is currently no unitary or paradigmatic analysis. Some works on process metaphysics (Whitehead 1929/1978; Weber and Basile 2007) seem to be overly speculative. To counteract this, some analytic philosophers have tried to develop process metaphysics rigorously in terms of symbolic systems and logic techniques (Rescher 1996; Seibt 2003; Heller and Herre 2003). Others have applied analytic process metaphysics to some special sciences (Needham 1999, 2003; Christiaens 2003). In a nutshell, current analytic process metaphysics uses a logical-semantic analysis of key terms as a means

---

[5] For a detailed introduction, see Rescher (2008) or Seibt (2012).

to explore general everyday cases and special scientific cases. Does the concept of process relate to the concepts of individuals and individuation?

According to Rescher (2008: 1), the term "process" refers to "a sequentially structured sequence of successive stages or phases." He tries to define "process" in terms of the following three statements:

1. A process is a complex—a unity of distinct stages or phases. A process is always a matter of now this, now that.
2. This complex has a certain temporal coherence and unity, and processes accordingly have an ineliminably temporal dimension.
3. A process has a structure, a formal generic format in virtue of which every concrete process is equipped with a shape or format (Rescher 2008: 1).

One can obviously find that these characterizations harbor the problem of the individuation of processes (i.e., how to individuate *a* process?), and they seem to provide a criterion for the individuation of processes in general. For philosophical naturalists, the next step is to examine this metaphysical theory in terms of scientific cases. Pemberton (Chapter 3 in this volume) uses cases across different sciences to explore the individuation of processes.

Process philosopher Johanna Seibt also connects individuals and processes in developing her concept of free process, claiming that: " . . . anything which is conceived of as occurring in the activity mode is a concrete, dynamic, *non-particular* individual. Such individuals, which I call 'free processes,' may be used for the interpretation of much more than just common sense activities"[6] (Seibt 2003: 23). However, she does not address these questions of what individuals are, what individuation is, and what is the relation between individual and individuation. In contrast, Dupré (Chapter 2 in this volume) defends the claim that organisms are processes. A number of other chapters in this volume understand individuation as a process as well and explore the close relation between individual and individuation in terms of cases in diverse sciences.

Although noting the relation between process and individuation, current process metaphysics offers few resources to explore those central issues in this volume. Instead, our main focus of prioritizing questions of individuation and scientific practice offers materials and support for empirically based, naturalistic process philosophy.

[6] Seibt (2003) uses a semantic analysis of activities and a classification of verb types as a starting point to develop her theory of free process. She first identifies processes with activities, and then analyzes entities referred by nouns in terms of activity as she talks of an occurrence as activity and as an entity that fulfills conditions such as completeness, resumability, recurrence, and dynamicity. In a nutshell, Seibt tries to reduce entities to activities, which are fundamental processes.

1.4 The Problem of Individuation Based on Scientific Practices

Foregrounding the individuation process in effect reformulates the classical question of individuality: How do we individuate things in scientific practice and thus count them as individuals? Rather than seek a general, universal answer, our main approach is to focus on different aspects of this problem, which can be formulated as different, but related, subproblems and subquestions. Taken together, this plurality of formulations offers a distinctive characterization of *the problem of individuation*, which accommodates the diversity of scientific practices and their individuated results. Across the chapters comprising this volume, eight subproblems can be distinguished:

1. The conception-of-individuality problem: *Is the conception of individuality one or many?* Existing literature shows that many conceptions of individuality have been proposed. Still, many philosophers hold a monistic perspective to this issue, hoping to find a unified theory. However, is a pluralistic stance justified? Dupré (Chapter 2), Kaiser (Chapter 4), Waters (Chapter 5), and Love (Chapter 8) examine this issue.
2. The identification-of-individuals problem: *According to what criteria in scientific practices do we identify entities as individuals?* This problem looks like a traditional one, but, rather, asks what criteria for identifying entities as individuals are revealed by considering scientific practices rather than theories only. Waters (Chapter 5), Fagan (Chapter 6), Griesemer (Chapter 7), Love (Chapter 8), and Chen (Chapter 9) address this problem.
3. The nature-of-individuals problem: *Are individuals substances with a set of essential properties, or processes depending causally on a range of activities?* This problem asks which—substance ontology or process ontology—is better supported by scientific practices. Dupré (Chapter 2) and Pemberton (Chapter 3) explicitly address this problem, arguing for process ontology. Many of the other chapters implicitly commit to the position as well.
4. The emergence-of-individuality problem: *Under what conditions do entities become or emerge as individuals?* Answers to this problem propose criteria for emerging individuality (individuation as a process), which in turn presuppose some theory of emergence. Guay and Sartenaer (Chapter 10) deal with this problem.
5. The composition-of-individuals problem: *What conditions make all components compose a whole individual? Or, what is the compositional relation that holds between a (complex) individual and all of its components?* Pemberton's, Kaiser's, and Chen's chapters address this problem.
6. The demarcation-of-individuals problem: *What criteria demarcate the boundary of an individual from its environment (including other*

individuals)? In other words, what conditions demarcate all parts of an individual from the surrounding environment? This problem is the other face of the composition problem of individuals. Pemberton's and Kaiser's chapters deal with this problem as well.

7. The individuation-of-parthood problem: *Under what conditions is something a part of the other thing?* In addition to the composition problem of individuals, Pemberton's and Kaiser's chapters also attempt to find the condition that can identify something as a part of an individual.
8. The maintenance-of-individuality (or the continuity-of-individuals) problem: *How does an individual maintain its identity through changes?* Understanding individuals as processes logically implies this problem. Dupré and Pemberton's chapters attend to this question. In addition, Chen's chapter explores the maintenance conditions for some experimental individuals.

In addition to the eight subproblems of individuation, related problems of individuation and reality are explored in this volume. They are formulated by a composite question: *How do individuation and individuals relate to reality, to their environment, and to our actions?* Bueno (Chapter 11) and Hricko (Chapter 12) connect the issue of individuation with the realism/anti-realism debates. Millstein (Chapter 13) argues that the issue of individuality matters for environmental science and land ethics. All ground their answers on scientific practices.

On the approach taken in this volume, insights about criteria of individuality emerge from piecemeal investigation of these problems and questions. For the most part, results of these investigations are based on individuation as that process is discovered in scientific practice, rather than on a single dominant theory or more abstract metaphysical speculations. Collectively, the investigations of various contributors to this volume tend to support the metaphysical view of individuals as processes. We conclude this introductory chapter with a more detailed summary of the authors' contributions.

1.5 Overview of Chapters

The eleven contributed chapters examine the individuation of scientific entities, explore different aspects of individuation, highlight individuation in experimental practices, and extend the issue of individuation to wider contexts. Two chapters (Guay and Sartenaer, Millstein) do not engage as directly with individuation, focusing instead on emergence and individuality, respectively. Although departing from the volume's main approach, these chapters enrich the collection by connecting individuality issues across different domains of inquiry: particle physics and general metaphysics; philosophy of biology, ecology, and environmental ethics. Overall, the volume's chapters are arranged into three parts: Part I, Aspects of Individuation: Metaphysical and Processual; Part II, Experimental Practices of Individuation; and Part III, Individuation in Philosophical Approaches to Science: Realism, Anti-realism, Environmentalism. We briefly introduce each in turn.

1.5.1 PART I: ASPECTS OF INDIVIDUATION: METAPHYSICAL AND PROCESSUAL

Authors in the first part of this book offer a general account of different aspects of individuation: the relationship between individuals and individuation, the nature of individuals in general, the relation of individuals to kinds, and the individuation criteria of parthood. Although the authors in this section each aim to provide a general account, similar to those of traditional metaphysicians, their arguments are developed by focusing on scientific cases and practices.

In his previous work, John Dupré has argued that the entities we think of as biological individuals are more appropriately treated as processes than as substances. In Chapter 2, he extends his process view to the relation of individual to kind, arguing that standard assumptions about kinds and individuals can break down in the classifications applied to plastic, process-like biological individuals. He continues to argue how these considerations provide a deeper ground for the classificatory pluralism.

John Pemberton, in Chapter 3, deals with the problem of individuation of processes from a consideration of their treatment in science. He views that processes involve parts acting together (at each stage) to bring about the next stage—they are commonly characterized by a rich range of criteria including the nature and arrangement of parts at each stage, how the process exhibits change and self-maintains across stages, and perhaps the role of the process within its larger context.

Marie I. Kaiser focuses on the biological version of the individuation-of-parthood problem, which is involved in the problems of identification, demarcation, and composition of individuals: Are there any general criteria that guide the individuation of part–whole relations in the biological realm? In Chapter 4, Kaiser develops a set of new criteria of spatial and temporal inclusion for biological parthood. At the end of the work, she discusses in what way a monistic approach to biological parthood still can account for the plurality of cases found in the biological realm.

1.5.2 PART II: EXPERIMENTAL PRACTICES OF INDIVIDUATION

Authors in the second part of the book explore experimental and theoretical practices of individuation through particular case studies in the sciences.

C. Kenneth Waters (Chapter 5) urges philosophers to shift attention from individuals to individuation. In his chapter, Waters addresses the conception-of-individuality and the identification-of-individuals problems. By taking the gene concept and the debate about holobionts (assemblages of a host with microbial symbionts) as examples, Waters argues for a pragmatic/contextual/pluralistic stance on the issue of individuals. Treating a thing (especially a biological one) as an individual depends on the purpose, the means, and the practical context of individuating it.

In exploring the problem of biological individuation, Melinda Bonnie Fagan focuses on both cells and organisms. These two kinds of entity occupy central places in investigating biological individuality because they are located at the key levels of the ontological hierarchy in the living world. In Chapter 6, Fagan focuses on the identification problem of a special biological individual: pluripotent stem cells. She draws an analogy between pluripotent stem cell colonies and multicellular organisms and argues that the former objects should be considered biological individuals analogous to the latter. Fagan's argument makes stem cells' developmental capacities central to questions of individuality. Her chapter thus bridges between the individuality problem of cells and organisms and the problem of developmental continuity, which is addressed in the following two chapters.

James Griesemer addresses mainly the demarcation-of-individuals problem that is involved in the context of the continuity-of-individuals problem. He asks: What distinguishes developmental stages that demarcate phases of a single biological individual from reproductive generations that demarcate distinct biological individuals? By focusing on the case of developmental systems, Griesemer argues in Chapter 7 that successful tracking of developmental systems embeds pragmatic individuation criteria into theoretical perspectives. This argument commits him to a pragmatic/pluralistic stance similar to Waters's on the conception of biological individuality.

Similarly, Alan Love defends a pluralistic perspective on individuality in terms of investigating individuation practices in experimental contexts of developmental biology. In Chapter 8, he argues that molecular and morphological practices used for tracking individuals and their components through ontogeny are not dependent on evolutionary theorizing. This thesis that experimental practices of individuation may be independent of theorizing is also defended in the context of physics in the following chapter.

Ruey-Lin Chen argues in Chapter 9 that experiments may offer a conception of individuality independent of theories by producing and maintaining individuals in well-designed mechanisms. He shows that some experiments in physics can create new individuals (bosonic and fermionic condensates) out of other materials, and some experiments in the biotechnology of gene transplantation can verify and present posited entities (genes) as individuals. He therefore identifies creation as the ontological mode of experimental individuation and presentation as the epistemological mode.

The last chapter of this part, coauthored by Alexandre Guay and Olivier Sartenaer, explores the emergence-of-individuality problem of quasiparticles. In Chapter 10, they argue for the way in which emergence can shed light on the ontological status of some quantum entities—more particularly, so-called quasiparticles. They take Laughlin's emergentism seriously, arguing that some quasiparticles, for example, anyons, would come into being on the occasion of certain physical transformation.

1.5.3 PART III: INDIVIDUATION IN PHILOSOPHICAL APPROACHES TO SCIENCE: REALISM, ANTI-REALISM, ENVIRONMENTALISM

Unlike the foregoing chapters, which examine the meanings and scientific practices of individuation, authors in the third part of this volume explore the roles that individuation and individuality play in other philosophical inquiries, including the relationship between empirical evidence and theories, the debate over realism, the construction of scientific fields, and the relevance of science to ethics.

Experimental individuation of a scientific entity may be regarded as the strongest evidence for a scientific theory's commitment to that entity. In Chapter 11, Otávio Bueno examines three physical experiments in which scientists allegedly manipulated and positioned a single atom or trapped an ion or a positron. These experiments seem to raise a challenge to constructive empiricism. Bueno offers an empiricist response to the challenge.

Jonathon Hricko also discusses the individuation of theoretical entities in relation to the realism debate in Chapter 12. Retail realists have argued that we ought to settle this debate at the level of specific theoretical entities, adopting realism toward some and anti-realism toward others. Thus, they require some means of individuating such entities. Hricko uses a case from chemistry to argue that retail realists ought to presuppose that such entities are individuated rather finely (i.e., that distinct theories involve distinct entities).

Finally, in Chapter 13, Roberta L. Millstein discusses the individuality of *biotic communities* in Aldo Leopold's sense. According to Leopold, a biotic community is "a fountain of energy flowing through a circuit of soils, plants, and animals." But he also describes biotic communities in terms of interactions between species and the way in which changes in some species affect other species. Millstein thus suggests that Leopold's conception of "biotic community," the referent of which is purported to have intrinsic value, blends the concepts of "ecosystem" and "community" in some fashion. She explores the way by which one may integrate a community–individual and an ecosystem–individual and thus provide a solid basis for Leopold's Land Ethic.

References

Adams, Robert M. 1979. Primitive thisness and primitive identity. *Journal of Philosophy* 76: 5–26.

Bouchard, Frédéric, and Philippe Huneman, eds. 2013. *From groups to individuals: Evolution and emerging individuality*. Cambridge, MA: MIT Press.

Calcott, Brett, and Kim Sterelny, eds. 2011. *The major transitions in evolution revised*. Cambridge, MA: MIT Press.

Chen, Ruey-Lin. 2016. The experimental realization of individuality. In *Individuals across the sciences*, eds. Alexandre Guay and Thomas Pradeu, 348–370. Oxford: Oxford University Press.

Clarke, Ellen. 2010. The problem of biological individuality. *Biological Theory* 5: 312–325.

Clarke, Ellen. 2012. Plant individuality: A solution to the demographer's dilemma. *Biology and Philosophy* 27(3): 321–361.

Clarke, Ellen. 2013. The multiple realizability of biological individuals. *The Journal of Philosophy* 110(8): 413–435.

Christiaens, Wim. 2003. The EPR-experiment and free process theory. In *Process theories: Crossdisciplinary studies in dynamic categories*, ed. Johanna Seibt, 267–283. Dordrecht: Kluwer Academic Publishers.

Dupré, John. 2010. The polygenomic organism. In *Nature after the genome*, eds. Sarah Parry and John Dupré, 19–31. Oxford: Blackwell Press.

Dupré, John. 2012. *Processes of life: Essays in the philosophy of biology*. Oxford: Oxford University Press.

Dupré, John, and Maureen O'Malley. 2007. Mategenomics and biological ontology. *Studies in the History and Philosophy of the Biological and Biomedical Science* 38: 834–846.

Collected in John Dupré. 2012. *Processes of life: Essays in the philosophy of biology*, 188–205. Oxford: Oxford University Press.

Dupré, John, and Maureen O'Malley. 2009. Varieties of living things: Life at the intersection of lineage and metabolism. *Philosophy and Theory in Biology* 1: 1–25.

Collected in John Dupré. 2012. *Processes of life: Essays in the philosophy of biology*, 206–229. Oxford: Oxford University Press.

Ereshefsky, Marc, and Makmiller Pedroso. 2013. Biological individuality: The case of biofilms. *Biology & Philosophy* 28(2): 331–349.

Ereshefsky, Marc, and Makmiller Pedroso. 2016. What biofilms can teach us about individuality. In *Individuals across the sciences*, eds. Alexandre Guay and Thomas Pradeu, 103–121. Oxford: Oxford University Press.

Fagan, Melinda B. 2016. Cell and body: Individuals in stem cell biology. In *Individuals across the sciences*, eds. Alexandre Guay and Thomas Pradeu, 122–143. Oxford: Oxford University Press.

Folse, Henri J., and Joan Roughgarden. 2010. What is an individual organism? A multilevel selection perspective. *Quarterly Review of Biology* 85(4): 447–472.

French, Steven. 2014. *The structure of the world: Metaphysics and representation*. Oxford: Oxford University Press.

French, Steven. 2016. Eliminating objects across the sciences. In *Individuals across the sciences*, eds. Alexandre Guay and Thomas Pradeu, 371–394. Oxford: Oxford University Press.

French, Steven, and Décio Krause. 2006. *Identity physics: A historical, philosophical and formal analysis.* Oxford: Oxford University Press.

French, Steven, and Michael Redhead. 1988. Quantum physics and the identity of indiscernibles. *The British Journal for the Philosophy of Science* 39: 233–246.

Ghiselin, Michael T. 1974. A radical solution to the species problem. *Systematic Zoology* 23: 536–544.

Glick, David. 2016. Minimal structural essentialism: Why physics doesn't care which is which? In *Individuals across the sciences*, eds. Alexandre Guay and Thomas Pradeu, 207–225. Oxford: Oxford University Press.

Godfrey-Smith, Peter. 2009. *Darwinian populations and natural selection.* New York: Oxford University Press.

Godfrey-Smith, Peter. 2013. Darwinian individuals. In *From groups to individuals: Evolution and emerging individuality*, eds. Frédéric Bouchard and Philippe Huneman, 17–36. Cambridge, MA: MIT Press.

Godfrey-Smith, Peter. 2016. Individuality and life circles. In *Individuals across the sciences*, eds. Alexandre Guay and Thomas Pradeu, 85–102. Oxford: Oxford University Press.

Gracia, Jorge J. E. 1988. *Individuality: An essay on the foundations of metaphysics.* New York: State University of New York Press.

Griesemer, James. 2000. The units of evolutionary transition. *Selection* 1: 67–80.

Guay, Alexandre, and Thomas Pradeu, eds. 2016. *Individuals across the sciences.* New York: Oxford University Press.

Haber, Matthew H. 2013. Colonies are individuals: Revisiting the superorganism revival. In *From groups to individuals: Evolution and emerging individuality*, eds. Frédéric Bouchard and Philippe Huneman, 195–217. Cambridge, MA: MIT Press.

Haber, Matthew H. 2016. The biological and the mereological: Metaphysical implications of the individuality thesis. In *Individuals across the sciences*, eds. Alexandre Guay and Thomas Pradeu, 295–316. Oxford: Oxford University Press.

Heller, Barbara, and Heinrich Herre. 2003. Ontological categories in GOL. *Process theories: Crossdisciplinary studies in dynamic categories*, ed. Johanna Seibt, 57–76. Dordrecht: Kluwer Academic Publishers.

Hull, David. 1978. A matter of individuality. *Philosophy of Science* 45(3): 335–360.

Hull, David. 1980. Individuality and selection. *Annual Review of Ecology and Systematics* 11: 311–332.

Hull, David. 1992. Individual. In *Keywords in evolutionary biology*, eds. E. Fox Keller and Elisabeth Lloyd, 181–187. Cambridge, MA: Harvard University Press.

Ladyman, James. 2016. Are there individuals in physics, and if so, what are they? In *Individuals across the sciences*, eds. Alexandre Guay and Thomas Pradeu, 193–206. Oxford: Oxford University Press.

Ladyman, James, and Tomasz Bigaj. 2010. The principle of the identity of indiscernibles and quantum mechanics. *Philosophy of Science* 77: 117–136.

Ladyman, James, and Don Ross. 2007. *Every thing must go: Metaphysics naturalized.* Oxford: Oxford University Press.

Lidgard, Scott, and Lynn K. Nyhart, eds. 2017. *Biological individuality.* Chicago: University of Chicago Press.

Lewontin, Richard. 2000. *The triple helix: Gene, organism, and environment.* Cambridge: Harvard University Press.

Loux, Michael J. 1978. *Substance and attribute: A study in ontology*. New York: Springer.
Love, Alan, and Ingo Brigandt. 2017. Philosophical dimensions of individuality. In *Biological individuality*, eds. Scott Lidgard and Lynn K. Nyhart, 318–348. Chicago: University of Chicago Press.
Lowe, E. Jonathan. 1989. *Kinds of being*. Oxford: Basil Blackwell Ltd.
Lowe, E. Jonathan. 1998. *The possibility of metaphysics: Substance, identity, and time*. Oxford: Oxford University Press.
Lowe, E. Jonathan 2005. Individuation. *The Oxford handbook of metaphysics*, eds. Michael J. Loux and Dean W. Zimmerman. Oxford: Oxford University Press.
Lowe, E. Jonathan. 2009. *More kinds of being*. Oxford: Wiley-Blackwell.
Mackie, Penelope. 2006. *How things might have been: Individuals, kinds, and essential properties*. Oxford: Clarendon Press.
Millstein, Roberta. 2009. Populations as individuals. *Biological Theory* 4(3): 267–273.
Morganti, Matteo. 2009. Inherent properties and statistics with individual particles in quantum mechanics. *Studies in the History and Philosophy of Modern Physics* 40: 223–231.
Morganti, Matteo. 2013. *Combining science and metaphysics. Contemporary physics, conceptual revision and common sense*. Houndmills, Basingstoke: Palgrave Macmillan.
Morganti, Matteo. 2016. The metaphysics of individuality and the sciences. In *Individuals across the sciences*, eds. Alexandre Guay and Thomas Pradeu, 273–294. Oxford: Oxford University Press.
Needham, Paul. 1999. Macroscopic processes. *Philosophy of Science* 66(2): 310–331.
Needham, Paul. 2003. Continuants and processes in macroscopic chemistry. In *Process theories: Cross-disciplinary studies in dynamic categories*, ed. Johanna Seibt, 237–265. Dordrecht: Kluwer.
Paternotte, Cédric. 2016. Collective individuals: Parallels between joint action and biological individuality. In *Individuals across the sciences*, eds. Alexandre Guay and Thomas Pradeu, 144–164. Oxford: Oxford University Press.
Pradeu, Thomas. 2010. What is an organism? An immunological answer. *History and Philosophy of Life Sciences* 32(2–3): 247–267.
Pradeu, Thomas. 2012. *The limits of the self: Immunology and biological identity*. Oxford: Oxford University Press.
Pradeu, Thomas. 2013. Immunity and the emergence of individuality. In *From groups to individuals: Evolution and emerging individuality*, eds. Frédéric Bouchard and Philippe Huneman, 77–96. Cambridge, MA: MIT Press.
Pylkkänen, Paavo, Basil J. Hiley, and Ilkka Pättiniemi. 2016. Bohm's approach and individuality. In *Individuals across the sciences*, eds. Alexandre Guay and Thomas Pradeu, 226–249. Oxford: Oxford University Press.
Rescher, Nicholas. 1996. Process metaphysics: An introduction to process philosophy. *Transactions of the Charles S. Peirce Society* 32(4): 689–697.
Rescher, Nicolas. 2008. Process philosophy (revised version). *The Stanford encyclopedia of philosophy* (spring 2018 edition), ed. Edward N. Zalta. https://plato.stanford.edu/archives/sum2012/entries/process-philosophy/.
Russell, Bertrand. 1940. *An inquiry into meaning and truth*. London: George Allen and Unwin.
Saunders, Simon. 2006. Are quantum particles objects? *Analysis* 66: 52–63.

Saunders, Simon. 2016. On the emergence of individuals in physics. In *Individuals across the sciences*, eds. Alexandre Guay and Thomas Pradeu, 165–192. Oxford: Oxford University Press.

Seibt, Johanna. 2003. Free process theory: Towards a typology of occurrings. In *Process theories: Crossdisciplinary studies in dynamic categories*, ed. Johanna Seibt, 23–55. Dordrecht: Kluwer Academic Publishers.

Seibt, Johanna. 2018. Process philosophy. *The Stanford encyclopedia of philosophy* (spring 2018 edition), ed. Edward N. Zalta. https://plato.stanford.edu/archives/spr2018/entries/process-philosophy/.

Sellars, Wilfred. 1952. Particulars. *Philosophy and Phenomenological Research* 13(2): 184–199.

Sober, Elliott. 1991. Organisms, individuals, and units of selection. In *Organism and the origins of self*, ed. Alfred I. Tauber, 275–296. Dordrecht: Springer Netherlands.

Strawson, Peter F. 1959. *Individuals: An essay in descriptive metaphysics*. London: Metheun & Co., Ltd.

Tooley, Michael. Ed. 1999. *Particulars, actuality, and identity over time*. New York and London: Garland Publishing, Inc.

van Cleve, James. 1985. Three versions of the bundle theory. *Philosophical Studies* 47: 95–107.

van Inwagen, Peter. 1990. *Material beings*. Ithaca: Cornell University Press.

Weber, Michel. 2007. Abstraction and individuation in Whitehead and Wiehl. In *Subjectivity, process, and rationality*, eds. Michel Weber and Pierfrancesco Basile, 39–119. Heusenstamm: Ontos Verlag.

Weber, Michel, and Pierfrancesco Basile, eds. 2007. *Subjectivity, process, and rationality*. Heusenstamm: Ontos Verlag.

Whitehead, Alfred. 1929/1978. *Process and reality*. New York: The Free Press.

Wiggins, David. 1980. *Sameness and substance*. Cambridge: Harvard University Press.

Wiggins, David. 2001. *Sameness and substance renewed*. Cambridge: Cambridge University Press.

Wilson, Robert. 1999. The individual in biology and psychology. In *Where biology meets psychology: Philosophical essays*, ed. Valerie G. Hardcastle, 357–374. Cambridge, MA: MIT Press.

Wilson, Robert, and Matthew Barker. 2013. The biological notion of individual. *The Stanford encyclopedia of philosophy* (spring 2018 edition), ed. Edward N. Zalta. https://plato.stanford.edu/entries/biology-individual/.

{ PART I }

Aspects of Individuation: Metaphysical and Processual

{2}

Processes, Organisms, Kinds, and the Inevitability of Pluralism

John Dupré

2.1 Introduction

This chapter begins with the proposal that we should treat organisms not, as is traditional, as a kind of thing or substance, but as a kind of process. I shall begin by explaining this idea a bit further and outlining some of the reasons in its favor. I shall then consider some of the implications of this position on how we should understand the classification of organisms. Finally I shall show how these considerations provide a deeper ground for the classificatory pluralism I have advocated for many years.

2.2 Organisms as Processes

One view of ontology, of what there is, has dominated Western philosophy since the Greeks: the most basic furnishings of the world are things or, in more technical philosophical terms, substances.[1] Without attempting to summarize two millennia of philosophical reflection on the nature of these primary things, we can say that they are thought of as integrated, persisting through time, not dependent on anything external for their existence, and as the bearers of properties. They are also the subjects of change though, as I shall briefly mention later, this role is problematic. For many philosophers, there are, among the properties exhibited by

[1] There are actually two main philosophical uses of the term "substance" (Robinson 2014). In one such usage, suggested by the Latin meaning of "standing under," substances are whatever are the basic constituents of reality. In the second, which is what is relevant here, substances are objects, or things, typically contrasted with the properties that they bear. In the first sense of substance, my (confusingly if thus phrased) claim is that substances are processes.

a thing, some essential ones, the possession of which is the necessary and sufficient condition for the existence of the things of the particular kind of which they are the essence. Starting with this ontological foundation, other familiar categories, processes, and events, can be understood as consisting in changes of the properties of things or the interactions between things. Paradigmatic things include rocks, cars, and pieces of furniture; in some philosophical schemes, a particularly important category are atoms as eternal and unchanging things. And, at least since Aristotle, among the most paradigmatic of things have been organisms.

Nevertheless an important minority of philosophers has defended a quite different vision.[2] For them what is most fundamental to the world is change, or process. Nothing truly stands still, and what we are tempted to see as stable things, only contingently changing their accidental properties, are in reality no more than partial stabilities in the surrounding flux, eddies in the flow of process. I am myself a subscriber to this metaphysical minority view. Here, however, I do not propose to defend the view in full or in detail.[3] In this section, I will argue only that a process ontology provides a much better way of understanding organisms, at least aiming thereby to colonize one of the prime exemplars of substance ontology. In the following section, I shall explore the implications of this claim for biological taxonomy—the classification of organisms. A main conclusion will be that a process view of organisms makes it clear why we should be pluralists rather than monists about biological classification. Happily, there are well-known independent grounds for this conclusion, so this result might be taken as offering abductive support for the process perspective.

Why might one better think of an organism as a process? We might usefully begin by comparing an organism—say, a cat—to something fully inanimate—say, a rock. A rock, we suppose, can sit quietly for years or centuries doing nothing. Its default state is inactivity. A cat, on the other hand, that does nothing is a dead cat. For the cat to stay alive, countless processes must occur. At a relatively gross level, its heart must pump and circulate its blood, its lungs must move to inhale oxygen and exhale carbon dioxide, and so on. The maintenance of its organs requires the constant division, retirement, and replacement of their constituent cells. Within the cell, now at a scale of microseconds and milliseconds, vast numbers of chemical reactions must occur to sustain the health of the cell and eventually its timely division or death (apoptosis). What might seem at first sight a stable thing is actually sustained at the finely articulated intersection of countless processes at multiple time scales. Of course, if I were here defending a full-blown process ontology I might remark that, even for the rock, the cohesion of its parts might best be seen as the upshot of trillions of dynamic quantum mechanical processes. But the

[2] The origin of this idea in the Western tradition is generally credited to Heraclitus. The *locus classicus* in contemporary philosophy is Whitehead (1929). A good recent overview is Seibt (2016).

[3] For such a defence, see Dupré and Nicholson (2018).

dynamic stabilization of matter in the form of a cat is at least a more obvious and complex, multilayered matter.

One way of seeing the difference between a process and a substance ontology is in terms of the explanatory problems they face. Whereas a substance ontology has to provide explanations of change, a process ontology instead faces the problem of explaining stability or the appearance of stability. A major difference between these problems, at least in the biological context currently under consideration, is that while the problem for substance ontology is a philosophical one (i.e., in what sense is it possible at all for the very same thing to have different properties at different times?), the problem for a process ontology is an empirical one (i.e., how is the [relative] stability of a process in fact maintained?). This difference will need some further elaboration.

The philosophical problem for substance theorists derives from a common understanding of identity. If *a* is identical to *b*, they are the same thing: there are not two things, somehow related to each other, but one thing, referred to in different ways. A thing has the same properties however it is referred to, so, if *a* is identical to *b, a* and *b* have all the same properties; this is an informal statement of Leibniz's law. But then, how is it possible for a thing to change and yet be the same thing. In 1980, my hair was all brown; now it is all gray. If I am identical to the relevant organism in 1980, there is a thing that both has only brown hair and only gray hair, which is impossible. The main attempts to solve this problem either claim that I am composed of many temporal parts, some of which are brown-haired and some of which are gray-haired, or that I do not have properties simpliciter, but only relative to particular times: being brown-haired is a transient property, not strictly suitable for characterizing a persisting entity, but being brown-haired in 1980 is timelessly true of me.[4] The philosophical problem is a classic and very difficult one, given the wide acceptance of Leibniz's law.[5] I shall make no attempt to resolve it here.

In what way is a process ontology better able to make sense of this problem? First, it is uncontroversial that processes contain temporal parts. Second, since change is intrinsic to a process, no one could suppose that all the parts of a process had identical properties. However one understands the relation between temporal parts of a process, it will not involve anything like Leibniz's law. But, third, there is an obvious way of addressing this problem through the causal continuity of a process, an idea that has been discussed under the rubric of *genidentity*.[6] My relation to the infant I once was, for example, is not a question of any interesting shared

[4] The locus classicus for these approaches to the problem of identity over time is Lewis (1986). There has been a great deal of subsequent discussion, but engaging with this would be beyond the scope of this essay.

[5] For recent discussion of these positions and an argument that they cannot ultimately make sense of change over time for substances, see Meincke (manuscript).

[6] The concept traces to Kurt Lewin in the 1920s. See Guay and Pradeu (2016).

properties but of an integrated nexus of causal processes involving the gradual transformation of the latter into the former.

This perspective profoundly transforms questions about identity over time. First, the persisting individual is not a metaphysical given, in which the individual is identified and investigated to detect the conditions under which it might persist. Rather, it is something the individual does: persisting is, indeed, hard work. Individuation, moreover—the distinction of a discrete individual from its material context—is not just something we do by pointing to a self-evident individual, but something the individual itself does by working to distinguish itself from the flux of its surroundings. The empirical problem of explaining the more or less limited stability of biological entities, finally, is central to the agenda of the biological sciences.[7] Biomedical science, for example, is concerned with exploring the conditions for stability of human organisms and possible interventions to maintain this stability. These are, needless to say, very hard problems—but there is no reason to think they are in principle impossible.

The severity of the problem for a substance ontology is immediately apparent when we turn to a second reason favoring a process ontology for living systems, the problem of ontogeny. We tend to associate with the word "cat" an adult member of that species. But cats were once kittens and before that zygotes and embryos. Is there some point in development when the embryo or the fetus becomes a cat (or kitten)? The progression between these developmental stages seems continuous, and any fine line between them seems arbitrary. Perhaps a cat only comes into being at birth, although this seems to be more a change in the external conditions of the being than any sudden change in its nature. This kind of problem is even more pressing when one considers organisms such as insects that go through major metamorphoses. There seems a clear and familiar sense in which the egg, caterpillar, pupa, and butterfly are all stages in the career of the very same individual. Yet the amount of change that this individual has passed through from egg to butterfly is such as to make it difficult indeed to argue that these stages are all part of the history of the same individual substance. In fact, there may be no interesting properties at all that they share. Some have been tempted to nominate the genome—or, better, its nucleotide sequence—for this role. But in fact genomes themselves are highly dynamic entities and subject to many kinds of changes throughout development. And there is even a question whether there is any unique entity that qualifies as the genome of an organism (see Dupré 2012: chapt. 7).

---

[7] Note that there are two relevant kinds of stability: stability of a particular property, such as temperature, and stability of a directional process, such as ontogeny, or the cell cycle. Typically, processes of the latter kind are involved in the stabilization of higher level states, as the cell cycle is vital to the health of an organism. C. H. Waddington (1957) distinguished the stabilization of properties and of sequences of states with the terms *homeostasis* and *homeorhesis*

From a process perspective, we can see a path toward a solution. There are no particular philosophical restrictions on the degrees of similarity and difference that are admissible for the temporal stages of a process. The conditions on being part of the same process have to do with continuity and causal connection between these temporal stages. And while it may be very difficult to provide a general account of the relevant conditions, it is easy to see that they may be realized by a sequence such as the developmental stages of an organism, even in cases as complex and diverse as that of the butterfly. If this sounds rather vague, it is because it is intended to. A source of pluralism about kinds is just the uncertainty about when a process continues, bifurcates, or transforms into a process of a different kind.

I assumed in discussing the problem of ontogeny that there is at least some clear phenomenon—the persistence of a particular individual organism—of which the competing ontological accounts are supposed to make sense. But, in a process ontology, this is far from being a clear general requirement. Think of the river, an entity that Heraclitus made exemplary of the process view. What is a river, and what is a tributary? The Missouri–Mississippi system is longer than what is commonly referred to as the Mississippi, but the latter carries more water. Either factor might perfectly well have been taken as a defining characteristic, though in fact it seems more likely that the decision reflects historical contingency. From the Treaty of Paris in 1763, the Mississippi served as the boundary between the British and Spanish empires, whereas Meriwether Lewis and William Clark in 1804 were the first Europeans to travel the length of the Missouri. The Mississippi was surely more salient to European settlers at the time that these nomenclatural decisions became established. There is, at any rate, surely no difference of kind between the Missouri River, now taken to start at its confluence with the Mississippi, plus the part of the Mississippi downstream from that confluence on the one hand and the full length of the Mississippi on the other. No fact in nature, beyond the facts that determine accordance with an arbitrary definition, determines that one is a river and the other part of a river and a tributary. And probably no one should care very much, though it is no doubt possible for momentous political consequences to be attached to such matters. At any rate, problems of this kind abound in living systems, problems that present no special difficulty of principle from a process perspective but that are difficult or impossible to make sense of in the context of a substance ontology in which fully distinct individuals are taken as given at the outset.

As is now well-known, the human body coexists with trillions of resident microbes. Some of these are essential for the well-being of the human, playing vital roles in digestion, development, the immune system, and more. Others are less necessary and still others harmful. Which, if any, of these are parts of the human organism? If one is tempted to say none of them, then what of the mitochondria, the long-captive bacteria that process the energy requirements of every cell in our bodies? One intuitively plausible criterion for excluding symbiotic bacteria from the human system might be the fact that they have separate lines of inheritance;

but this would exclude the mitochondria, which no one, as far as I know, would be willing to do. Similarly (and for other reasons; see Dupré 2012: chapt. 7), it will not work to limit the human to cells containing a particular genome sequence.

For reasons of this sort I have argued elsewhere (Dupré and O'Malley 2009, reprinted in Dupré 2012) that the individuals we should be most interested in distinguishing are metabolically connected segments of lineages. The human microbiome includes entities belonging to very different lineages from the human lineage, but these distantly related entities are nevertheless deeply intertwined in chemical and physical interactions. These assemblies are the functional wholes that interact more or less effectively with their biotic and abiotic environments. Human and microbial lineages intersect in the human body to provide a stable and highly organized nexus of living process. This is, moreover, absolutely typical. All or almost all multicellular organisms are similarly symbiotic, and even unicellular organisms typically exist in multispecies communities such as biofilms. The important point here is that what should be included as part of such a nexus is underdetermined. The spectrum from passengers and pathogens to essential symbionts has not been disposed of but instead put in a conceptual frame in which it is unproblematic.

One promising solution that has been suggested for this problem is that the boundaries of the individual are determined by the immune system (Pradeu 2012). What is or is not part of the system is not decided by our conceptual legislation but by the complex processes that determine which elements are accepted and which rejected by the system itself. I have considerable sympathy with this suggestion. Such a discriminatory process is essential for the survival, the stability, of a complex living system. (Or even, if there are any such, a simple living system. The immune systems of bacteria, regulating the flow of nucleic acids into the bacterial cell, are an increasingly lively area of study.) All I would insist is that immune response is not an unambiguous demarcator of the boundaries of a system. It is, rather, a complex web of processes that stabilize a nexus of biological activity by regulating access to it. Much scope remains for different decisions about organismal boundaries. On a process understanding of life, this is just what is to be expected; for a substance theory, it is, at the least, a serious problem.

It is best, I suggest, to distinguish two senses of "organism." The first, commonly assumed in theoretical contexts, takes an organism to be part of a cell lineage. On this view, the human organism is just the set of cells derived from an initiating zygote. This is a concept that has significant theoretical uses in genetics and in genetic models of evolution.[8] Call these organisms_1. But the more central and useful concept of an organism is as a living individual, and here there is no avoiding the

[8] Such a concept is at least an intelligible way of distinguishing living entities and one that has been assumed in a lot of important theoretical work. It is not without serious difficulties in application deriving from twinning, vegetative reproduction, and similar phenomena. It will not be necessary for present purposes to decide whether these can be resolved.

problems posed by the near universality of symbiosis. These are the organisms that develop, behave, engage in ecological interactions, and so on. I shall call these organisms$_2$. As these different functions of the two concepts suggest, an important moral of this discussion is that how we should define the boundaries of organisms depends on the theoretical purposes we have in view.

2.3 Classification

This bifurcation of the organism concept presents a puzzling dilemma for taxonomy, the classification of organisms. The dominant contemporary view sees taxonomy as tracking phylogeny. The history of life is generally represented as a tree, with branching points representing events of speciation. Cladistic taxonomy insists on the principle of monophyly: any classificatory unit, or taxon, should include all and only the descendants of an ancestral group. Thinking of the tree of life as a real tree, a monophyletic group is what falls off when one saws through any branch. Species are then the smallest twigs that we choose to distinguish.[9]

It will be clear that what this model classifies are organisms$_1$. Moreover, this classification of organisms$_1$ fits perfectly with a methodology: genetic comparison. Whereas the rise of ever faster genome sequencing and genomic marker recognition technologies have not delivered quite the benefits that have sometimes been advertised, one thing they are unquestionably good for is comparison and classification. The ability to discern similarity and difference between genome sequences underlies such successful technologies as forensic genomics and paternity testing, and also something of a revolution in taxonomic practice.

It should be noted in passing that the revolution engendered by these genomic technologies for tracing phylogenetic relations has had some unexpected results. Specifically, they have made possible the tracking not only of vertical relations of inheritance through reproduction, but also horizontal relations of lateral gene transfer (LGT). The prevalence of LGT in prokaryotes (bacteria and archaea) has been such as to lead many to conclude that there is no unique tree of life to trace (Bapteste et al. 2009). Using different genes will lead to a range of different trees. However, at least for multicellular eukaryotes such as ourselves, these techniques have led to an increasing consensus on the phylogenetic relations between organisms$_1$.

This leads to something of a problem. The organisms$_1$ that we classify are not quite the same entities as the organisms$_2$ that figure in much of our functional biology. The difficulty for our classification schemes in dealing with symbiosis is not

[9] The actual definition of a species within this model is difficult to specify and must ultimately be somewhat arbitrary. We may choose to recognize the branching nature of the species itself by distinguishing subspecies, varieties, or races. For various perspectives on the species problem, see Wilson (1999).

a new one. Consider one of the most familiar examples of symbiosis, the lichens, symbiotic collaborations between a fungus and a photosynthetic microbe, either an alga or a cyanobacterium. Twenty percent of fungus species form lichens, and these include a wide variety of kinds, though most commonly asomyocetes (cup fungi).

One might first want to ask what exactly is the relationship between these partners. In most cases the photobiont can survive independently of the lichen relationship, while the fungus usually cannot or, if it can, will fail to develop normally. Is this a pure mutualism (benefit to both parties), or is one exploiting the other, a controlled parasitism? Given the diversity of such relations and the fact that, in many cases of symbiosis, the question of benefit or harm will depend on further details of the circumstances (Méthot and Alizon 2014), it may be better not to worry too much about such questions. In Dupré and O'Malley (2009) we suggested that a neutral term, "collaboration," would most usefully refer to this range of interactions. Life, then, is massively collaborative.[10]

How are lichens fitted into the tree of life? In fact the official practice is to give the lichen the same name as the fungus. This has some curious consequences. Consider the following from a government site on Australian lichens:

> *Sticta filix* is a foliose lichen with an algal photobiont and *Dendriscocaulon* was initially described as a fruticose genus with a cyanobacterial photobiont and in each genus fungal tissue dominates the thallus. However, you can find lobes of *Sticta filix* growing from branched *Dendriscocaulon* thalli and microscopic examination has shown that in such cases there is continuity between the fungal partner. That is, it is the same fungus in both the fruticose and foliose thalli, rather than the foliose *Sticta* thalli being parasitic or epiphytic on the fruticose *Dendriscocaulon*. Since lichens are classified by the fungal partner what we have is the same species of lichen showing two strikingly different growth forms, depending on the photobiont. Of course, since a species may have only one name this lichen cannot be both a species of *Sticta* and a species of *Dendriscocaulon*. The genus *Sticta* was introduced in 1803 and *Dendriscocaulon* in 1885. Given the dramatic difference in thallus form it's not surprising that two separate genera were erected.[11]

Moreover, it appears that the different partners provide different ecological advantages. The website referenced in the preceding text illustrates a case where the lichens are growing on the surfaces of boulders that provide a narrow gap above a stream of water experiencing spray from a waterfall. Nearest the water the fungus is all of the "*Dendriscocaulon*" form (the quotes indicate that this is no longer a valid name), and, as the boulders separate further from the water, only

[10] Collaboration is very common within as well as between species, a point of considerable relevance to our own species (see Dupré 2018). This will not be of concern in this chapter.

[11] http://www.cpbr.gov.au/lichen/form-structure-sticta.html. Accessed February 25, 2016.

the *Sticta* form is found. In between are fungi parts which harbor each of the two photobionts.

According to standard taxonomic practice, only one species is present here. On the other hand, from a phenotypic perspective, the organisms at each end of this microcline are so different as to have made their placement in separate genera "not surprising." And, of course, the photobionts that are somewhat arbitrarily excluded from the classificatory scheme are at vast distances in the traditional tree of life. The cyanobacterium is a bacterium and the alga is a eukaryote. These are two of the three most fundamental and distinct branches of the tree.

A similar problem is posed by optionally lichenized fungi. For example, a species of *Stictis* (a saprophytic fungus) appears to be indistinguishable from the fungus species in the lichen formerly know as *Contrema*, a lichen. The two are found in close association, the lichen growing on living plants of the tree, the unlichenized fungus on decaying parts (Wedin, Döring, and Gilenstam 2004). From a taxonomic point of view it turns out that "*Contrema*" doesn't exist, or at least it is not a valid name for anything. Its instances are individuals of the *Stictis* species. Nonetheless *Stictis* and "*Contrema*" are morphologically and ecologically distinct, and "*Contrema*" has a whole additional genome not found in *Stictis*.

There is no great mystery here. The lichen "species" isn't a branch of the tree of life but two intertwined branches. The evolutionary histories of the two partners in the lichen will be quite different, at least for their more or less distant parts. Phylogenetic taxonomists consider the function of taxonomy to be, or at least to include, the tracing of evolutionary history. Given this aim there is never going to be a satisfactory way of classifying lichens because they do not have an evolutionary history; they have two.

It is important to note that this is only an extreme case. As I noted earlier, it seems increasingly clear that the vast majority of functionally distinguished organisms are symbiotic systems. For many cases, notably large eukaryotes, this is not too serious a problem. Although a human individual is host to trillions of symbiotic microbes, it seems unproblematic to classify the whole system with reference to the dominant partner—dominant at least in terms of mass. But it should not make us forget that the functional human individual is, just as much as the lichen, not a part of a branch of the tree of life but the intertwining of a large number of branches or parts of branches (assuming again, and probably counterfactually, that the microbial partners can properly be arranged on a tree at all [Bapteste et al. 2009]).

2.4 Pluralism

The necessity for multiple overlapping and cross-cutting classifications of biological individuals has been increasingly recognized by philosophers of biology (Dupré 1993; Ereshefsky 1992), and the preceding examples provide a particularly striking illustration of the unavoidability of such pluralism, a pluralism that recognizes the dependence of a classificatory scheme on the theoretical aims in view. If one is interested in tracing phylogeny, the assimilation of lichens to their fungal partner makes sense; but, if one is concerned with morphology, physiology, or ecology, then a fungus and a lichen are very different kinds of things.

I want here to suggest more systematically that a processual perspective on life makes pluralism more generally to be expected. Organized arrays of living matter participate in evolving lineages, develop, and interact. They are constantly changing both intrinsically and in their relations to other living entities. When we classify an entity at some stage in these processes we may be interested either in identifying the process or one of the processes of which it is a temporal stage or the properties and relations of the entity as it functions at the stage we are encountering.

The conflict between functional and phylogenetic classification in lichens just discussed illustrates the point. Phylogenetic taxonomy identifies an organism by reference to the evolutionary process of which it is a part and functional taxonomy by its current properties and relations. One might make a similar argument about development. We are very comfortable with the idea that a caterpillar and a butterfly might be members of the same kind, merely at different temporal stages in the same developmental process. However, it is also true that they are physiologically, morphologically, and ecologically quite different. Ecologically, for example, whereas caterpillars are typically voracious herbivores, butterflies are reproducers and often pollinators. Because the prevalence of caterpillars and of butterflies of a particular species are naturally related, it will sometimes be appropriate to treat them as of the same kind in an ecological model. But, for many models, one life cycle stage will be pivotal while the other will be irrelevant. And, of course, the relationship between caterpillar numbers and butterfly numbers will itself be an output of ecological processes, requiring that they be distinguished in any model in which both their numbers are parameters.

Parallel to this point about development we should remember that even within a purely phylogenetic taxonomy a core activity is to distinguish between stages in a process. In this case, the relevant processes are evolving lineages. Speciation is then the analogue of metamorphosis in the developmental example. Whether every species distinction represents a natural kind of event or process at its origin is a more difficult question. The currently dominant cladistic taxonomy requires by fiat a process of lineage bifurcation, though the prevalence of hybridization can make this difficult to define except retrospectively. Due particularly to the influence of Ernst Mayr (1942), it has become common to understand speciation as stemming from geographic isolation. However, it now seems increasingly plausible that sympatric speciation—speciation without any physical obstacle to interbreeding—may be equally important. This is known to occur frequently in plants through the process of polyploidy, and various more gradual processes have been proposed for animal speciation (e.g., Maynard Smith 1966; for a review, see Bolnick and Fitzpatrick 2007).

All of these processes result in a split in a lineage. But it is also possible for a great deal of change to occur within a coherent lineage, and pragmatically it may be sensible to make taxonomic distinctions between temporal stages of such an evolving lineage. This is anagenetic (as opposed to cladogenetic, lineage splitting) speciation. A problem with this is that, as opposed to the processes of cladogenesis mentioned earlier, anagenesis does not in general include anything that could count as a mechanism of speciation. Classificatory distinctions between stages of a lineage are to some extent unavoidably arbitrary. Nonetheless, if one is interested in classification in relation to biological function, such arbitrary divisions may be well motivated.

The general point is this. Once one sees life as a deeply intertwined set of processes, evolutionary, developmental, ecological, and symbiotic, classification faces a series of dilemmas. First, should one classify processes or stages of processes? Phylogenetic taxonomies and standard descriptions of kinds of organisms as, for example, found in a field guide, both represent process classifications. The problem then is that processes, by their nature, undergo change.[12] So the same category may encompass very diverse entities: caterpillars and butterflies, fungi and lichens, different stages in anagenetically evolving lineages. If, on the other hand, we classify in terms of similarity and present function, which means classifying specific stages of the underlying processes, not only do we abstract from the temporarily extended processes in a way that can be hard to motivate theoretically, we confront the problem of intertwining.

I noted above my earlier account of the biological individual as lying at the metabolic intersection of multiple lineages (Dupré and O'Malley 2009, reprinted in Dupré 2012).[13] Metabolism here encompasses broadly the processes through which segments of multiple eukaryotic and prokaryotic lineages interact to provide the stable systems we recognize as organisms. If we extend our view to the life cycle of an organism, we see different such lineage segments engaging and disengaging at different times. The gut bacteria of an infant, for instance, are very different from those of an adult human. To describe this life cycle, we must have the resources to refer to the various lineages that compose it and to various stages through which the life cycle progresses. These lineage segments will, as already noted, have different evolutionary histories and often will have the capacity to

[12] In some simple cases, such as a whirlpool, such changes are no more than the activity required to sustain the process. In living systems, there are also more or less (homeorhetically) stabilized developmental trajectories that are in turn subject to more or less adaptive variability.

[13] To be clear, I do not understand this account as picking out one unique set of entities or as capturing the only concept that could legitimately be applied to distinguish biological individuals. The proposal is just that this is an important, probably the most important, kind of entity that functions in an integrated way in many significant biological contexts. The idea must be seen in the context of the broader pluralism discussed further later.

participate in a range of different organism life cycles. The need for multiple classificatory systems becomes even clearer.

I have not even mentioned yet the entanglement of the organism and its environment (Barker, Desjardins, and Pearce 2014). This is a phenomenon the importance of which has long been recognized within areas of research such as developmental systems theory (Oyama, Griffiths, and Gray 2001), niche construction (Odling-Smee, Laland, and Feldman 2003), and cultural evolution (Richerson and Boyd 2005). All of these developments explore ways in which the nature of a biological individual cannot be abstracted from the context in which it lives. Developmental systems theory describes the ways in which features of the environment are co-opted as parts of the processes by which organisms develop; niche construction, closely related, is the process by which organisms shape their environments in ways that not only mesh with their biological needs, but also direct their further evolution. Cultural evolution, which again may intersect with niche construction for species like our own, in which there is an important material component of culture, is a process somewhat analogous to natural selection, but in which inheritance is accomplished by such processes as imitation and social learning. Here, the social environment is a vital contributor to the development of the phenotype.

These phenomena give rise to the possibility that purely biological features may be insufficient to provide the distinctions between organisms that are needed for some purposes. For a species such as our own, in which cultural evolution and niche construction play fundamental roles, the mere identification of an organism as *Homo sapiens* tells us, for many purposes, very little. Different cultures, and different material cultures, shape the developing organism in very diverse ways, leading to differences not only in behavior, but even in physiology. As one from countless possible examples, one might consider the diverse ways in which the high demands for water and labor imposed by the cultivation of rice, interacting with the peculiarities of diverse social systems, have shaped the many cultures for which this crop has been a central element of the food supply (Bray et al. 2015).

I have argued elsewhere (Dupré 2001) that there is a strong case to be made for a general purpose taxonomy that is as little theory-driven as possible and hence as resistant as possible to theoretically driven change. The argument is simply that we need to store and communicate biological information, and, to do so, we need to be able to identify what part of the living nexus we are talking about. But far from being a concession to any kind of classificatory monism, the idea that there is a unique and uniquely correct taxonomy out there to be discovered, this argument is precisely a recognition that there is no such thing. The multiple possible and actual theoretical taxonomies that serve specific investigative purposes are just what create the need for a *lingua franca* that is as nearly theory-free as possible.

Earlier, I alluded in passing to a doubt as to whether phylogenetic classification was even in principle applicable to prokaryotes (bacteria and archaea). Many microbiologists believe that it is not applicable, on the grounds of the prevalence of LGT (Bapteste et al. 2009). It is now widely agreed that LGT has occurred frequently in the history of prokaryote evolution. It might be supposed that this was merely an epistemological problem. If genes have entered a cell from various different lineages, then the phylogenetic history inferred from comparison of gene sequences will depend on which genes are used for comparison.

But I believe the problem is ontological. There must be, assuming that all cells originate through the division of an ancestral cell, a tree of cells, if an extremely lush and bushy one. Two questions arise: first, what kind of tree could provide a basis for phylogenetic classification? And, second, would such a classification be a useful one? One reason for skepticism as to the classificatory potential of the tree of cells is just that it has far too many divisions; in fact, every organism (cell) originates from a splitting. Sexual reproduction provides us with lineages with many individual members by creating gene flow within a significant but fairly well circumscribed population. It is imaginable that LGT could provide such gene flow for unicellular microbes. Unfortunately, it is both too rare and too promiscuous (i.e., insufficiently constrained by phylogenetic distance) to do this.

In theory, any microbial cell could be taken as the founder of a new lineage. It would be reasonable to take seriously only the claims of cells with a somewhat interesting innovation. There is a glimpse of a simple picture here: we identify the important innovations in microbial evolution and identify significant clades as those originating with the first organism to display this innovative feature. This picture, however, is wholly undermined by LGT.

Any mutation or any instance of LGT will provide a possible starting point for defining a new species. Assuming that the quantity of material in a genetic transfer is small and that the phylogenetic distance between donor and recipient is quite large, it may be unproblematic to identify the event as the appearance of a new kind of some sort within the recipient clade. However, if the transfer occurs within a clade above—but not much above—the level of species, the problem is more difficult. Suppose the donor and recipient species are differentiated by only one important innovation X. Call the transferred material Y. We now have three related groups possessing, respectively, traits X, Y, and X + Y. As neither X nor Y is a novel feature, it appears that the new lineage defined by X + Y is a member of both the species with X and the species with Y, which is unacceptable on any standard view of taxonomy. The only solution at this point would be to call the combination, X + Y, an innovation. A problem with this already somewhat problematic solution is that subsequent transfers of either X to the lineage Y or Y to the lineage X or either X or Y from the lineage X + Y to the lineages Y or X will produce lineages that would be assigned genetically to the same species as the original X + Y. Thus we would have species that could become indefinitely paraphyletic (i.e., with multiple origins, or nonmonphyletic), contradicting the original hypothesis about how the system was to be set up.

If this sounds a minor problem, consider that, within what is, if one takes seriously its Linnean nomenclature, the species *Escherichia coli* is in fact a large number of distinct strains. Any particular stain will contain only one-third to one-quarter of the full set of genes found in *E. coli* strains, and only about 20% of the genome is common to all known strains. This is indicative of the fact that the majority of genes in any particular part of this lineage arrived by lateral transfer. The problem abstractly described in the previous paragraph will concretely permeate such a taxon.[14]

A rather different type of case is exemplified by the much-discussed case of *Thermatoga maritima* (Nelson et al. 1999; Zhaxybayeva et al. 2009). This bacterium is notable for having derived about one-quarter of its genes from an archaeon, which is to say, from a different domain of life. Other major elements of gene transfer combined to make the placement of the organism in the generally accepted tree of life highly disputed, not just in detail, but over huge phylogenetic differences. Such cases challenge the assumption earlier in this discussion that the amounts of genetic material transferred are small compared to the overall genome size. Note that contradiction of this assumption doesn't require that all the material be transferred at the same time. A series of small transfers is, at least from the point of view of the taxonomist, indistinguishable from a large transfer. This is not just a theoretical possibility. Part of the explanation of the issue proposed by Zhaxybayeva et al. (2009: 5869) is that two of the groups that contributed to the genome in question have been much thrown together; they are the only two groups found in hot oil reservoirs. They have had great stretches of time in which to interact and sample one another's genetic menus.

The last case perfectly illustrates my general thesis. If life is, rather than a diverging tree of fully distinct lineages, a system of intertwined lineages, sometimes separating intact, sometimes fully merging, and sometimes transferring parts within the intertwined system, there is no reason to expect any unique and definitive classification into kinds; *a fortiori*, phylogeny cannot provide such a classification. In the realm of prokaryotes, where not only functional intertwining but exchange of parts is endemic, this problem seems fully to rule out a unique phylogenetic classification or, for that matter, a unique phylogeny. The presence of large multicellular organisms providing obvious candidates for anchoring the phylogeny makes the problem much less severe for many parts of eukaryote taxonomy. Nonetheless, cases such as the lichens remind us that this solution is to some degree pragmatic and only contingently available.

2.5 Conclusion

For many years I have argued, together with a significant and growing number of philosophers of biology, that there is no future in seeking a unique way of

[14] For more details on the diversity of *E. coli* strains, see Monk et al. (2013).

classifying the living world, somehow provided for us by Nature herself. The quest for a Natural System of classification is a wild goose chase. However, this thesis does naturally raise the question whether this pluralism is itself something amenable to explanation. Is it just a brute fact that nature declined to organize herself into a single scheme of kinds waiting for us to discover? Or is it rather something following from some even deeper fact about the natural—or at least the living—world?

For most of the time that I have been defending pluralism I have tended to assume something like the brute fact view. Monism, the denial of pluralism, did indeed strike me as a part of a wider system of ideas—essentialism, the universal reign of natural law, or, more specifically, the tyranny of the microphysical—that I took to be false (Dupré 1993). Nonetheless, the ingredients of this picture I took to be only contingently false, and, although natural bedfellows, I did not take them to be necessarily connected.

In this chapter, I have proposed a more fundamental grounding for pluralism. Monism, I want to suggest, is inevitably connected to a substance ontology. In a world composed of things, we naturally ask whether these things have essences (perhaps they must have if they are even to be persisting things). If they do, then there is only one kind that any thing can belong to, the kind of things with which it shares its essence. An obvious next step is to look for patterns among these essences. The paradigm for such a pattern is the periodical table of the elements, and cladistic taxonomy might seem to aspire to something broadly similar.

In a substance world, then, both essentialism and monism have some clear appeal. But in a process world, monism, at least, has no appeal. Certainly there may be optimal ways of distinguishing individuals from the processual nexus for particular purposes, but it is not plausible that nature provides us with only one set of such modes of distinction. Processes are not defined as entities by their essential (or even nonessential) properties, but by relations, notably causal relations, between their temporal stages. Their persistence is something they enact, not something that needs to be sought in the metaphysical recesses of their property inventories. But then, given the tendency of processes to split and merge, to contain elements that are doing different things over different time scales and so on, it would be improbable indeed that we could expect to inspect a process and decide on some unique kind to which it belonged. It may be that such tendencies are no more than contingent features of the processes observed in our world, in which case one might imagine a monistic world of processes. At the least, a process ontology points to an explanation of pluralism not available in a world of substances.

So, process metaphysics does, I think, provide that deeper explanation of pluralism. Is it a contingent fact—if it is a fact—that the world is a world of processes rather than things? I suspect not, though I have no confidence that I can prove it. It is, at the least, a deep fact if it is a fact; perhaps as deep as a metaphysical fact could aspire to be.

I conclude with a confession: I would like the world to be a process world. Things have fixed properties, and, except when they run into accidents (their path to the marketplace intersects someone else's, for instance), what they do is determined by their being the kinds of things they are. Processes are not, or need not be, like that. They change in all kinds of ways, including, in the context of life, evolution and development. Evolution is, of course, an open-ended process despite the remarkable and perverse attempts of some thinkers to constrain it within the grip of physical law. The realization that development is also open-ended has been instrumental in shaking evolutionary thinking free of this moribund grip. Processes can, and in biology do, adapt to their environments and change in ways that make their persistence in a changing and often hostile environment more probable. How either individual organisms or evolving lineages do this is a hugely complex and difficult set of scientific questions; but, once we have placed it in the context of a process metaphysics, it should no longer be a philosophical problem.

I, at any rate, am very pleased to be a process.

Acknowledgments

I would like to thank two anonymous referees and the participants in a meeting of the Egenis Biological Interest Group for very helpful comments. I am especially grateful to my colleagues on the ProBio project, Stephan Guttinger, Anne Sophie Meincke, and Dan Nicholson, for extensive and rewarding discussion of many issues here addressed. The research leading to this article has received funding from the European Research Council under the European Union's Seventh Framework Programme (FP7/2007-2013)/ERC Grant Agreement 324186.

References

Bapteste, Eric, Maureen A. O'Malley, Robert G. Beiko, Marc Ereshefsky, Johann P. Gogarten, Laura Franklin-Hall, et al. 2009. Prokaryotic evolution and the tree of life are two different things. *Biology Direct* 4(34).

Barker, Gillian, Eric Desjardins, and Trevor Pearce, eds. 2014. *Entangled life: Organism and environment in the biological and social sciences*. Dordrecht: Springer.

Bolnick, Daniel I., and Ben M. Fitzpatrick. 2007. Sympatric speciation: Models and empirical evidence. *Annual Review of Ecology, Evolution, and Systematics* 38: 459–487.

Bray, Francesca, Peter A. Coclanis, Edda L. Fields Black, and Dagmar Schafer, eds. 2015. *RICE—Global networks and new histories*. New York: Cambridge University Press.

Dupré, John. 1993. *The disorder of things: Metaphysical foundations of the disunity of science*. Cambridge, MA: Harvard University Press.

Dupré, John. 2001. In defense of classification. *Studies in the History and Philosophy of the Biological and Biomedical Sciences* 32: 203–219.

Dupré, John. 2012. *Processes of life: Essays in the philosophy of biology.* Oxford: Oxford University Press.

Dupré, John. 2018. Human nature: A process perspective. In *Why we disagree about human nature*, ed. Tim Lewens. Cambridge: Cambridge University Press.

Dupré, John, and Maureen A. O'Malley. 2009. Varieties of living things: Life at the intersection of lineage and metabolism. *Philosophy and Theory in Biology.* http://hdl.handle.net/2027/spo.6959004.0001.003.

Dupré, John, and Daniel J. Nicholson. 2018. A manifesto for a processual philosophy of biology. In *Everything flows: Towards a processual philosophy of biology*, eds. Daniel J. Nicholson and John Dupré, 3–45. Oxford: Oxford University Press.

Ereshefsky, Marc. 1992. Eliminative pluralism. *Philosophy of Science* 59: 671–690.

Guay, Alexandre, and Thomas Pradeu. 2016. To be continued: The genidentity of physical and biological processes. In *Individuals across the sciences*, eds. Alexandre Guay and Thomas Pradeu, 317–347. Oxford: Oxford University Press.

Lewis, David K. 1986. *On the plurality of worlds.* Oxford: Blackwell.

Maynard Smith, John. 1966. Sympatric speciation. *American Naturalist* 100: 637–650.

Mayr, Ernst. 1942. *Systematics and the origin of species.* New York: Columbia University Press.

Meincke, Anne Sophie. The disappearance of change. Manuscript.

Methot, Pierre-Olivier, and Samuel Alizon. 2014. What is a pathogen? Toward a process view of host–parasite interactions. *Virulence* 5(8): 775–785.

Monk, Jonathan M., Pep Charusanti, Ramy K. Aziz, Joshua A. Lerman, Ned Premyodhin, Jeffrey D. Orth, et al. 2013. Genome-scale metabolic reconstructions of multiple *Escherichia coli* strains highlight strain-specific adaptations to nutritional environments. *Proceedings of the National Academy of Sciences, USA* 110(50): 20338–20343.

Nelson, Karen E., Rebecca A. Clayton, Steven R. Gill, et al. 1999. Evidence for lateral gene transfer between Archaea and Bacteria from genome sequence of *Thermotoga maritima. Nature* 399: 323–329.

Odling-Smee, F. John, Kevin N. Laland, and Marcus W. Feldman. 2003. *Niche construction: The neglected process in evolution.* Princeton: Princeton University Press.

Oyama, Susan, Paul E. Griffiths, and Russell D. Gray. 2001. *Cycles of contingency: Developmental systems and evolution.* Cambridge, MA: MIT Press.

Pradeu, Thomas. 2012. *The limits of the self: Immunology and biological identity.* New York: Oxford University Press.

Richerson Peter J., and Robert Boyd. 2005. *Not by genes alone: How culture transformed human evolution.* Chicago: University of Chicago Press

Robinson, Howard. 2014. Substance. *The Stanford encyclopedia of philosophy* (spring 2014 edition), ed. Edward N. Zalta. http://plato.stanford.edu/archives/spr2014/entries/substance/.

Seibt, Johanna. 2016. Process philosophy. *The Stanford encyclopedia of philosophy* (spring 2016 edition), ed. Edward N. Zalta. http://plato.stanford.edu/archives/spr2016/entries/process-philosophy/.

Waddington, Conrad H. 1957. *The strategy of the genes: A discussion of some aspects of theoretical biology.* London: Allen and Unwin.

Wedin, Mats, Heidi Döring, and Gunnar Gilenstam. 2004. Saprotrophy and lichenization as options for the same fungal species on different substrata: Environmental plasticity and fungal lifestyles in the Stictis–Conotrema complex. *New Phytologist* 164: 459–465.

Whitehead, Alfred N. 1929. *Process and reality. An essay in cosmology*. New York: Macmillan.
Wilson, Robert A. 1999. *Species: New interdisciplinary essays*. Cambridge, MA: MIT Press.
Zhaxybayeva, Olga, Kristen S. Swithers, Pascal Lapierre, Gregory P. Fournier, Derek M. Nickhart, Robert T. DeBoy, et al. 2009. On the chimeric nature, thermophilic origin, and phylogenetic placement of the Thermotogales. *Proceedings of the National Academy of Sciences USA* 106: 5865–5870.

{3}

Individuating Processes

John Pemberton

3.1 Introduction

Recent years have seen the development of a view that at least some entities that are widely taken to be individuals are processes, perhaps most notably biological organisms (See, e.g., Dupré 2013; Bapteste and Dupré 2013; Dupré, Chapter 2 this volume). But what are processes? How should we individuate them?

One way forward to address these questions would be to appeal to an existing metaphysical view of individuals and to seek to impose this on processes. But, as the brief sketch in Section 3.4.1 shows, metaphysics encompasses a divergent and complex range of view concerning what it is to be an individual. To choose any one of these views is to make a choice and commitment and, for many, this may seem too quick.

I propose proceeding in more modest steps. In the first step, I set aside metaphysics to explore the precise knowledge of processes achieved by practitioners and philosophers of science focusing on successful practice areas (e.g., biology, chemistry, engineering, physics) with the aim of distilling and articulating a prevailing view of processes adopted within these areas. In the second step, I sketch some of the accounts of individuals proposed by metaphysics. By juxtaposing these metaphysical views of individuals with the view of processes distilled from science, this chapter aims to identify the tensions between these perspectives and hence to cast light on the nature of processes and how they might be individuated.[1]

Section 3.2 undertakes this first step: distilling a characterization of processes employed in the sciences. A consideration of many examples of the careful and specific accounts of processes developed in the sciences suggests that processes

[1] This approach aims to follow the recommendations of Guay and Pradeu (2016: chapt. 1) and the long-standing principles of the Stanford School and is congruent with the program to develop a "*metaphysics of science*" (see, e.g., Mumford and Tugby 2013).

are taken to involve parts acting together (at each stage) to bring about the next stage of the process. In addition to how the process is at each stage (e.g., the spatial arrangement of parts at that stage), we may characterize how the process changes across stages and how the process acts to self-maintain across each stage; that is, an account of how it survives and hence exists (its existence is not taken to be just a brute fact). Moreover, many processes may often be intimately connected with larger processes of which they are a part and have roles in maintaining this larger contextual process, so that the nature of the context in which it occurs and the role(s) it plays within this larger context may often provide further useful characterizing criteria of a process. There is thus a rich range of criteria that may be used to characterize processes.

The metaphysical discussion of individuals which seems potentially most helpful here is not focused on processes directly, but on individual things (sometimes termed *substances* or *objects*).[2] In order to compare processes with things, I restrict (within this initial investigation) the focus to entities which are plausibly both processes, from the perspective of science, and things, from the perspective of metaphysics. As discussed in Section 3.3, this chapter supposes that at least some entities fall into this overlap and suggests that perhaps many may do (e.g., perhaps organisms, parts of organisms, atoms, molecules, bundles, artefacts).

Section 3.4.1 adumbrates the broad range of metaphysical views concerning what it is to be an individual (thing) in order to make explicit the divergence of views and the key themes involved. This divergence of view presents a challenge for undertaking the required comparison with processes. To facilitate progress on this comparison, I articulate a single simple exemplar of these metaphysical views: an individual is taken to satisfy E. J. Lowe's *identity* and *unity conditions* (Lowe 2016: 59) and to be wholly characterized by a list of (admissible) properties. This simple exemplar view captures central aspects of many popular metaphysical account of individuals, such as bundle accounts and bare particular accounts. Comparison of this exemplar view with the view of processes distilled from science (in the context of the key metaphysical themes concerning what it is to be an individual, which have been identified) is a first achievable step undertaken by this chapter. This initial comparison, undertaken in Section 3.4, identifies interesting tensions between these perspectives.

Section 3.5 draws some provisional conclusions: until tensions with the metaphysical account of individuals are better resolved, it seems reasonable to choose to use methods for individuating processes that have respect for the rich account of processes reflected in the practices of science. Respecting such a processual view may license recognizing both processes and certain of their parts (e.g., organism/

[2] I have not found salient material on individuation within the process metaphysics literature (see, e.g., works by Whitehead, Rescher, Seibt)—I do not here seek to employ insights from this domain.

cell, biofilm/bacterium, species/organism) as being valid unities (e.g., sufficiently robust processes).

3.2 Processes in Science

A review of textbooks, articles, and research papers across the sciences reveals the widespread use of diagrams that represent the spatial layout of the parts of things or processes which are of interest. Often these diagrams feature arrows indicating how the layout changes over time or perhaps a sequence of such diagrams that denote the stages of such change. Typically, accompanying text describes this changing of the arrangement and the properties of the parts through time and how it is that the parts bring about this changing.

As an example, Figure 3.1[3] shows the layout of some of the key parts involved in synaptic transmission, which is similar to diagrams used in many biology textbooks, such as Bear et al. (Bear, Connors, and Paradiso 2007: figure 2.15), whose accompanying text explains:

> When a nerve impulse arrives in the presynaptic axon terminal, neurotransmitter molecules are released from synaptic vesicles into the synaptic cleft. Neurotransmitter then binds to specific receptor proteins causing the generation of electrical or chemical signals in the post-synaptic cell. (Bear et al. 2007: 39)

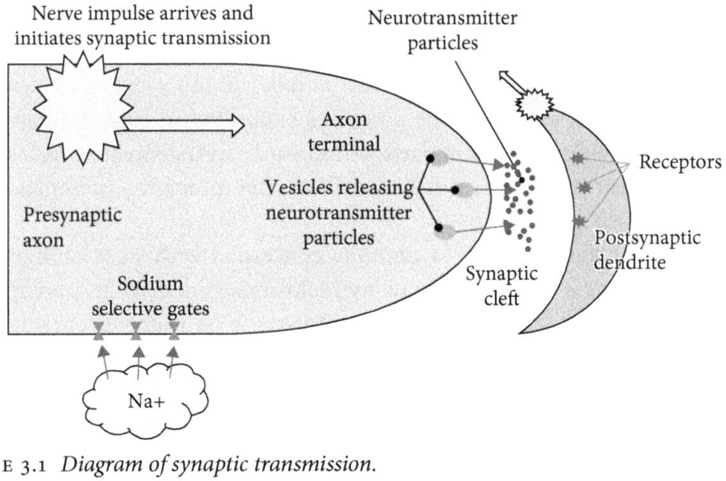

FIGURE 3.1 *Diagram of synaptic transmission.*

[3] All figures in this paper are by the author.

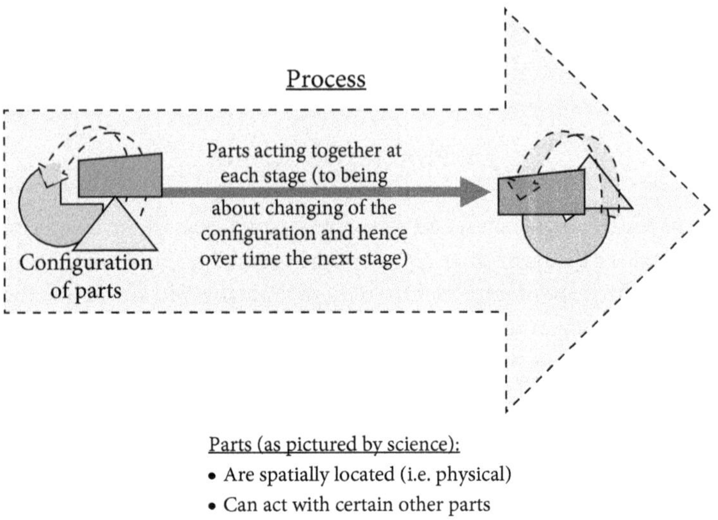

FIGURE 3.2 *A process.*

Other works focus on certain of the parts shown in Figure 3.1 and depict diagrammatically the acting together of their more detailed parts—Südhof and Rizo, for example, use a series of spatial diagrams to explicate the vesicles and the processes in which they engage (Südhof and Rizo 2011: figures 1–5). Other texts focus on numerical measurements of aspects of such processes; for example, Eijkelkamp et al. present alongside a spatial diagram representing the operation of a sodium gate a graph depicting the magnitude of the potential across the gate over time during the passage of a nerve impulse (i.e., an action potential; Eijkelkamp et al. 2012: figure 1).

This example is representative of a vast number of other examples across the sciences. A common feature of the processes explicated in such diagram-based accounts is the acting together of parts within some arrangement at each stage to bring about a next stage—this is consistent with the minimal representation of a process shown in Figure 3.2.

The view that much science is centrally concerned with such configurations of interacting parts is supported by many philosophers of science, perhaps most notably the *new mechanists*, who take *mechanisms* to be widely occurring.[4] Such mechanisms are taken to involve parts that are spatially organized acting together to exhibit patterns of change (sometimes referred to as *activities* or *behaviors* of the mechanism; Machamer, Darden, and Craver 2000; Glennan 2002; Bechtel and Abrahamsen 2005; Illari and Williamson 2012). The views of such philosophers of

[4] Some new mechanists take mechanisms to be widely occurring throughout the sciences (e.g., Glennan 2017), others focus on narrower areas and take mechanisms to occur widely within their areas of focus (e.g., Darden 2006; Craver 2007).

science have particular salience as they are typically based on careful and extensive observation of examples from the practice of science. These examples, which generally feature processes in the sense of Figure 3.2, are taken to exhibit patterns of change associated with a mechanism and are often explicitly recognized as processes. [5]

I take Figure 3.2 as representing at least many of the processes studied by science, and I adopt it as the starting point for the scientific picture of processes.

3.2.1 CHARACTERIZING PROCESSES

This section builds on the characterization of a process in Figure 3.2 by identifying the various characteristics that processes are commonly taken to have by those who make use of processes across the various fields of scientific practice. The characterizing criteria identified are not intended as a definitive list but as a reasonable starting point which others might choose to amend according to their differing views or purposes.

Figure 3.2 makes clear that the processes posited in the sciences are time-extended: they involve a changing (or perhaps stasis) of a configuration over time. The first three of the following criteria follow straightforwardly from Figure 3.2.

3.2.1.1 Criterion A: The Nature of the Parts at Each Stage

The elements of the spatial diagrams used in science, noted earlier, are typically shapes or pictures representing the parts of the process—that is, the spatially located entities which act together with the other parts at each stage to give rise to the process. The diagram typically represents some stage of the process (although arrows or other pictorial devices may be used to suggest the pattern of change over time)—and hence the parts which are present and perhaps acting at that stage. The nature of the parts varies from science to science: in biology these may be neurons, vesicles, neurotransmitter particles, as in Figure 3.1. In astronomy, familiar spatial diagrams include those with force arrows indicating the mutual gravitational attracting (acting together) of stars (planets) and their planets (moons) and the resultant elliptic orbits—here parts include stars, planets, and moons. In chemistry, parts may include electrons, protons, molecules; in engineering, perhaps cogs, spark plugs, cylinders. In each case the parts are typically entities familiar to the science concerned.

---

[5] Although many accounts of mechanism do explicitly embrace processual characteristics, they fall short of providing an adequate account of the link between the arrangement of parts and the change which occurs; e.g., in the Machamer et al. account the link between *entities* and *activities* (see Pemberton 2011). Cartwright (recently with Pemberton) by embracing powers does provide a basis for such a link within *nomological machines*, which may be understood as a version of mechanisms (Cartwright and Pemberton 2012; Pemberton and Cartwright 2014; Chen 2017).

3.2.1.2 Criterion B: The Spatial Arrangement of the Parts at Each Stage

By choosing the duration of a salient stage as sufficiently brief, we may typically be able to limit the change of the configuration during that stage so as to be able to characterize the arrangement of parts (i.e., their spatial locations, orientations, and velocities) at that stage using a single snapshot spatial diagram. Sometimes a diagram may show schematically the spatial layout of differing parts at differing times, perhaps using arrows to indicate the time sequencing (as in Figure 3.1). In engineering, the spatial arrangement of parts may typically be specified by use of a blueprint (which may also generally specify the nature the parts), which may be used as the basis for constructing the configuration and hence instantiating the relevant process. Similarly, for certain laboratory experiments, the initial spatial layout of pieces of laboratory equipment and entities under study may be represented in a diagram—when the layout is instantiated, the aim may typically be to produce some repeatable process involving the entity under study (i.e., a run of the experiment). In other sciences (such as astronomy, biology, geology), the processes in focus are typically found in nature—the diagrams may show the spatial layout of typical processes of the relevant kind at salient stages.

3.2.1.3 Criterion C: The Nature of the Change a Process Exhibits Across Certain Stages

The swinging of a pendulum might be described as U-shaped or perhaps as exhibiting a stopped–slow–fast–slow–stopped pattern. Other processes may be smooth, intermittent, or explosive or perhaps give rise to spherical, circular, or linear spatial patterns of change. In Figure 3.1, the rough pattern of change over time is indicated by the arrows and explicated further by the text.

We may note that the temporally extended nature of processes (represented in Figure 3.2) leads to two differing senses in which the term "process" is often used. Sometimes the changing that occurs may be in focus, such as the swinging of a pendulum, the beating of a heart, a run of some experimental set-up in the laboratory. This changing occurs over time, and this sense of the term "process" might be thought of as somewhat abstract. Alternatively, we might have in focus the configuration of parts acting at each stage (e.g., the swinging pendulum, the beating heart, the acting parts of the experimental set-up).[6] I take a process to embrace both these notions—this is reflected in the recognition of criteria A, B, and C.

3.2.1.4 Criterion D: Beginning-to-End/Homogeneous Ongoing/Heterogeneous Ongoing

Some types of processes, *beginning-to-end* processes say, run through from a certain type of starting configuration to a certain type of ending configuration via a series of well-defined stages—perhaps over some rough given timescale. The

[6] In Carl Craver's terms, "the ψ-ing S" rather than "the ψ-ing of S" (Craver 2007: 7).

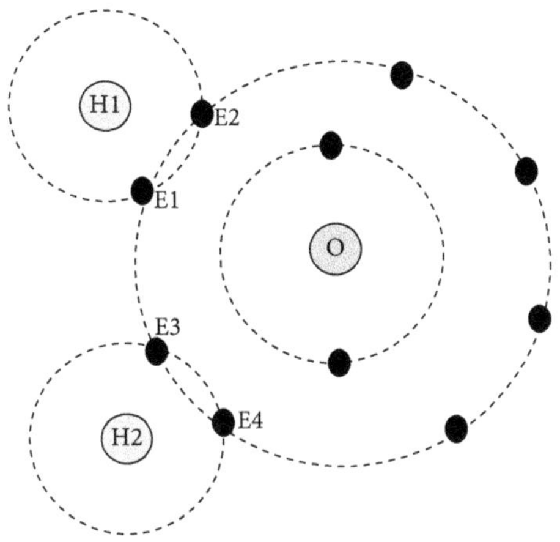

FIGURE 3.3 *A water molecule: an example of a process.*

firing of a neuron, for example, may start with neurotransmitter particles binding with ligand-gated neuroreceptors, hence opening certain ion channels. The movement of ions across the neuronal membrane which this licences may result in an increased potential across the membrane and hence the opening of nearby voltage-gated ion channels, which may in turn lead to a cascade of voltage-gated ion channels opening along the neuron and then to the opening of vesicles which release neurotransmitter particles.

Again, when a coin is inserted in a drink vending machine, a process of well-defined stages may ensue which results in a drink in the output bin.

Beginning-to-end processes exhibit heterogeneity: perhaps differing parts act at differing stages, or perhaps parts act in differing ways at differing stages. The ligand-gated neuroreceptors feature at the start of the process of the neuron firing, whereas voltage-gated ion channels and vesicles feature later on. The coin inserted in the vending machine perhaps depresses a lever which turns various cogs leading on to the acting of different parts which release the cup and liquid.

Other processes (although they do generally have a beginning and an end) may be thought of as ongoing. This is typically the case with processes which exhibit a high degree of homogeneity. As an example of such a process, consider a water molecule.

On the standard scientific account, the parts of the water molecule are protons, an oxygen nucleus, and electrons arranged along the lines shown in Figure 3.3 (see, e.g., Keller 2013: cover diagram). Within this configuration, the parts exert (and experience) basic forces of physics on (due to) other parts—perhaps most notably electrostatic forces associated with charge.[7] This acting together of the

[7] This chapter does not attempt to address the complex and divergent views prevalent within the metaphysics of quantum mechanics—the classical view suffices for our purposes here.

parts[8] in this way may be understood as a process which continues through time in a way which is self-sustaining. This process seems to exhibit a high degree of homogeneity: the same parts are acting at each stage and acting in the same sort of way (although perhaps the strength and direction of the basic forces may vary over time).

Bundles, as characterized by Aristotle for example; e.g., *Metaphysics V.6.*, may also be taken to be examples of homogeneous ongoing processes. Consider, for example, a bundle of sticks bound by some binding such as a string. When roughly static, the binding gives rise to forces which hold the bundle together, whereas the sticks play their role by pushing against each other and the binding to form a stable physical unit—the same parts and roughly the same forces act homogeneously at each stage. When external forces act—for example, when the bundle of sticks is kicked—the binding forces within the bundle react: they increase to prevent the sticks flying apart, for example.[9] In this reactive way, bundles are effective at self-maintaining and may be viewed as a common, if somewhat prosaic, form of process. At the micro level, bindings may be atomic or molecular bonds, so that the bundles may then be graphite rods, diamonds, lumps of copper, rocks, drops of water, and the like. At the macro level, forms of binding may include tying, gluing, nailing, bolting, interlocking of congruent shapes, containing by a wall, or linking by tissue or fiber. Examples of macro bundles may include bundles of sticks, packets of biscuits, broken (or perhaps nonfunctioning) mechanisms, and dead organisms.

By contrast, consider an example from engineering: a process which is exhibited by an artifact such as a motorbike. This process transitions through differing stages involving the acting of differing parts at each stage: the sparking of the cylinder by the spark plug, the exploding of hydrocarbons, the driving down of the piston, the opening of the exhaust valve, the emitting of the exhaust gasses, the closing of the exhaust valve, the moving upward of the piston, and the admission of new hydrocarbons. In this sense, the process exhibits a high degree of heterogeneity. By ensuring that the process arrives at roughly the same configuration as it started, the process is rendered cyclical and hence ongoing—a *heterogeneous ongoing process* we may say.

Organisms may involve a processual transition through developmental stages (e.g., seed, young and mature organism) where each stage exhibits multiple life processes (e.g., mitosis, respiration, digestion, photosynthesis, self-repair, etc.). The view has therefore developed that organisms are processes—if so, then these processes also exhibit heterogeneity but in a vastly more complex way than machines. The differing parts of the organism may act in differing ways at differing

[8] The exerting of a force and the experiencing of a force are how the actings of the parts (e.g., masses, charged bodies) may be understood in this context.

[9] Of course, a sufficiently hard kick may cause the bindings to break or sticks to escape the bindings, so that the bundle ceases to be.

stages to self-maintain the organism and perform desired functions. Many of the subprocesses of the organism (e.g., the firing of a neuron, the contraction of a muscle, the digestion of a meal) might be taken to have a beginning-to-end character. But the appropriate coordination of such subprocesses ensures that the organism as a whole is ongoing—it is another form of heterogeneous ongoing process.

This threefold classification of processes into beginning-to-end/homogeneous ongoing/heterogeneous ongoing might be further refined (e.g., to capture distinctions between mechanical and organic heterogeneous ongoing processes) or developed—it is intended as indicating a starting point for a classificatory criterion.

3.2.1.5 Criterion E: Originating-Things

Sometimes we may characterize a process according to the things which came together to give rise to that process, the *originating-things* say. Originating-things may or may not remain as identifiable unities within the new process.

For example, we might take the cogs, cylinders, and spark plugs used to build a motorbike as the originating-things of that bike—these originating-things do typically remain as identifiable parts of the motorbike. In biology, we might take the originating-things of a certain mule to be the gametes from the donkey and the horse which fused to form the zygote which was the earliest stage of this mule. The story of these gametes after their fusion is complex—it does not seem that these originating-things remain as identifiable unities within the mule longer term. Certain eggs may be among the originating-things of a cake—these originating-things do not remain as unities within the cake.

3.2.1.6 Criterion F: The Nature of the Things/Parts that Join or Leave the Process at Each Stage

Things may both enter and leave a process at each stage (as illustrated in Figure 3.4). For example, hydrocarbon fuel (motorbike) or nutrients (organism) may enter a process, whereas exhaust gasses or urea may leave a process. Processes are often characterized by the nature of the things which join or leave; for example, organisms may be characterized as herbivores or carnivores (according to the nature of food taken in by the organism), and similarly, engines as diesel or petrol—or perhaps according to the rate of emission of certain polluting gases.

3.2.1.7 Criterion G: Residual-Things

When a process comes to an end, there may typically be some things which are left, *residual-things* say. Following the death of an organism (i.e., the ending of its life process), a corpse remains (initially at least). When an experimental setup is run in the laboratory, perhaps a chemical experiment, the process which occurs may give rise to characteristic products and residues.

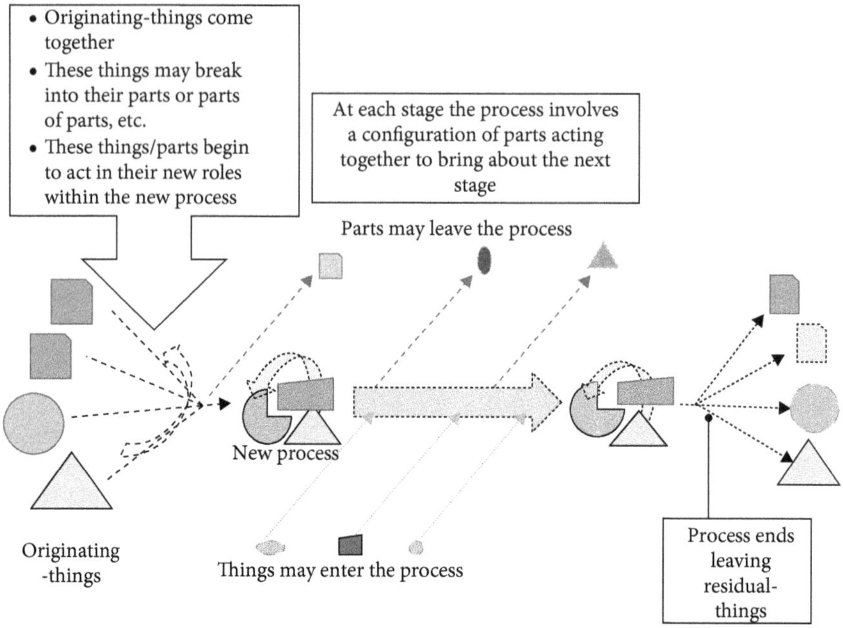

FIGURE 3.4 *Originating-things come together to form a new process, and then things may join and leave the process.*

3.2.1.8 Criterion H: How the Process Self-Maintains at Each Stage

The widespread supposition of science is not that the sequence of stage configurations of a process is inexplicable, but rather that the nature of the parts and their configuration at each stage brings about the next stage. (Where the process exhibits a single ongoing configuration, it may be supposed that the acting of the parts at each stage maintains that configuration.) Typically, an account of how the parts act together at each stage to bring about changing, and hence the next stage of the process, is available for each of the spatial diagrams of the parts illustrated and discussed earlier. For example, the approaching nerve impulse (action potential) of the neuron in Figure 3.1 opens sodium-selective gates, allowing movement of sodium ions and hence the local changing of potential difference across the wall of the neuron, which underwrites the movement of the action potential along this part of the neuron. At the axon terminal, the action potential opens the vesicles, releasing neurotransmitter particles which cross the synaptic cleft and dock with receptors, and so on. It is supposed that a detailed story is available (even if not yet discovered) about the acting of parts at each stage and how these lead on to the next stage.

How a process self-maintains across each stage may often be a useful criterion for classification. In the case of homogeneous processes, the process may self-maintain in largely the same way at each stage—a hydrogen atom, for example, self-maintains at each stage through the acting together of a proton and an electron (i.e., the mutual exertion of basic forces of physics). A bundle self-maintains through binding forces which are similar at each stage (although these forces may increase reactively when the bundle is stressed).

For heterogeneous processes, the account may typically vary across stages. A fire self-maintains at each stage by the oxidation of hydrocarbons releasing energy to maintain a high localized temperature and hence the release of more flammable gasses—but the parts (e.g., firelighter, kindling, logs, coal) acting at each stage may vary. For an engineered machine which transitions through a dynamic locus of configurations before perhaps returning (more or less exactly) to some *starting*[10] configuration (such as motorbikes, cisterns, toasters, pendulums), an adequate account might reference how the process self-maintains through each different stage of the locus of configurations.

3.2.1.9 Criterion I: Role in a Larger Process

A key (which might be viewed as a shaped-bundle–type process) can engage in a process of lock opening when it is in a suitable configuration with a lock and a person; for example, it is inserted in the lock and rotated by the person (a *person-turning-key-in-lock process,* say). We might ascribe to the key (qua shaped-bundle process) the role of opening the lock. We might also ascribe this role to the person-turning-key-in-lock process but in a different sense: this process may end with the lock being open.

In the biological context, many biochemical structures, such as proteins, have key-like (or lock-like) characteristics.[11] A protein that is an enzyme, for example, may have a role as catalyst in some particular biochemical reaction—typically, the reactants bind with the protein's active site whose spatial molecular structure and charge pattern lowers the energy requirement for the reaction (see, e.g., Copeland 2000, especially 1.4 and 1.5). Other possible roles for proteins include transporter, inhibitor, and binding agent for specific molecules or reactions. The role, or function, of a protein has been embraced as a crucial taxonomic criterion within extensive recent work in this area (see e.g., Wu et al. 2004).

Many other parts of organisms (and the processes in which such parts engage) may also be classified according to their role. For example, a heart pumps blood, a kidney cleans blood. In engineering, a cog in the context of a bicycle might be classified as a *drive sprocket*—that is, it has the role of transmitting drive via a chain.

In some cases, the role performed by a part is central to the identification of that part and the salient process in which it engages. When a neuron is transmitting an action potential, certain configurations of polypeptide strings located in and

---

[10] As the locus of configurations exhibits a cycle, we might choose to ascribe any point in the cycle as the *starting* configuration, but where such mechanisms require a human action (e.g., pulling a chain, pushing a button, depressing a lever) to initiate a cycle, it is most natural to choose the rest configuration before this human action as the *starting* configuration.

[11] This is not to advocate simplistic key-in-lock theories of, e.g., enzyme operation.

around the walls of the neuron act together to open or close a pore in the lining of the neuron, hence allowing or inhibiting the passage of sodium ions across the neuron wall. It is this role within the higher level process of neuron firing that helps to license recognition of the unity of this process (i.e., the acting together of this configuration of polypeptide strings), which may be referred to as the opening/closing of a sodium-selective gate, and hence the unity of the sodium-selective gate itself.

3.2.1.10 Criterion J: The Context in Which the Process Occurs

Processes may also be classified according to the context in which they occur (e.g., *stellar nucleosynthesis* as occurring within a star, certain biochemical process as intracellular, and action potential transmission as neuronal).

One reason a process may be classified according to such a context is that the process requires a configuration of certain types of parts, and such configurations only obtain in certain restricted types of context. Another reason is that a process is recognized as performing a role in a larger context (as in Criterion I)—the larger context must then obtain: for a heart to pump blood it must be located within a suitable circulatory system. The (pumping) heart may be classified as part of this circulatory system.

On other occasions, certain features of a context may facilitate a process in other ways. For example, the operation of many proteins within intracellular processes depends critically on their being correctly folded and spatially configured (and not merely having a certain topological structure). The achievement of such suitable folding and spatial configuration may rely upon *chaperone* proteins, so that the context must feature such proteins for the relevant processes to occur.[12]

3.2.1.11 Criterion K: The *Phase-Stages* of a Process

As noted in Criterion D, processes vary considerably in their degree of heterogeneity. Some processes exhibit such a high degree of heterogeneity across sequential stages that it may be unclear whether they should be treated as one process or a series of sequential processes. For example, an astronomic feature may have the stages of diffuse nebula, main-sequence star, red giant, and white dwarf; an organism may have stages of egg, caterpillar, chrysalis, and butterfly. I dub such radically differing stages *phase-stages*. Where a process is taken to exhibit such phase-stages, the nature of these phase-stages and how they lead on from each other may provide a further useful classificatory criterion.

This consideration of processes that are referenced within science identifies a rich framework of characterizing criteria. Which of these types of criteria are useful in any given case will depend on the nature of the process in focus. Typically, a number of criteria of differing types may be helpful in establishing a suitable

---

[12] See Dupré (2010) for a discussion of the importance of context for biological entities.

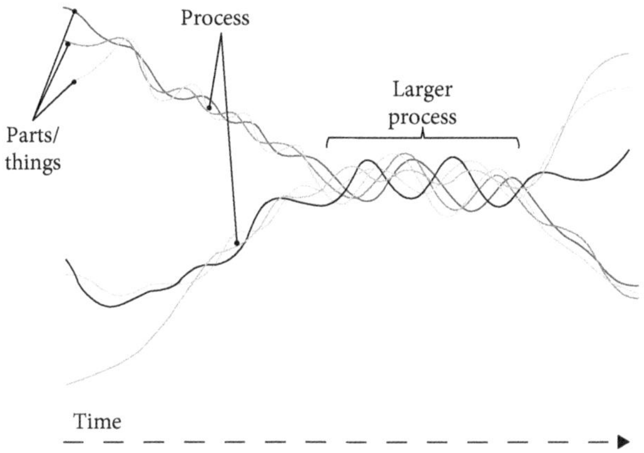

FIGURE 3.5 *A picture of processes drawn from science.*

characterization. Where a process may be taken to involve phase-stages, it may be appropriate to characterize each phase-stage using criteria of types A through J.

Processes, on this characterization, involve the acting together of parts at each stage. As noted in the discussion of Figure 3.4, the coming in to being of a process may involve the coming together of things which then act together as parts in the new process, as in the construction of a motorbike. The vastly more complex processes studied in biology may come into being via the union of gametes, themselves vastly complex processes, and then prolonged development stages. Astronomic processes, such as stars, may involve the incremental coming together of massive bodies in a way which is different again.

Figure 3.5 develops Figure 3.4 to illustrate this pictorially: things, perhaps themselves processes, may come together to form processes (represented by intertwining strands), which may in turn combine (perhaps via incremental development) to form larger processes. Each of the parts (represented in the picture by a single strand) may itself be a process of smaller parts (an intertwining of smaller strands). As discussed earlier, originating things may or may not survive as unities within a new larger process.

3.3 Individuating Processes: Preliminary Comments

An initial thought might be that careful characterization of processes along the lines just outlined, together with a specification of time and place, might be sufficient for individuating processes. In the case of some well-understood sufficiently macroscopic processes, such as a beating heart or firing neuron, this may often be so. For example, we may be able to pick out and keep track of a particular neuron and identify particular processes (e.g., neuron firings) as those processes of neuron-firing type which occur in this particular neuron at certain particular times. Each firing might be recorded as having a certain action potential intensity and duration, and a plurality of firings might then allow, for example, for the typical range of intensities and durations of firings of this neuron to be estimated.

In other cases, individuation of processes is more challenging: If we take a leaf mold to be processual, is it one process or many processes? Is a certain butterfly the same process as its chrysalis? Is the forest fire the same fire as the flame of Harry's match which lit the cigarette whose glowing butt caused the fire?

It is consideration of these more challenging questions that suggests enlisting the expertise of metaphysics concerning the nature of individuals and individuation. Metaphysics, however, is not typically focused on the individuation of processes (at least not explicitly), but its more common focus is individual substances or things. In order to bring to bear the insights of metaphysics concerning the nature of individual things, I propose to focus in this initial investigation on entities that may be viewed by science as processes and that are widely considered by metaphysics to be individual things—a plausible preliminary view of this overlap is indicated in Figure 3.6.

If we take a string and attach one end to a bob and the other end to the top of a bench stand, and then set the bob swinging, we may have a simple pendulum, which we may take to be a process. If we now lie the stand-string-bob on its side on the bench it no longer swings—perhaps this apparatus is now a different type of process, maybe an organized bundle. If these are two differing processes, are they the same thing?

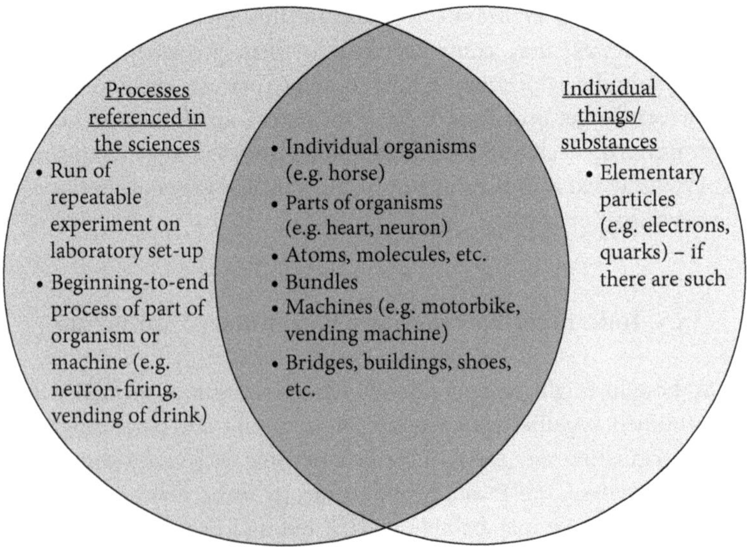

FIGURE 3.6 *Entities that may be viewed by science as processes and by metaphysics as individual things—one possible view.*

Similarly, we may ask: Is a beating heart understood as the process of the parts of the heart acting together over time, the same as a heart taken to be a thing, perhaps some compresent plurality of properties? Aristotle, for example, draws a clear ontological distinction between a heart acting in its bodily context and a disconnected heart, viewed as only homonymously a heart (Aristotle *Metaphysics Z*).

These are difficult questions that seem highly relevant to the topic of individuating processes. The strategy of this chapter, as indicated in Section 3.1, is not to address such questions directly from metaphysics, but rather to juxtapose a scientific view of processes and a metaphysical view of things as the basis for an initial exploration of the tensions between these views (this exploration is undertaken in Section 3.4). It is important to be clear that Figure 3.6 is not presupposing an answer to such questions but is merely identifying entities that are plausibly viewed by science as processes and by metaphysics as individual things.

It is a supposition of this chapter that at least some entities fall into this overlap—individual organisms, such as horses, are perhaps the most cited example. We noted in Section 3.2 many processes referred to by the sciences and indicated the sense in which they may be taken to be processes. Many of these are also often considered to be individual things, such as parts of animals (e.g., hearts), atoms, molecules, bundles, fires.

Some processes are less plausibly things. For example, although an experimental setup in the laboratory running from some starting configuration through to termination may generally be considered a well-defined process, it might not generally be taken to be a thing. Similarly, an action potential (neuron-firing), perhaps a beginning-to-end process involving certain parts of a neuron cell, does not seem like a thing. Rather the neuron cell as a whole, a coordinated and complex multitude of processes that perhaps self-maintains in a way more similar to an organism, might more commonly be viewed as a thing. Machines seem similar in this respect to neurons: it is the vending machine which we might typically take to be a thing, rather than the beginning-to-end process of vending a drink.

Many seemingly static artifacts which are taken to be things (e.g., shoes, bridges, buildings) self-maintain their organized structure under conditions of use in a way that is similar to, but more precise than, bundles. Tall buildings, for example, must be designed to sway in the wind and return to equilibrium, bridges must contort as they accept a load and then return to equilibrium as the load is removed. Each of these artifacts involves the self-maintaining acting together of its parts and might reasonably be considered a process.

Are there things that are clearly not processes? If physics is right that there are some elementary particles—for example, the particles of the Standard Model, such as electrons, muons, and quarks—then it seems they do not involve the acting together of parts and hence they would be things which are not processes (in the sense explored here).

Although this discussion suggests that it is plausible that many (but not all) processes and many (but not all) things may fall into the overlap (as indicated in

Figure 3.6), opinions on this may differ. The exact extent of the overlap is not crucial to the argument in the next section (although those who take it to be much narrower may need to disregard some of the examples).

3.4 Comparing Processes in Science with Individuals in Metaphysics

There are many widely differing views of individuals within metaphysics—it is not plausible to compare each of these against the picture of processes sketched in Section 3.2. This chapter therefore adopts the following pragmatic approach. First, some of the leading views of individuals in metaphysics are briefly recapitulated to make explicit a rough adumbration of the range of views concerned and the key themes involved. Second, a single simplified exemplar metaphysical view of individuals from within this range is outlined. Although this is just one view, it reflects a consensus of many leading contemporary metaphysical views of individuals. Comparing this exemplar metaphysical view with the view of processes in science (focusing on processes which are plausibly individuals) provides an initial indication of the tensions between these perspectives. This is, of course, merely a first step to determining whether and how metaphysics may help with the individuation of processes achievable within this brief chapter.

3.4.1 INDIVIDUALS IN METAPHYSICS: THE DIVERGENCE AND COMPLEXITY OF VIEWS

Aristotle supposes that metaphysics is the study of *being qua being* (*Metaphysics*, Book Γ) and, at the very center of this project, is the explication of discrete individuals, *substances*, which have ontological priority. Aristotle rejects Plato's account of individuals as the co-instantiation of universals at spatiotemporal locations due, inter alia, to its failure to solve the *nature-feature problem* (i.e., to distinguish what a thing is from its accidental properties; see Scaltsas 1994, especially chapt. 3)—and hence to provide an account of continuity through change. Individual organisms, such as horses or humans, are identified by Aristotle as paradigm examples of substances—and these are understood as involving both matter and form. But to achieve ontological priority, the substantial whole must be an unqualified unity of its matter and form—establishing this unity presents Aristotle with a challenge. His answer is the hylomorphism, according to which matter is reidentified according to the substantial form so as to become unified as a whole that has no actual parts.[13]

[13] See Marmodoro (2013) for a lucid account and further references to the recent debate concerning Aristotle's hylomorphism.

Aristotle's project to identify substances is taken up by many generations of metaphysicians who are often cautious of Aristotelian forms. Medieval philosophers, perhaps most notably Duns Scotus, posit *haecceities*: nonqualitative properties which are responsible for individuation and identity (Duns Scotus 1639). Leibniz posits *monads* as fundamental particulars which exhibit discrete individuality (see, e.g., McCullough 1996). A long tradition posits bare particulars as the basis of ontological individuality (see, e.g., Sider 2006).

More recently, linguistic analytic philosophy has a differing motivation for positing discrete individuals, as Ruth Barcan Marcus explains:

> The notion of an individual object or thing is an indispensable primitive for theories of meaning grounded in standard model theoretic semantics. One begins with a domain of individuals, and there are no prima facie constraints as to what counts as an individual except [each] must be distinct from every other and identical to itself. (Marcus 1975: 39)

This logico-linguistic concept of individual perhaps finds its most well-known expression in Quine's dictum, often expressed as "*to be is to be the value of a bound variable*" (see, e.g., Quine 1948). The notion of individuality in play here does not appeal to any ontological criteria for demarcating individuals, such as spatial continuity, boundedness, or other physical criteria which may often be salient within scientific contexts.

Central to a consideration of metaphysical accounts of individuals is their differing motivations for individuation:

1. One focus is picking out individuals at a particular time to separately identify one individual (Tom, say) from another (Dick, say). An individual that can be picked out as distinct from a sibling in this way satisfies what Lowe calls the *identity requirement* on individuality (Lowe 2016: 59). An electron within an entangled pairing may be an example of a putative individual which fails this requirement. It is often suggested (controversially) that Aristotle adopts matter as a principle of individuation: Tom is then distinct from Dick on account of differing constituent matter. Spatiotemporal location is another approach proposed to achieve such individuation, notably in the neo-platonic or Humean traditions.
2. Another issue is picking out the same individual (or different individuals) at differing times. Now it seems that matter is problematic as a principle of individuation: the matter of organisms (the leading contenders for substances) may change over time. For macroscopic individuals, in biology for example, spatiotemporal continuity may provide a practical criterion for resolving this challenge, but it is less

clear that this will do for the sorts of particles posited by contemporary physics.

3. A further important focus is determining whether a putative plurality of parts obtaining at some time constitute a single whole—that is, a single individual constituted by the parts—or not (van Inwagen 1990). This requirement to be one entity is what Lowe terms the *unity requirement* on individuality (Lowe 2016: 59). Parts are held to present a challenge for many metaphysical accounts of individuals as they are often deemed to threaten the unity of the whole. Aristotle, for example, supposes that there can be no actual parts within a substantial whole, only *potential* parts.

Even this brief sketch of a handful of views makes clear that there are divergent metaphysical opinions on what it is to be an individual and that picking out a single metaphysical view involves a choice and commitment.

3.4.2 EXEMPLAR METAPHYSICAL VIEW OF INDIVIDUALS

The exemplar view supposes that an individual satisfies E. J. Lowe's *unity* and *identity* requirements for individuality and that individuals can be wholly characterized by a list of admissible properties at each time (so that an individual may perhaps be taken to be a series of such property-defined entities at successive times). This exemplar view captures central aspects of metaphysical accounts according to which individuals are co-instantiations of platonic or other universals, bundles of properties, bare particulars with properties, collections of tropes, or co-locations of powers.

Figure 3.7 pictures this exemplar view: each box, which is wholly characterized by a list of properties, represents an individual at some time. The horizontal dashed arrows represent the passage through time of these individuals. The boxes are pictured as separate and distinct—this reflects the individuals satisfying Lowe's identity requirement. No box has other boxes as parts—this reflects the individuals satisfying Lowe's unity requirement. It is supposed that the context of an individual can be represented by external relations (e.g., spatial or causal relations): these are represented by the double-headed arrows.

3.4.3 EXEMPLAR METAPHYSICAL VIEW OF INDIVIDUALS VERSUS PROCESSES FOUND IN SCIENCE

How well can this exemplar metaphysical view of individuals capture processes found in science, including the range of characterizing criteria of such processes identified in Section 3.2? As outlined in Section 3.3, to facilitate consideration of this question, the focus is on entities which might be viewed both as processes and as individual things.

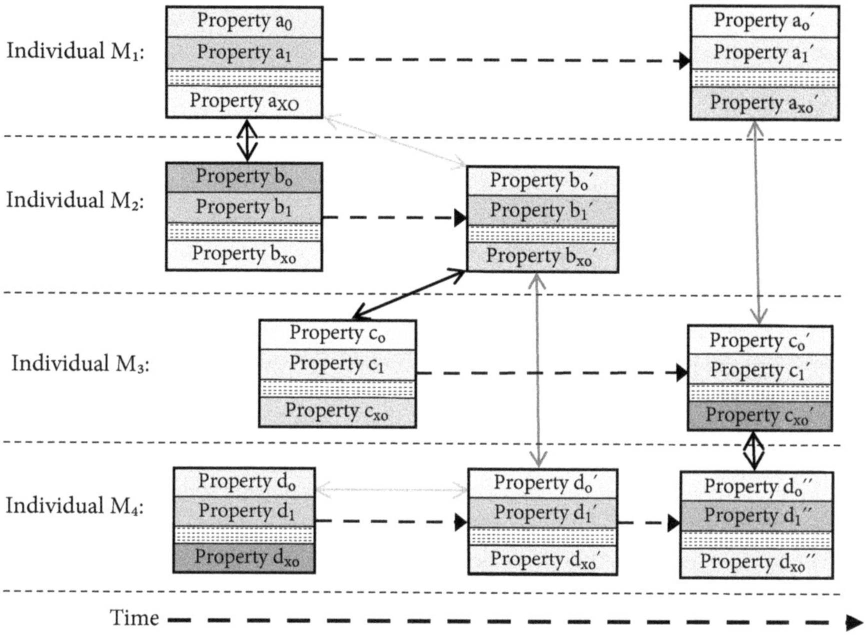

FIGURE 3.7 *A picture of the exemplar metaphysical view of individuals.*

The exemplar view seems prima facie well-placed to capture how a process is at each stage. Perhaps the having of a part of some type (Criterion A) and/or the obtaining of a spatial arrangement of these parts (Criterion B) may be taken to be admissible properties. Lewis, however, rejects such *structural properties* (Lewis 1986), and although Armstrong does accept structural properties in a limited sense (Armstrong 1986), his account of such properties cannot capture a distinction between butane and isobutane, for example, which are both C^4H^{10}. So, even in capturing the characteristics of a single stage of a process, the exemplar view faces challenges.

How might the exemplar view capture the nature of the change that a process exhibits across certain stages (Criterion C)? For processes that are not fully homogeneous (Criterion D), doing so will typically require, inter alia, positing a property which captures changing from one state to another state. Consider, for example, a neuron (which may be viewed from a process perspective as a heterogeneous ongoing process) that at some stage has certain parts acting to transmit an action potential. Viewed as an individual thing, it seems we would want to ascribe to this neuron some such property as *is-firing*. The neuron firing entails, we may suppose, the receipt of neurotransmitter particles (state A, say) initiating a process which features as a later stage the release of further neurotransmitter particles (state B, say). The putative property *is-firing* then has the form *is-changing-from-state-A-to-state-B* (i.e., *is-firing* entails the neuron has the property of being about to release neurotransmitter particles).[14] But such a property does not obtain before or after such a change, nor can it apply during the change because the change is then incomplete and may not be completed—it seems there is no time at which such a property might obtain. Perhaps we could make do with the less specific property: *being-such-as-to-change-from-state-A-to-state-B.* But here it seems that how the individual changes across stages is ontologically prior to the being of the individual at each stage (at which this putative property obtains), and this might be deemed by some as incompatible with the notion of an individual on their version of the exemplar view.[15] Another approach might be to associate an individual with its characteristic processual behavior via a power; for example, an individual is ascribed the power *can-change-from-state-A-to-state-B.* Of course, an adoption of powers would be anathema to many metaphysicians sympathetic to exemplar-like views. In any case, admitting powers may not resolve the problem: if the manifestation of the power is a process, then it seems that processes must be admitted to the ontology after all; and if the manifestation is not a process, then it is not clear that the characteristic of exhibiting processual change has been captured. It seems controversial whether the exemplar view can capture how a process changes across stages.

Kripke, among others, proposes that certain properties associated with origin are admissible properties of an individual (Kripke 1980). Perhaps, then, the association of an individual with certain originating-things (Criterion E) might count as admissible properties on the exemplar view. But Kripke's view is controversial. And further liberalism would be required to license the inclusion of putative properties that capture the nature of things that leave or join during a process (Criterion F) or that remain at the end of a process (Criterion G).

Recognizing an entity as a process licenses inclusion within our characterization criteria of how it self-maintains at each stage. By contrast, the representation of change on the exemplar view would seem to be limited to the obtaining of one property at one time and then the obtaining of another different property at a later time. It is not clear how to capture the characteristic form of self-maintenance exhibited by a process (Criterion H) within the exemplar view.

How might we capture the role of a part in a larger process (Criterion I) and/or the context in which that part occurs (Criterion J). If this part is taken to be an individual, then the process of which it is a part cannot also be an individual on the exemplar view (due to the *unity requirement*). It seems the roles and context of a part must be captured by external relations between

[14] I am not considering here behaviors which may be thought continuous or ongoing (e.g., rolling, expanding, cooling)—these might perhaps plausibly be taken as characterizing how a thing is at some point (or perhaps brief period) of time and hence be allowed as properties of the thing at that time.

[15] This point echoes the insight of E. J. Lowe that instantaneous velocity might plausibly be a power (and hence a property), but changing location from one place to another cannot (Lowe 2006: 139–140).

that part and other parts. These external relations would often need to capture multiple subtle and perhaps intricate roles and highly complex contexts which are changing over time. Whether sufficient relations of the required kind are admissible is doubtful—even the admissibility of relatively simple spatial and causal relations is controversial.

Again, the exemplar view seems to face challenges in capturing the unity of an individual that has differing phase-stages (Criterion K)—what is it on this view that unites the differing stages with widely differing properties?

The more liberally we interpret the characteristics of an individual which we can capture within the exemplar list of admissible properties, the more we compromise the parsimonious metaphysical intuitions which motivate such a view initially. Even on the most liberal interpretations of such properties, it is not clear how important aspects of processes can be adequately captured.

This brief investigation suggests that metaphysical views of individuals face challenges in capturing the wide range of characteristics of processes we find reflected in the practice of science.

The contrast between Figures 3.5 and 3.7 highlights pictorially a further important tension between the view of processes in science and the exemplar metaphysical view of individuals: it seems that parts may remain present in the processes of which they are a part so that both the part and the process (of which it is a part) may then obtain as valid unities (the latter, and perhaps the former, as a process). This contrasts with the exemplar metaphysical view according to which an individual cannot have a part which is also an individual (as represented by the discreteness of the boxes in Figure 3.7).

It may be that the metaphysical view that either a (possibly complex) whole or a part of that whole may be an individual, but not both, has influenced current debates in the sciences. Questions concerning individuals are often couched in terms of exclusive alternatives within these debates: Is an organism or a species an individual (Hull 1976; Rieppel 2009)? Is the eukaryotic part of an organism or the symbiotic consortium (with prokaryotic parts) an individual (Dupré 2012)? Is a biofilm or a constituent bacterium an individual (Ereshefsky and Pedroso 2016)? Is a cell or the organism of which it is a part an individual (see, e.g., Fagan 2016, especially section 7.2)? Figure 3.5 suggests the way in which a processual view may license a more accommodative answer: a processual unity at one level (pictured as an entwining of strands) does not preclude another differing processual unity at a higher or a lower level.

3.5 Conclusion

A consideration of the scientific treatment of processes suggests that the adequate characterization of processes requires a rich framework of criteria which may include (in addition to how the process is at each stage) how the process changes and self-maintains across stages, the nature of the context in which the process typically acts, and the role(s) it plays in this context.

A preliminary and limited comparison of this view of processes distilled from science with metaphysical views of individuals suggests a tension between these views. Metaphysical views often suppose individuals can be wholly characterized by a list of admissible properties, where tight admissibility may be adopted on grounds of parsimony. There is a tension between the adequate characterization of processes and parsimony of admissible properties. Furthermore, metaphysics often supposes that individuals cannot have parts that are themselves individuals, whereas no such limitation obtains for processes: processes may have parts that are themselves processes. Further articulating, understanding, and perhaps resolving these tensions may be some of the work required to develop an adequate account of the individuation of processes. [16]

In the meantime, it seems reasonable to choose to individuate processes in ways which have respect for the rich view of processes reflected in the practice of science, including the wide range of characterizing criteria employed. Using more limited criteria for individuation (e.g., a list of admissible properties), perhaps in deference to a particular metaphysical account of individuals, may be detrimental to the task of individuating processes—perhaps this is one source of the problems concerning individuation which this volume seeks to address. If the suggestion in Section 3.3 is right that many things may be processes, then the individuation of processes has wide application.

Respecting such a processual view may license recognizing both processes and certain of their parts (e.g., organism/cell, biofilm/bacterium, species/organism) as both being valid unities—perhaps robustly self-maintaining processes which can be picked out sufficiently reliably from their natural context. This may provide a basis for the resolution of some important contemporary debates concerning individuation.

Acknowledgments

For helpful comments and advice on earlier drafts of this chapter, I should like to thank Nancy Cartwright, Anna Marmodoro, discussants at a presentation to the Powers Structuralism programme in Oxford (November 2014), various participants at the Taiwan conference on Scientific Individuation (December 2014), and two anonymous referees.

---

[16] I take up this task in Pemberton (forthcoming).

References

Armstrong, David M. 1986. In defence of structural universals. *Australasian Journal of Philosophy* 64(1): 85–88.

Bapteste, Eric, and John Dupré. 2013. Towards a processual microbial ontology. *Biology and Philosophy* 28: 379–404.

Bear, Mark F., Barry W. Connnors, and M. A. Paradiso. 2007. *Neuroscience: Exploring the brain. Third edition.* Baltimore MD: Lippincott, Williams and Wilkins.

Bechtel, William, and Adele Abrahamsen. 2005. Explanation: A mechanist alternative. *Studies in the History and Philosophy of the Biological and Biomedical Sciences* 36: 421–441.

Cartwright, Nancy, and John M. Pemberton. 2012. Aristotelian powers: without them, what would modern science do? In *Powers and capacities in philosophy: The new Aristotelianism*, eds. John Greco and Ruth Groff, 93–112. London: Routledge.

Chen, Ruey-Lin. 2017. Mechanisms, capacities, and nomological machines: integrating Cartwright's account of nomological machines and Machamer, Darden and Craver's account of mechanisms. In *Philosophy of science in practice*, eds. Hsiang-Ke Chao and Julian Reiss, 127–145. Switzerland: Springer. Synthese Library 379.

Copeland, Robert A. 2000. *Enzymes* (2nd edition). New York: Wiley-VCH.

Craver, Carl. 2007. *Explaining the brain.* Oxford: Clarendon Press.

Darden, Lindley. 2006. *Reasoning in biological discoveries: Essays on mechanisms, interfield relations, and anomaly resolution.* Cambridge: Cambridge University Press.

Duns Scotus. 1639. *Opera omnia*, ed. Luke Wadding. 12 volumes, Lyon, reprinted Hildesheim: Georg Olms Verlagsbuchhandlung, 1968.

Dupré, John. 2010. It is not possible to reduce biological explanations to explanations in chemistry and/or physics. In *Contemporary debates in philosophy of biology*, eds. Francisco José Ayala and Robert Arp, 32–47. Malden, MA: Wiley-Blackwell Publications.

Dupré, John. 2012. *Processes of life. Essays in the philosophy of biology.* Oxford and New York: Oxford University Press.

Dupré, John. 2013. Living causes. In *Mechanisms and causation in biology. Proceedings of the Aristotelian Society* Suppl 87: 19–37.

Eijkelkamp, Niels, John E. Linley, Mark, D. Baker, Michael S. Minett, Roman Cregg, Robert Werdehausen et al. 2012. Neurological perspectives on voltage-gated sodium channels. *Brain* 135(9): 2585–2612.

Ereshefsky, Mark, and Makmiller Pedroso. 2016. What biofilms can teach us about individuality. In *Individuals across the sciences*, eds. Alexandre Guay and Thomas Pradeu, 103–121. New York: Oxford University Press.

Fagan, Melinda B. 2016. Cell and body. In *Individuals across the sciences*, eds. Alexandre Guay and Thomas Pradeu, 122–143. New York: Oxford University Press.

Glennan, Stuart. 2002. Rethinking mechanistic explanation. *Philosophy of Science* 69: S342–S353.

Glennan, Stuart. 2017. *The new mechanical philosophy.* New York: Oxford University Press.

Guay, Alexandre, and Thomas Pradeu. 2016. *Individuals across the sciences.* New York: Oxford University Press.

Hull, David L. 1976. Are species really individuals? *Systematic Zoology* 25(2): 174–191.

Illari, Phyllis, and Jon Williamson. 2012. What is a mechanism? Thinking about mechanisms across the sciences. *European Journal of Philosophy of Science* 2: 119–135.

Keller, Rebecca W. 2013. *Middle school chemistry grades 5–8*. Albuquerque, NM: Gravitas Publications, Inc.

Kripke, Saul A. 1980. *Naming and necessity*. Cambridge, MA: Harvard University Press.

Lewis, David K. 1986. Against structural universals. *Australasian Journal of Philosophy* 64: 25–46.

Lowe, E. Jonathan. 2006. *The four-category ontology*. New York: Oxford University Press.

Lowe, E. Jonathan. 2016. Non-individuals. In *Individuals across the sciences*, eds. Alexandre Guay and Thomas Pradeu, 46–60. New York: Oxford University Press.

Machamer, Peter, Lindley Darden, and Carl F. Craver. 2000. Thinking about mechanisms. *Philosophy of science* 67: 1–25.

Marcus, Ruth B. 1975. Dispensing with possibilia. *Proceedings and Addresses of the American Philosophical Association* 49: 39–51.

Marmodoro, Anna. 2013. Aristotle's hylomorphism without reconditioning. *Philosophical Inquiry* 37(1–2): 5–22.

McCullough, Laurence B. 1996. *Leibniz on individuals and individuation*. Dordrecht: Springer Science and Business Media, B.V.

Mumford, Stephen, and Matthew Tugby, eds. 2013. *Metaphysics and science*. Oxford: Oxford University Press.

Pemberton, John M. 2011. Integrating mechanist and nomological machine ontologies to make sense of what-how-that evidence (manuscript). http://personal.lse.ac.uk/pemberto.

Pemberton, John M. Forthcoming. *Actor-process ontology*.

Pemberton, John M., and Nancy Cartwright. 2014. Ceteris paribus laws need machines to generate them. *Erkenntnis* 79: 1745–1758.

Quine, Willard V. O. 1948. On what there is. Reprinted in Quine 1953. *From a logical point of view*. Cambridge, MA: Harvard University Press.

Rieppel, Olivier. 2009. Species as process. *Acta Biotheoretica* 57(1–2): 33–49.

Scaltsas, Theodore. 1994. *Substances and universals in Aristotle's Metaphysics*. Ithaca and London: Cornell University Press.

Sider, Theodore. 2006. Bare particulars. *Philosophical Perspectives* 20: 387–397.

Südhof, Thomas C., and Josep Rizo. 2011. Synaptic vesicle exocytosis. *Cold Spring Harbor Perspectives in Biology* 3(12): a005637.

van Inwagen, Peter. 1990. *Material beings*. Ithaca and London: Cornell University Press.

Wu, Cathy H., Hongzhan Huang, Anastasia Nikolskaya, Zhangzhi Hu, and Winona C. Barker. 2004. The iProClass integrated database for protein functional analysis. *Computational Biology and Chemistry* 28(1): 87–96.

{4}

Individuating Part–Whole Relations in the Biological World

Marie I. Kaiser

4.1 Introduction

The view that nature is divided into part–whole hierarchies is deeply embedded in the sciences, for instance, in the biological sciences. Biologists represent objects (e.g., cells) as being constituted of a certain collection of organized parts (e.g., DNA, mitochondria, proteins, etc.). Assumptions about part–whole relations are involved in classifications of biological objects into kinds (e.g., the assumption that fish have gills as parts, whereas mammals have lungs as parts). Moreover, the methodological principle that one can understand the behavior of a whole by decomposing it into its parts and studying the behavior of the parts is central to generating knowledge in the biological sciences (for the limitations of reductive methods, see Kaiser 2011, 2015). The importance of part–whole relations to the biological sciences raises the question under which conditions something is a part of a biological whole and what it means that the parts constitute the whole. For instance, when does a vesicle that is transported from the endoplasmic reticulum to the Golgi apparatus in a eukaryotic cell cease to be a part of the endoplasmic reticulum and become a part of the Golgi apparatus? Is the case that is attached to the body of a caddisfly larva and that promotes the larva's survival a part of the larva, or does it belong to the larva's environment? Under which conditions is a particular species or abiotic factor (e.g., nitrogen) a part of an ecosystem? Does a virus that enters a host cell and uses the cell's machinery and metabolism to replicate itself become a part of the host cell?

The goal of this chapter is to answer questions like these by developing a general, systematic account of biological parthood. Such an account specifies general criteria that guide the individuation of wholes and parts in the biological sciences, and it reveals the conditions under which biological objects stand in a part–whole relation to each other. The central question of the chapter is: under which conditions is one biological object *x* a part of another biological object *y*?[1] If two biological objects stand in a part–whole relation to each other, I will speak of one being a *biological part* of the other. Hence, biological parthood applies to biological objects only.[2] The question under which conditions *x* is a biological part of *y* is a special version of what van Inwagen has called the "special composition question" (van Inwagen 1990: 21). It is a special version because it concerns only part–whole relations in the biological world, rather than the concept of a part in general (which is the object of mereology; see, e.g., Simons 1987). This chapter is about biological parthood, not about parthood simpliciter.[3]

Questions about biological parts are intertwined with questions about biological individuality. Typically, the concept of a biological individual is centered on but not confined to organisms. In addition to organisms, also parts of organisms (e.g., genes and cells) and groups of organisms (e.g., populations) are discussed as being biological individuals (e.g., Clarke 2011; Wilson and Barker 2017). Accordingly, the relata of many part–whole relations in the biological realm will be biological individuals. Rather than focusing on the question of what biological individuals are and how they are individuated, this chapter examines how biological individuals *relate* to each other; in particular, which conditions must be satisfied so that biological individuals relate to each other as parts and wholes. One of my central claims will be that the individuation and characterization of a biological individual as a whole constrains the individuation of its parts in several ways. This chapter uncovers the various kinds of constraints by formulating different criteria for biological parthood.

Methodologically, my account of biological parthood arises from and is sustained by an analysis of a wide variety of paradigmatic examples of part–whole relations from the biological sciences. My analysis takes into account actual cases of biological part–whole relations as well as the explanatory and investigative strategies that biologists employ when studying part–whole relations. The account of biological parthood that I develop here is thus an exercise in what I call a *metaphysics of biological practice*. First, it is a metaphysical account because it aims at describing a feature of reality; namely, part–whole relations that exist out there in the biological world. It provides an understanding of the characteristics of biological part–whole relations and of the conditions under which they hold. Second, I agree with naturalistic metaphysics (e.g., Ladyman and Ross 2007; Chakravartty

[1] I take biological objects to be those objects that belong to the domain of the biological sciences.

[2] For reasons of simplicity, I will often speak of one object being a *biological part* of another object. This must be read synonymously with the claim that one biological object is a part of another biological object.

[3] My focus on biological parthood is not driven by the conviction that biological part–whole relations are fundamentally different from other kinds of part–whole relations. In fact, my analysis yields parthood criteria that might also be applied to nonbiological objects, for instance, in the chemical realm.

2013) in that we should consult the sciences to develop metaphysical views about which kinds of entities exist and how they are like. But instead of considering physical theories only, a metaphysics of biological practice draws our attention to the metaphysical underpinnings of the nonphysical sciences, and it develops metaphysical claims that take into account also the nontheoretical aspects of biological practice (e.g., scientific activities, epistemic values, reasoning strategies; Chang 2011; Waters 2014).[4] Because of the relatively broad scope of my account (i.e., parts and wholes in the biological world, in general), I focus on analyzing examples of part-whole relations that are paradigmatic and figure in successful biological explanations. In addition, my analysis is backed up by philosophical case studies that are concerned with more specific kinds of part-whole relations (e.g., DNA sequences being parts of the human genome) and that analyze the concrete investigative practices and reasoning strategies that biologists employ (see, e.g., Kaiser forthcoming).

A pluralistically inclined philosopher might object that my goal of paying attention to actual biological practice conflicts with my other goal of developing a general, systematic account of biological parthood. According to the pluralist, a monistic account that specifies a single set of criteria supposed to apply to all part-whole relations in the biological world will fail to capture the diversity of biological part-whole relations and of individuation practices in the biological sciences (cf. Kellert, Longino, and Waters 2006). I agree with the pluralist that it might turn out that only a pluralistic notion of biological parthood that recognizes different, perhaps conflicting, parthood criteria (such as Wimsatt 1972, 2007) accounts for the diversity of biological practice. However, diversity and difference is only one aspect of biological practice that philosophers can and should account for (Kaiser 2015: chapt. 2). From a philosophical perspective, it is also interesting to zoom out and look for generalities and for similarities between different practices of individuating parts and wholes. When traditional metaphysicians argue about the general structure of reality, they adopt such a general perspective and develop philosophical views that are supposed to apply universally. I think we should not reject the monistic aspirations of traditional metaphysics too hastily because we can learn a lot from striving for monism and from analyzing the obstacles we meet. Hence, the aim of my chapter is twofold. First, I develop a monistic account of biological parthood that proposes two general, necessary conditions for

[4] I am aware of the fact that a metaphysics of biological practice understood in that way presupposes scientific realism and requires an account of which elements of biological practice allow for drawing metaphysical conclusions (and which do not). A promising approach is to focus on successful and stable practices. However, it is a controversial question whether scientific success provides us with access to the world as it really is. Those skeptical about this might prefer to adopt a weaker view of metaphysics, according to which metaphysics makes general claims about our conceptual apparatus, rather than about the world (e.g., Strawson 1959). I agree that these are urgent and very interesting questions, but they lie beyond the scope of this chapter. Fortunately, my analysis of biological parthood is compatible with both views of metaphysics.

the existence of biological part–whole relations, and I show why these criteria are preferable to alternative criteria. Second, my analysis enables me to explain why the search for a monistic account that specifies individually necessary and jointly sufficient criteria for biological parthood is so difficult. Understanding the reasons for these difficulties delivers valuable insights into the nature of biological part–whole relations.

This chapter is organized as follows. In Section 4.2, I distinguish different kinds of questions that one can ask concerning biological parthood, and I specify the relata of biological part–whole relations. Section 4.3 develops spatial inclusion as the first criterion for biological parthood and introduces the notion of a natural boundary, which is central to the spatial inclusion criterion. In Section 4.4, I develop the second criterion for biological parthood, compositional relevance, which refers to the characteristic behaviors of the whole and specifies the relevance condition in terms of an insufficient but necessary part of a condition that is unnecessary but sufficient (INUS-condition). In Section 4.5, I discuss in how far my account of biological parthood can deal with hard cases, such as redundant parts or collective parts. I conclude in Section 4.6 by revealing the merits and limits of a monistic account of biological parthood.

4.2 Preliminaries

4.2.1 RELATING DIFFERENT PART–WHOLE QUESTIONS

I shall call the question of whether a particular biological object x (e.g., a vesicle) is a part of another biological object y (e.g., the Golgi apparatus) the *parthood question*. It is closely related to but less demanding than the *decomposition question*, which asks for not only one but for all parts into which a whole is partitioned. My account of biological parthood focuses on the relation between *one* biological part and its whole (i.e., on the parthood question) and specifies the conditions under which this one-to-one relation holds. From this focus, it does not follow that the other parts of the whole are irrelevant to the existence of a part–whole relation. On the contrary, several authors have drawn attention to the fact that a biological part always is a member of a team of parts that interact with each other in very "intense" (Simon 1962; Haugeland 1998; McShea 2000) or "productive" ways (Machamer, Darden, and Craver 2000; Gillett 2013) and that exhibit a special "jointness" (Fagan 2012). How, exactly, the existence of a particular part–whole relation depends on other parts of the same whole will be examined in Section 4.4. This dependency does not imply that answering the parthood question presupposes answering the decomposition question. We can individuate one particular biological object as a part of a whole without having decomposed the whole into all of its parts (but not without knowing some of the other parts).

Whether a particular object is a biological part of another object, the whole, does not only depend on the other parts but also, and in particular, on how the whole is individuated and by which properties and behaviors it is characterized. I shall call the question "What is the whole, how can it be individuated and characterized?" the *individuation-of-the-whole question*. A central idea that I will elaborate in the following sections is that answering the individuation-of-the-whole question constrains answering the parthood question. In other words, which object I pick out as the whole and how I characterize its properties, typical behaviors, and spatial boundaries constrains what counts as a biological part of this whole. For instance, biologists do not treat a green alga that is spatially included in a fungus as a part of the fungus. However, if the whole is referred to as a lichen, the green alga will be a part of the lichen. Likewise, if biologists conceive of a genome as having a specific chemical structure (i.e., as consisting of DNA only), transcription factors and histones fail to be biological parts of the genome because they are proteins. By contrast, if biologists characterize a genome in purely functional terms (e.g., as guiding development), transcription factors and histones turn out to be biological parts of the genome because they are central to gene regulation. The criteria for biological parthood that I develop in this chapter specify the different ways in which the individuation of the whole constrains the individuation of its parts.

The individuation-of-the-whole question must be distinguished from the *demarcation question*, which concerns the outer boundary of the whole and asks how the whole can be demarcated from its context. One might claim that answering the demarcation question boils down to answering the decomposition question because to demarcate an object from its context is nothing but identifying all of its parts and classifying all non-parts as context. In Section 4.3.1, I shall argue that this view invites a circularity objection, which can only be avoided by a substantial notion of a spatial boundary that allows us to demarcate an object from its environment independently of individuating all of its parts. In other words, I will argue that answering the demarcation question is independent of answering the decomposition question. Identifying what I will refer to as the "natural boundary" of a biological object may be one aspect of individuating the object as a whole.[5] Hence, answering the demarcation question may be part of answering the individuation-of-the-whole question.

In sum, we must distinguish four questions concerning part–whole relations:

PART–WHOLE QUESTIONS

1. *Individuation-of-the-whole question:* What is the whole, how can it be identified?

[5] For instance, biologists spatially characterize a cell as being surrounded by a cell membrane and insects as being surrounded by an exoskeleton with a particular shape and structure.

2. *Demarcation question:* Where does the outer boundary of the whole run? How can the whole be demarcated from its context?
3. *Decomposition question:* Into which collection of parts can the whole be decomposed?
4. *Parthood question:* Is a given biological object a part of another biological object, the whole?

4.2.2 THE RELATA OF BIOLOGICAL PART–WHOLE RELATIONS

Claims about part–whole relations in the biological sciences typically concern types or kinds of objects, not tokens. For instance, all individuals of the kind lichen are composed of green algae and fungi, not only a particular lichen. The fact that science is often concerned with kinds of part–whole relations, however, does not imply that the part–whole relation itself holds between types or kinds.[6] Depending on one's ontology, kinds might be viewed as abstract entities (e.g., universals) that do not exist in space and time, or they might not be said to exist at all—at least not independently of our classification practices. It seems to me that the more parsimonious and less controversial assumption to start with is that part–whole relations in the biological realm relate token objects that exist in space and time. The account of biological parthood that I develop in this chapter is thus a singularist account.[7] According to a singularist account of biological parthood, part–whole relations exist between individual biological objects, and claims about kinds of biological part–whole relations are generalizations that arise from investigating particular part–whole relations.

Even if kinds are not the relata of biological part–whole relations, they are still relevant to the conditions under which biological part–whole relations exist. Whether two particular biological objects are related as part and whole depends also on the kinds to which they belong. To see this, consider the question of whether a virus that has infected a host cell is a part of it. Answering this question depends not only on how biologists individuate the whole but also on how they classify it. Viruses are surely not among the parts that cells have, in general. But they may be parts of a subtype of cells, namely infected host cells. A virus normally does not contribute to but often hinders the behaviors that are typical for objects of the kind "cell," such as growth, DNA replication, protein synthesis, and cell division. A token object that belongs to the kind "host cell," by contrast, is characterized by different typical behaviors, such as reduced cell defenses, viral replication, and release of viruses. The virus is a biological part of a cell of the

[6] This is not a trivial point because some authors claim that part–whole or constitutive relations exist between properties or types of events (e.g., Harbecke 2010).

[7] My singularist account is in accordance with the fact that the part–whole relation in the classical, mereological sense is conceived of as a first-order relation between individuals.

kind host cell because the virus is essential to the characteristic behaviors of host cells.[8] I will pick up this point in Section 4.4.1 when I explain the notion of a characteristic behavior in more detail and incorporate it into my second criterion for biological parthood. To be clear, my claim that kinds are relevant to the conditions under which biological part-whole relations exist does not imply that the existence of biological part-whole relations depends on classification preferences of individual biologists. Not any biological kind that one might think of is scientifically legitimate and thus can be said to be real (in a promiscuous way; Dupré 1993: 36). Hence, from the fact that biological part-whole relations exist only relative to specific kinds, it does neither follow that biological parthood is subjective nor does it follow that my account of biological parthood is epistemological rather than metaphysical in character.

Some philosophers might agree that part-whole relations exist between tokens but reject the view that material objects—understood in accordance with endurantism[9]—are the appropriate relata of part-whole relations. Process ontologists argue that an object-bias or "substance paradigm" (Seibt 2016) fails to acknowledge the processual nature of the living world (e.g., Whitehead 1929; Rescher 1996, 2000; Seibt 2003; Dupré 2012; O'Malley 2014). I agree that the ontological category of processes (or occurrents, which include processes, events, and states) is important to capture the constantly changing, complex biological world (see Kaiser and Krickel 2017). However, this does not commit one to the radical claim that "everything is process" (Bickhard 2011: 95)[10] and that part-whole relations must exist between processes only. It is possible to identify material objects as the relata of part-whole relations and yet to acknowledge the importance of processes. As my analysis will show, the processes in which biological objects are involved co-determine whether these objects are related as parts and wholes. A static view of biological parthood that considers only the properties of and relations between material objects at a certain time is thus highly implausible. Biological part-whole relations essentially involve processes, even though processes are not the relata of part-whole relations.[11]

---

[8] There are several other examples that support this claim. For instance, the F_1 subunit seems not to be a biological part of a protein of the kind "transmembrane protein," but rather a biological part of a protein of the kind "ATP synthase."

[9] Endurantists believe that material objects have spatial parts but no temporal parts and that material objects are wholly present whenever they exist. By contrast, perdurantists claim that material objects are four-dimensional space-time worms that are extended over time and that have also temporal parts.

[10] It is unclear whether there is any philosopher of biology who subscribes to the radical view that only processes exist. Dupré claims that biological entities, such as organisms and genomes, are dynamical, constantly changing entities that cannot be understood in terms of "properties of and relations between their structural constituents" (Dupré 2012: 8).

[11] Along these lines, one might add that a part-whole relation between biological objects does not only depend on the processes they implement but also on the properties and powers they possess (Gillett 2013).

To conclude, I assume that part–whole relations in the biological world each exist between particular biological objects. My heart, for example, is a biological part of my circulatory system. Despite this focus on particulars and on material objects, I argue that we can only understand the conditions under which a certain part–whole relation exists if we broaden our perspective and take into account also other parts of the same whole, the processes in which the objects are involved, and the kinds to which they belong. The criteria for biological parthood that I develop in the following sections explicate how other parts, processes, and kinds determine the existence of biological part–whole relations.

4.3 Spatial Inclusion

4.3.1 WHY A PRIMITIVE NOTION OF SPATIAL INCLUSION FAILS

A widespread and intuitively plausible view is that the parts of material objects are spatial parts. There is a dispute in metaphysics about whether material objects have temporal parts as well, as the perdurantist believes, but everybody agrees that material objects have at least spatial parts. To be a spatial part of a whole means to be spatially included or contained in the whole; that is, to be located inside the spatial boundary that continuously surrounds the whole. The view that something is a part if it is spatially included in the whole seems to apply to the biological world as well (see Craver 2007; Clarke 2011; Gillett 2013). Chloroplasts are parts of cells because they are located inside cells. My liver is a part of me because it is spatially included in me. An individual black-headed gull is a part of a certain population because it is located in the spatial distribution of that population. A green alga is a part of a lichen because it is contained in the lichen. Based on these intuitions, you might suggest specifying biological parthood as primitive spatial inclusion:

> Primitive spatial inclusion (PSI)
> An object x is a biological part of an object y if and only if
> x is spatially included in y; that is, x is located inside the continuous spatial boundary of y.[12]

Despite its initial plausibility, paradigmatic examples of part–whole relations from biological practice show that spatial inclusion is neither sufficient nor necessary for biological parthood. Consider first why it is not sufficient. In the case of lichens, green algae are spatially included in the fungus but they are not conceived of as parts of the fungus (rather, they are parts of the composite organism, the lichen). Similarly, if a doctor leaves a cotton ball inside of my stomach during

[12] Note that spatial inclusion does not require spatial contact or proximity among the objects that are spatially included in y. Spatial contact or proximity may have an influence on part–whole relations (e.g., by allowing for causal interactions), but it is not presupposed by spatial inclusion.

surgery we would not say that the cotton ball became a part of me just because it is spatially located inside of me. Another example is the individuation of the parts of the human genome. According to ENCODE (2012), not any arbitrary DNA sequence contained in the human genome is a biological part of it (Kaiser forthcoming). Examples like these are widespread in the biological sciences. They show that there is more to biological parthood than spatial inclusion and that PSI specifies a criterion for biological parthood that is not sufficient.

Other cases show that spatial inclusion is not even necessary for biological parthood. Take the example of a population of black-headed gulls. As an individual, it might have a specific spatial distribution (e.g., at the North Sea coast around Büsum), but this does not preclude that some members of that population are located outside of this region (if they potentially interbreed with members located in that region). In general, populations seem not to be the kind of objects whose identity depends on having a specific spatial boundary, contrary to, for instance, cells or mitochondria that are necessarily surrounded by a membrane. Gene regulatory networks and ecosystems are further examples of biological objects whose parts seem not to be held together by what I will call a natural boundary (see next section). PSI thus fails to provide even a necessary criterion for the existence of part–whole relations in the biological world.

Finally, PSI is problematic because it runs the risk of being circular (or trivial). PSI says that biological parts of a whole must be spatially included in the whole, which means that they must be located inside the spatial boundary of the whole. This raises the question of what the spatial boundary of the whole is and how it can be identified (the demarcation question; recall Section 4.2.1). The simplest answer is that we demarcate an object from its context by identifying all of its parts and drawing a three-dimensional boundary that encompasses all parts and a minimal set of other objects (this is similar to what Kaplan [2012] proposes). This suggestion, however, renders PSI circular because whether x is a biological part of y would depend on whether x is spatially included inside the boundary around y and y's other parts. In other words, the individuation of one part of a whole would presuppose having individuated all of its parts. For example, whether the case of a caddisfly larva is a biological part of the larva would depend on whether it is located inside the spatial boundary that surrounds all parts of the larva. To avoid the circularity of the spatial inclusion criterion, we must find a way to individuate the spatial boundary of the whole *independently* of individuating its parts. The next section introduces the notion of a natural boundary, which provides us with such an independent demarcation of biological objects.

4.3.2 NATURAL BOUNDARIES

The main idea behind the notion of a natural boundary is that there exist boundaries in the biological world that are of particular importance to the identity of many biological objects and to individuating their parts. Paradigmatic cases of natural boundaries include the skin of mammals, the exoskeleton of insects, the cell wall of plant cells, other membranes such as the blood–brain barrier or the alveolar–capillary membrane, and also things such as rivers or thick lines of scrub. I refer to these boundaries as "natural" to emphasize that they exist in reality rather than being introduced solely through human demarcation. In short, they are "bona fide boundaries" not "fiat boundaries" (Smith and Varzi 2000: 401).[13] Not all boundaries that biologists draw—for instance, when identifying the target of their investigation—correspond to natural boundaries. Sometimes biologists are interested in questions that require studying an object together with parts of its environment. At other times, they ignore parts of an object because they are irrelevant to the behavior they investigate.

Demarcating a biological object, the whole, from its context by identifying its natural boundary is independent in two respects. First, it does not require identifying all of the biological parts of the whole (i.e., answering the demarcation question does not require answering the decomposition question). Second, demarcating a biological object by identifying its natural boundary is independent of characterizing the natural boundary, itself, as a biological part (i.e., answering the demarcation question does not require answering the parthood question about the natural boundary). These independencies arise from the fact that natural boundaries typically are identified by their function as selective barriers and by the material discontinuities or structural differences that they involve. Let me explain this in more detail.

First, natural boundaries that demarcate individuals function as selective barriers. They bind together the objects that are located inside of them and separate them from what is outside. In most cases, this separation will not be complete but rather selective. For instance, the nuclear membrane ensures that ribosomes stay outside the nucleus, but its nuclear pore complexes allow some molecules to pass (e.g., mRNAs with specific signal sequences). As a result, natural boundaries reduce the causal interactions between objects inside and objects outside of them (e.g., between ribosomes and intranuclear mRNAs) and allow specific kinds of causal interactions among the objects they encompass (e.g., between mRNAs and enzymes during mRNA processing). The fact that interactions between the parts of a whole and parts of its environment are generally fewer and weaker than the interactions among the parts of the whole is the guiding idea of so-called intensity-of-interactions approaches to parthood (e.g., Simon 1962; Wimsatt 1972, 2007; Haugeland 1998; McShea 2000; McShea and Venit 2001).

Second, natural boundaries usually involve material discontinuities or structural differences (e.g., differences in the chemical structure, texture, or material constitution). The cell membrane, for instance, is composed of a lipid bilayer and

[13] Artificially produced biological objects (e.g., in synthetic biology) can have natural boundaries as well because the term "natural" denotes the independent existence of these boundaries in the natural world, not their natural origin.

transmembrane proteins and, as such, forms a unit that has a different material constitution and chemical structure than what is located outside and inside of it. In the case of the membrane of blood cells, for instance, the cytoplasm (inside) consists of a complex mixture of cell organelles, cytoskeleton filaments, dissolved molecules, and water, whereas the blood plasma (outside) consists mainly of water.[14]

In sum, natural boundaries exist in the natural world (independent of human demarcation), and they are individuated by their functions as selective barriers and by the structural differences and material discontinuities they involve. Natural boundaries play a crucial role in determining what is a part of a biological individual and what is not. In the next section, I use the notion of a natural boundary to substantiate the condition of spatial inclusion and to turn it into a plausible criterion for biological parthood, which avoids the three objections that a primitive notion of spatial inclusion faces (recall Section 4.3.1). The substantial criterion of spatial inclusion that I develop in the next section relies on the idea that natural boundaries constrain relations between biological parts and biological wholes. However, it does neither imply that *any* natural boundary demarcates biological individuals (because other necessary conditions might not be satisfied), nor does it imply that *any* biological individual must be demarcated by a natural boundary (because there can be individuals without natural boundaries).

4.3.3 A SUBSTANTIAL CRITERION OF SPATIAL INCLUSION

The leading idea of developing a substantial spatial inclusion criterion is that natural boundaries of biological objects constrain the individuation of their parts insofar as their biological parts must be spatially included in or inside the natural boundary of the object. The dung ball that a dorbeetle rolls is not a biological part of the beetle because it is located outside its exoskeleton; the neurotransmitter molecule is not a part of the neuron because it is located in the synaptic cleft outside the cell membrane of the neuron; and a green alga is discussed to be a biological part of the lichen because it is located inside the fungal membrane. Paradigmatic cases like these give rise to the following substantive notion of spatial inclusion as a criterion for biological parthood.

> Substantial spatial inclusion (SSI)
> An object x is a biological part of an object y only if
> 1. x must be spatially located inside or in the region that the natural boundary occupies if y has a natural boundary.

[14] This is compatible with the fact that some natural boundaries are scattered (e.g., of the Golgi apparatus).

This criterion explicates one way that the nature of the whole—in this case, its spatial nature, that is, its feature of possessing a specific natural boundary—constrains the individuation of its biological parts. The disjunction "inside or in the region of" is important because it also captures cases where a biological part is located in the natural boundary itself, such as a receptor molecule or an ion channel that are located in the region that the cell membrane occupies.

SSI avoids all three problems that a primitive notion of spatial inclusion faces (recall Section 4.3.1). First, SSI is not formulated as a sufficient condition for biological parthood. In the next sections, I will supplement SSI by an additional criterion that allows distinguishing cases of mere spatial containment from cases of genuine biological parthood (cf. Jansen and Schulz 2014). Second, even though natural boundaries are of great importance to the identity of biological objects and co-determine their parts, this is not true for all cases. Some biological objects, such as gene regulatory networks, immune systems, certain populations, and certain ecosystems, do not possess natural boundaries. Nevertheless, we can retain SSI as a necessary condition if we formulate it conditionally; that is, if we require biological parts to be spatially included in their wholes only if the wholes possess natural boundaries. The conditional form of SSI accounts for cases in which spatial inclusion is irrelevant to biological part–whole relations. Third, in the previous section, I characterized natural boundaries as involving material discontinuities or structural differences and as functioning as selective barriers. These structural and functional features ensure that natural boundaries of biological objects can be individuated independently from identifying their parts. Hence, by referring to the spatial boundary of a biological object as a natural boundary, SSI avoids the circularity objection because where the natural boundary of the whole runs is independent of what its parts are.

4.4 Compositional Relevance

Paradigmatic examples of biological part–whole relations show that biological parts are bound together to a whole not only spatially but also causally-functionally. To get an idea of what this means, consider the following reasoning strategies for why something is a part of a biological object that can be found in biological practice. A vesicle in the cytoplasm of a cell is a part of the Golgi apparatus (rather than, e.g., of the endoplasmic reticulum) because it contributes to the processing and transport of proteins. A particular DNA sequence out of the billions of possible sequences is a part of the human genome because it plays a causal role in the human genome as a whole; that is, because it contributes to gene expression or regulation (Kaiser forthcoming). Glia cells are not parts of the central nervous system because they do not directly participate in synaptic interactions and electrical signaling; rather, glia cells are parts of the brain because they provide physical support for neurons and regulate the internal environment of the brain. An amino acid sequence or protein region is a part of adenosine triphosphate (ATP) synthase because it is necessary to the protein's function of synthesizing ATP through transporting protons. A leukocyte or antibody is a part of an immune system because it is crucial to the protection of an organism against diseases. A cotton ball left inside the stomach during surgery is not a part of the human because it does not contribute to the survival of the human but rather impedes it if an infection occurs. Harmless strains of *Escherichia coli* bacteria are parts of the human gut because they facilitate the survival of humans by producing vitamin K_2 and preventing colonization of the intestine with pathogenic bacteria.

These paradigmatic examples reveal that we cannot understand the conditions under which biological part–whole relations hold if we consider biological objects and their properties only. We must also take into account the processes in which parts and wholes are involved and how these processes relate to each other, such as how the binding of a particular DNA sequence to a transcription factor contributes to the regulation of genes of the human genome.[15] My account of biological parthood thus stands in the tradition of philosophies that emphasize the importance of processes and of activities (which I take to be subtypes of processes; Kaiser 2018) to the biological world (Machamer, Darden, and Craver 2000; Dupré 2012; O'Malley 2014).[16]

The preceding examples suggest that biological parts must be involved in processes (which I will refer to as "parts-processes") that are in a certain sense relevant to—that is, contribute to, are necessary to, play a causal role in, are crucial to—one or more processes that the whole engages in (which I will refer to as "whole-processes"). In the following sections, I turn this rough idea into a specific, clear criterion for biological parthood. I argue that we should think of the whole-processes to which parts-processes must be relevant as the characteristic behaviors of biological objects, such as the characteristic behaviors of a Golgi apparatus to process and transport proteins (Section 4.4.1). I then draw attention to the parts-processes and explain why we cannot understand biological parthood by considering parts-processes alone (Section 4.4.2). Finally, I use Mackie's (1965) idea of INUS-conditions to specify the relevance relation that must hold between parts- and whole-processes (Section 4.4.3).

4.4.1 CHARACTERISTIC BEHAVIORS OF BIOLOGICAL WHOLES

Plausibly, not any arbitrary process of the whole should determine what its parts are because the whole might be involved in some processes only accidentally or exceptionally. For example, a lichen growing on the wall of a playground might

---

[15] Note that my focus is still on objects as the relata of part–whole relations. There might also be part–whole relations between processes, but this is not the focus of this chapter. My claim is that if we accept objects as the relata, we must take into account processes as well.

[16] With respect to part–whole relations, Gillett (2013) argues that we must adopt a dimensioned account that considers not only individual objects but also their properties, powers, and the processes they engage in.

be painted blue by school kids, but this is not a process in which a lichen is typically involved. HIV might damage the immune system of a person suffering from AIDS, but this process is not characteristic of human immune systems in general.

Processes of this kind can be excluded by introducing the notion of a characteristic behavior of a biological object. Characteristic behaviors are processes in which a biological object engages very generally, that is, under a wide range of contexts in which the object is naturally found. For instance, a sunflower grows and attracts bees and other insects to promote its reproduction whether it stands in my garden, on a sunflower farm, or at the edge of a grain field. Characteristic behaviors determine, at least in part, the character or nature of a biological object. For example, the functional nature of an ATP synthase is to use a proton gradient to synthesize ATP, which is why molecules of this kind have been named "ATP synthase." Contrary to mechanisms, which are always for a single behavior (Glennan 2002), most biological objects are characterized by more than one typical behavior. Cells, for example, divide, synthesize proteins, and grow. Which behaviors are characteristic of a particular biological object depends also on the kind to which it belongs (recall Section 4.2.2). Blood cells, for instance, exhibit characteristic behaviors different from those of muscle cells or of infected host cells.

Characteristic behaviors of biological objects will often be realizations of biological functions. The realization of the function of the heart to pump blood, for instance, is also a characteristic behavior of the heart. Likewise, the characteristic behaviors of a stem cell to differentiate into specific kinds of cells, to synthesize proteins, and to grow might be seen as realizations of functions of a stem cell. However, I prefer to speak about characteristic behaviors rather than about functions. Many etiological theories of biological function prioritize organisms and their characteristic behaviors to survive and to reproduce (e.g., Neander 1991). Such a restricted focus impedes our understanding of biological part–whole relations because biological wholes are not confined to the level of organisms and because even organisms display characteristic behaviors other than survival and reproduction (e.g., growth). Cummins's causal role theory of function (1975) seems to provide an understanding of the concept of a biological function that is much more adequate to the present purposes. However, Cummins's account also must be applied to biological part–whole relations with caution since it is misleading in some ways. Most importantly, Cummins argues that functions are dispositions or capacities, not processes (1975: 757). But the mere disposition to engage in a process seems not to be sufficient to an object being a biological part (more on this in Section 4.5). A mitochondrion, for instance, is a biological part of a cell because it produces ATP, not because it has the disposition to produce ATP, which may never be manifested.[17]

---

[17] Furthermore, Cummins (1975) states that functions are individuated with respect to the capacity of a containing system. Accordingly, Cummins's functions are features of parts which they possess in virtue of being parts of a whole (the containing system). Cummins's account thus applies very naturally

4.4.2 WORKING PARTS

Understanding the conditions under which part–whole relations in the biological realm hold requires considering not only the processes in which wholes engages but also the processes in which parts are involved. For instance, it is the process of a specific DNA sequence binding certain transcription factors that is relevant to the regulation of genes of the human genome, and it is the process of antibodies recognizing and neutralizing pathogens that is relevant to the protection of an organism against diseases. This is the point where causation enters the scene. Many of the parts-processes will be causal processes that involve not a single biological part but causal interactions between two or more biological parts of the same whole (e.g., the DNA sequence interacting with transcription factors, or the antibodies interacting with pathogens). In a similar vein, van Inwagen has argued that "parthood essentially involves causation" (1990: 81).[18]

The observation that causal interactions are central to biological parts and wholes has led several authors to defend what can be called the *intensity-of-interactions approach* to biological parthood (Simon 1962; Wimsatt 1972, 2007; Haugeland 1998; McShea 2000; McShea and Venit 2001). The basic idea of this approach is that we can decompose a system into parts according to the principle that interactions among parts are generally weaker and less frequent than interactions within parts. For instance, different molecules in a cell can be distinguished from another because the atoms inside of each molecule interact more frequently and stronger with themselves (e.g., they form certain kinds of chemical bindings) rather than, for instance, with atoms that belong to different molecules. The intensity-of-interactions approach has initial plausibility, but it encounters some serious objections. Most importantly, the approach overlooks that biological parts are not only determined by the intensity of causal interactions in and among parts but also by specific features of the whole.[19] In Section 4.3, I argued that biological wholes spatially constrain the individuation of their biological parts through the natural boundaries they possess. In the next section, I will argue that biological parts must be involved in causal interactions that are relevant at least to one of the characteristic behaviors of the whole. If we consider the strength of

---

to characterizing parts-processes as functions (if we ignore the capacity-process difference). By contrast, characterizing whole-processes (i.e., characteristic behaviors of wholes) as Cummins's functions would require positing an additional whole, which contains the original whole as a part.

[18] Van Inwagen thinks about causal interactions between simples, rather than between biological objects of various kinds.

[19] Other major objections are that the intensity-of-interactions approach requires specifying a threshold of number and intensity of interactions above which they are included and under which they are excluded. Such a threshold seems arbitrary or highly context-dependent (Wimsatt 1972). Furthermore, the intensity-of-interactions approach fails to exclude non-parts or background conditions that possess relatively many and strong interactions to other parts (e.g., Craver 2007: 143–144).

causal interactions only, we overlook this directedness of causal interactions to the behaviors of the whole.

To conclude, considering only the processes in which parts engage is not sufficient to understanding biological part–whole relations. Instead, we must analyze how parts-processes relate to whole-processes and what makes biological parts into "working parts" (Mellor 2008: 68) that work to bring about one of the characteristic behaviors of the whole.

4.4.3 SPECIFYING THE RELEVANCE CONDITION

Biologists conceive of a vesicle as a biological part of the Golgi apparatus if it contributes to the processing and transport of proteins. *E. coli* bacteria are regarded as biological parts of the human gut if they facilitate the survival of humans, for instance, by producing vitamin K_2 and preventing colonization of the intestine with pathogenic bacteria. Neurobiologists treat glia cells as biological parts of the brain rather than of the central nervous system because they play no direct role in electrical signaling but provide physical support for neurons and regulate the internal environment of the brain. What unifies all these paradigmatic cases of part–whole relations is that biological parts engage in processes (e.g., a vesicle budding off the Golgi apparatus and moving toward the cell membrane) that are relevant to at least one of the characteristic behaviors of the whole (e.g., the Golgi apparatus transporting proteins). Under which conditions, exactly, is the relation of relevance satisfied? What does it mean that a part-process must contribute to, play a role in, or be crucial to one of the characteristic behaviors of the whole (a whole-process)? The philosophical literature yields different ideas that one could pursue further. In this section, I briefly introduce the major ideas and point out their shortcomings before I then present my own relevance condition, which I refer to as *compositional relevance.*

One idea is that biological parts must have significant causal effects on the properties of the whole such that they give the whole a "causal unity" (Mellor 2008: 67). This is not an implausible idea, but it does not help to specify the relevance condition because it remains unclear what makes a causal effect significant and under which conditions causal unity is reached. Gillett argues that a part–whole relation between individual objects exists only if the part is a "member of a spatiotemporally related team of individuals many of whose members bear powerful and/or productive relations to each other" (2013: 321). Gillett nicely draws attention to the fact that biological parts are not isolated but *jointly* working parts (see also Fagan 2012). However, the parthood criterion he specifies—namely, that a biological part must bear powerful and/or productive relations to other parts of the same whole—runs the risk of becoming circular (at least it requires that other parts of the whole are already fixed). It also seems too weak because it does not restrict the kinds of dispositions that a biological part must have and the kinds of processes it must engage in (also, the cotton ball left inside my stomach during surgery, e.g., has some dispositions and engages in some processes). Moreover, to capture cases, such as inactive parts, Gillett backs away from formulating the parthood criterion as a necessary condition. His account thus fails to provide a satisfying answer to the central question under which conditions a biological object is a member of a jointly working team.

Craver's account of "constitutive relevance" (2007: 139) applies to the components of mechanisms, but one might suggest that its mutual manipulability condition can be used to specify also the sense in which biological parts must be relevant to their wholes. The mutual manipulability condition requires that the components of a mechanism (which are acting entities; i.e., material objects engaging in processes of a specific kind; Kaiser 2018) and the phenomenon that a mechanism is responsible for (e.g., an object-involving process; Kaiser and Krickel 2017) must be mutually manipulable. That is, some change in the component must change the phenomenon and vice versa. Craver's mutual manipulability condition is criticized, for instance, for being epistemic rather than metaphysical because it specifies how researchers get evidence about the components of mechanisms, rather than specifying which features in the world determine component-mechanism relations (Couch 2011).[20] Even more importantly, Craver's account of constitutive relevance requires components and mechanistic phenomena not only to be mutually manipulable but also to be related as parts and wholes.[21] Using the mutual manipulability criterion to specify biological parthood thus introduces a vicious circularity into Craver's account of constitutive relevance.

In the early literature on functional explanation and functional analysis, Hempel (1965) and Nagel (1961) put forward the idea that a biological part must be a bearer of a function in the sense that it must have certain effects that contribute to some activity of the containing system, the whole (cf. Cummins 1975: 741). Much of the discussion focuses on the question of whether the effects that biological parts have on their wholes can be understood as necessary conditions, such as the heart circulating blood might be a necessary condition for the proper working of the organism. However, counterexamples such as artificial pumps show that a heart pumping blood is not necessary for an organism's survival, and, even if one excludes things such as artificial pumps by adding the phrase "under normal conditions," one still has to face counterexamples such as redundant parts (e.g., the second kidney which is not necessary for the organism's survival but is a biological part of it) and relevant non-parts (e.g., the case of the caddisfly larva, which seems to be necessary for the larva's survival, but is not a biological part of it). Still, the idea that biological parts must in a certain sense be necessary to their wholes is

---

[20] Other objections can be found, for instance, in Harbecke (2010), Couch (2011), Leuridan (2012), Harinen (2018), and Baumgartner and Gebharter (2016).

[21] More specifically, the component-objects *X* must be parts of the system *S* that is involved in the phenomenon that the mechanism is responsible for (Craver 2007: 153; Kaiser and Krickel 2017).

appealing because it may give rise to a criterion of compositional relevance that is specific and strong enough.

Mackie (1965) claims that we should understand causes as INUS-conditions, that is, as being insufficient but necessary elements of overall conditions that are themselves unnecessary but sufficient for their effects. Mackie's idea provides us with a promising tool to maintain the idea that biological parts must be necessary to their wholes, while it allows us to weaken the relevance criterion in a way that it accounts for cases, such as redundant parts (e.g., the second kidney) and collective parts (e.g., the calcium ions in a muscle fiber that only collectively have a significant effect on muscle contraction).[22] I argued in the previous section that we cannot individuate biological parts by considering their causal interactions only, as the intensity-of-interactions approach proposes. Instead, the processes in which parts engage (which comprise causal interactions) must be examined with regard to how they affect the characteristic behaviors of the whole. The contribution that a part-process makes to one of the characteristic behaviors of the whole need not be necessary because it can be redundant (e.g., the second kidney filtering blood is not necessary for the human's survival) or only collectively significant (e.g., the transport of a single calcium ion into the sarcoplasmic reticulum does not raise the electrostatic potential). Instead, a part-process must be an INUS-condition for at least one of the behaviors that the whole characteristically displays. That is, biological parts must engage in processes that are necessary members of a minimal subset of parts-processes, which are jointly sufficient to a characteristic behavior of the whole.[23] Applying Mackie's idea of INUS-conditions to the question of biological parthood gives rise to the following second criterion for biological parthood.

> Compositional relevance (CR)
> An object *x*, which engages in biological process *p*, is a biological part of an object *y*, which shows characteristic behaviors $b_1, \ldots b_n$, only if
> 2. *p* is relevant to at least one of $b_1, \ldots b_n$, that is, *p* makes a necessary contribution to a condition that is minimally sufficient to one or more of $b_1, \ldots b_n$.

CR is a clearly ontological criterion (contrary to, e.g., Craver's mutually manipulability condition) because it specifies how the processes in which putative biological parts engage must relate to the characteristic behaviors of the whole. CR is also

[22] In a similar vein, Harbecke (2010) and Couch (2011) use Mackie's account to specify the conditions under which an acting entity is a component of a mechanism.

[23] Compositional relevance thus turns out to be a subtype of causal relevance (if one accepts Mackie's claim that causes are INUS-conditions for their effects). Note, however, that this does not imply that biological part–whole relations are special kinds of causal relations. Biological parts must satisfy further criteria (e.g., they must be spatially included in the whole in a substantial way) and, thereby, violate conditions that many authors regard as characteristic of causal relations (e.g., that cause and effect are distinct).

more specific than other approaches because it specifies precise conditions under which a (causal) process of a putative part is relevant to a whole. In other words, CR specifies the conditions under which causal effects of putative parts are significant (Mellor 2008) and the conditions under which putative parts bear powerful and/or productive relations to each other (Gillett 2013).

4.5 The Hard Cases Challenge to Monism

The goal of this chapter is to specify the general conditions under which particular biological objects stand in a part–whole relation to each other. To sum up, my analysis of paradigmatic examples from biological practice reveals two criteria for biological parthood:

> An account of biological parthood
> An object *x*, which engages in biological process *p*, is a biological part of an object *y*, which shows characteristic behaviors $b_1, \ldots b_n$, if and only if
> 1. *Substantial spatial inclusion (SSI)*: *x* must be spatially located inside or in the region that the natural boundary occupies if *y* has a natural boundary and
> 2. *Compositional Relevance (CR)*: *p* is relevant to at least one of $b_1, \ldots b_n$; that is, *p* makes a necessary contribution to a condition that is minimally sufficient to one or more of $b_1, \ldots b_n$.

This is a monistic account of biological parthood because it specifies two criteria that are individually necessary and jointly sufficient for a biological part–whole relation to exist.[24] That is, my account identifies a single set of criteria that is supposed to hold for any object in the biological world.

My analysis also shows that there exist certain kinds of part–whole relations that can be referred to as "hard cases" because they place requirements on a monistic account of biological parthood that pull in different directions. Understanding what these hard cases are and why it is so difficult to account for them at once sheds light on the merits and limits of monism.

Hard cases for an account of biological parthood
1. *Redundant parts* (e.g., kidney of humans removing waste products from the body): Some biological parts are redundant because there are other token objects that engage in the same kind of process, only one of which is necessary to a behavior of the whole.

---

[24] In what follows I provide support for the claim that these criteria are jointly sufficient. I consider and dismiss additional criteria (e.g., common origin/history or genetic identity) for biological parthood because they are not necessary.

2. *Irrelevant parts* (e.g., appendix of humans): Some biological parts engage in processes that are irrelevant to any characteristic behavior of the whole (e.g., because they have become nonfunctional during evolutionary history).
3. *Inactive parts* (e.g., members of a population not interbreeding): Some biological parts do not actually engage in relevant processes; they merely have the disposition to do so, which is not manifest.
4. *Collective parts* (e.g., calcium ion in a muscle fiber being released into the cytosol): Some biological parts only collectively have a significant effect on the behavior of the whole.
5. *Relevant non-parts* (e.g., case of a caddisfly larva that protects it from predators): Some biological objects engage in processes that are relevant to the behavior of the whole but are treated as background conditions or as parts of the context.

How does my account of biological parthood cope with these hard cases? The CR criterion accounts for redundant parts because the processes in which parts engage need not be necessary to one of the characteristic behaviors of the whole; they must only be necessary members of a set of biological parts, which are sufficient to one of the characteristic behaviors of the whole. Each kidney of a human, for example, is an INUS-condition for the human's survival—even though they are not necessary parts of the same condition. Cases of collective parts do not present a problem for CR because even if a biological part, such as a calcium ion, on its own lacks a significant effect on the contraction of a muscle fiber, it still belongs to an INUS-condition for muscle contraction. CR is only a necessary but not a sufficient condition for biological parthood. Cases of relevant non-parts, such as the blood that is transported to the heart and thus is an INUS-condition for the blood pumping of a heart, pass CR but are excluded by the SSI criterion because only the blood inside the ventricles and atria is located inside the natural boundary of the heart. This is a major reason that SSI is a separate criterion for biological parthood that cannot be reduced to CR.

Inactive parts present a much stronger challenge to CR because they only potentially engage in processes that are INUS-conditions for one of the characteristic behaviors of the whole and thus call into question that CR is a necessary condition. For example, not all organisms that belong to a particular population actually interbreed with other members of the population. The mere capacity to interbreed suffices to be a biological part of the population. One might argue that examples like this require that CR must be weakened, for instance, by replacing "*x* engages in biological process *p*" by "*x* has the disposition to engage in biological process *p*." A weakened CR criterion, however, invites more counterexamples than it copes with because several non-parts would turn out to be relevant. T-helper cells, for example, have the disposition to recognize antigens and thus would count as biological parts of our immune system. In addition, a dispositionally formulated relevance criterion neglects the fact that processes are central to biological parthood (recall Section 4.4).

Cases of irrelevant parts also challenge the status of CR as a necessary condition. For example, our appendix does not engage in processes that are INUS-conditions for one of our characteristic behaviors, such as our survival, reproduction, reasoning, or cooperation with others, but we typically consider our appendix to be a biological part of us. Again, one might argue that examples like this show that we must replace CR by a weaker relevance criterion. But if we weaken the relevance criterion, we will face many new counterexamples of biological objects which pass the weaker relevance criterion but which we do not regard as biological parts (i.e., relevant non-parts), such as the cotton ball that is left inside my stomach during surgery.

Alternatively, one might argue that the two criteria that I present in this chapter are not sufficient to biological parthood and must be supplemented by additional necessary criteria. In the case of our appendix, for example, referring to the common history or origin of the appendix and the human body as a whole or to the genetic identity of most of their cells might provide us with a criterion that individuates the appendix as a biological part, but not the cotton ball inside my stomach (cf. Jansen and Schulz 2014). The impression that spatial and causal-functional relations are only part of the story about biological part–whole relations is confirmed by examples that show how some biological wholes also structurally constrain their parts. For example, it follows from the chemical-structural nature of the human genome (i.e., from the fact that it is a nucleic acid) that its biological parts must be sequences of nucleotides rather than, for instance, proteins.[25] Each of these putative additional criteria, however, captures only a small subset of biological part–whole relations and thus does not constitute a necessary criterion that is universally applicable to the biological world. For instance, only some biological wholes have a structural nature that constrains their parts in a nontrivial manner.[26] The criterion of genetic identity is violated in many cases (e.g., genetic chimera, symbionts, ecosystems, etc.). In addition, the requirement that biological parts must originate together with or inside the whole does not hold for the majority of part–whole relations because most biological objects continuously lose and acquire particular parts.

[25] Likewise, the structure of the Golgi apparatus implies that its parts must be membrane-enclosed compartments or vesicles, and a population of black-headed gulls consists only of organisms of the species *Chroicocephalus ridibundus* (not of the beach or of other kinds of birds).

[26] One might argue, for example, that the heart can be structurally characterized as having two ventricles, two atria, connected by valves, etc. This structural description, however, is nothing but a description of the biological parts of the heart.

4.6 Conclusions: On the Merits and Limits of Monism

My analysis of hard cases reveals at which point monism reaches its limit. For a monistic account of biological parthood, it is impossible to cope with all hard cases at once. If you weaken the CR criterion to account for irrelevant parts and inactive parts, you will not be able to exclude several cases of relevant non-parts (and the other way around). You might solve the problem by introducing additional criteria that capture other kinds of constraints on biological part–whole relations (e.g., historical, genetic, or structural constraints). These additional criteria, however, will apply only to certain kinds of biological objects and will not hold universally. The goal of universality thus runs contrary to the goal of descriptive adequacy (Kaiser 2015: chapt 2), and it seems as if we need to make a principled decision at this point—in favor of more descriptive adequacy and pluralism or in favor of more universality and monism.

Among the virtues of my monistic account is that it gives a clear and precise answer to the general question under which conditions one biological object is a part of another biological object. Monism highlights the commonalities of part–whole relations from different levels of organization (e.g., a vesicle being a part of a Golgi apparatus and an organism being a part of a population) and it reveals how biological wholes in general constrain their parts. A monistic account of biological parthood thus provides unification. On the other hand, monism runs the risk of yielding an overly simple and descriptively inadequate view of biological parthood, one that fails to account for the diversity of biological part–whole relations reflected in the various approaches that biologists adopt to individuate part–whole relations (cf. Kellert, Longino, and Waters 2006).

The goal of this chapter is not to defend my monistic account against pluralistic alternatives. Rather, I presuppose the goal of monism, defend my account against monistic alternatives, and use my analysis to point out the limits that monism faces. Recognizing these limits allows us to distinguish general constraints on biological parthood from constraints that arise more locally, and it helps us to specify how far the individuation of part–whole relations in the biological world is diverse. Understanding the limits of monism thus allows us to draw general conclusions about the nature of biological parthood.

Acknowledgments

This work was supported by the Deutsche Forschungsgemeinschaft (DFG) [FOR-1063]. It also benefited much from my collaboration with the John Templeton Research Group "From Biological Practice to Scientific Metaphysics" [50191]. I would like to thank the members of Egenis at the University of Exeter, the participants of the Colloquium for Theoretical Philosophy at the University of Cologne, and the two anonymous referees for their helpful comments on earlier drafts.

References

Baumgartner, Michael, and Alexander Gebharter. 2016. Constitutive relevance, mutual manipulability, and fat-handedness. *British Journal for the Philosophy of Science* 67(3): 731–756.

Bickhard, Mark. 2011. Systems and process metaphysics. In *Handbook of philosophy of science. Philosophy of complex systems, vol. 10*, ed. Cliff Hooker, 91–104. Amsterdam: Elsevier.

Chakravartty, Anjan. 2013. On the prospects of naturalized metaphysics. In *Scientific metaphysics*, eds. Don Ross, James Ladyman, and Harold Kincaid, 27–50. Oxford: Oxford University Press.

Chang, Hasok. 2011. The philosophical grammar of scientific practice. *International Studies in the Philosophy of Science* 25(3): 205–221.

Clarke, Ellen. 2011. The problem of biological individuality. *Biological Theory* 5(4): 312–325.

Couch, Mark B. 2011. Mechanisms and constitutive relevance. *Synthese* 183: 375–388.

Craver, Carl F. 2007. *Explaining the brain. Mechanisms and the mosaic unity of neuroscience.* Oxford: Oxford University Press.

Cummins, Robert. 1975. Functional analysis. *The Journal of Philosophy* 72(20): 741–765.

Dupré, John. 1993. *The disorder of things. Metaphysical foundations of the disunity of science.* Cambridge, MA: Harvard University Press.

Dupré, John. 2012. *Processes of life: Essays in the philosophy of biology.* Oxford: Oxford University Press.

The ENCODE Project Consortium. 2012. An integrated encyclopedia of DNA elements in the human genome. *Nature* 489: 57–74.

Fagan, Melinda B. 2012. The joint account of mechanistic explanation. *Philosophy of Science* 79: 448–472.

Gillett, Carl. 2013. Constitution, and multiple constitution, in the sciences: Using the neuron to construct a starting framework. *Minds and Machines* 23: 301–337.

Glennan, Stuart S. 2002. Rethinking mechanistic explanation. *Philosophy of Science* 69: 342–353.

Harbecke, Jens. 2010. Mechanistic constitution in neurobiological explanations. *International Studies in the Philosophy of Science* 24: 267–285.

Harinen, Totte. 2018. Mutual manipulability and causal inbetweenness. *Synthese* 195: 35–54.

Haugeland, John. 1998. *Having thought.* Cambridge, MA: Harvard University Press.

Hempel, Carl G. 1965. *Aspects of scientific explanation. And other essays in the philosophy of science.* New York: Free Press.

Jansen, Ludger, and Stefan Schulz. 2014. Crisp islands in vague seas: Cases of determinate parthood relations in biological objects. In *Mereology in the sciences. Parts and wholes in contemporary scientific contexts*, eds. Claudio Calosi and Pierluigi Graziani, 163–188. Cham: Springer.

Kaiser, Marie I. 2011. Limits of reductionism in the life sciences. *History and Philosophy of the Life Sciences* 33: 453–476.

Kaiser, Marie I. 2015. *Reductive explanation in the biological sciences*. Cham: Springer.

Kaiser, Marie I. 2018. The components and boundaries of mechanisms. In *The Routledge handbook of mechanisms and mechanical philosophy*, eds. Stuart Glennan and Phyllis Illari, 116–130. New York: Routledge.

Kaiser, Marie I. Forthcoming. ENCODE and the parts of human genome.

Kaiser, Marie I., and Beate Krickel. 2017. The metaphysics of constitutive mechanistic phenomena. *British Journal for the Philosophy of Science* 68: 745–779.

Kaplan, David M. 2012. How to demarcate the boundaries of cognition. *Biology and Philosophy* 27: 545–570.

Kellert, Stephen H., Helen E. Longino, and C. Kenneth Waters, eds. 2006. *Scientific pluralism. Minnesota studies in the philosophy of science*. Minneapolis and London: University of Minnesota Press.

Ladyman, James, and Don Ross. 2007. *Every thing must go: Metaphysics naturalized*. Oxford: Oxford University Press.

Leuridan, Bert. 2012. Three problems for the mutual manipulability account of constitutive relevance in mechanisms. *British Journal for the Philosophy of Science* 63: 399–427.

Machamer, Peter, Lindley Darden, and Carl F. Craver. 2000. Thinking about mechanisms. *Philosophy of Science* 67: 1–25.

Mackie, John L. 1965. Causes and conditions. *American Philosophical Quarterly* 2(4): 245–264.

McShea, Daniel W. 2000. Functional complexity in organisms: Parts as proxies. *Biology and Philosophy* 15: 641–668.

McShea, Daniel W., and Edward Venit. 2001. What is a part? In *The character concept in evolutionary biology*, ed. Gunter P. Wagner, 259–284. London: Academic Press.

Mellor, D. Hugh. 2008. Micro-composition. *Royal Institute of Philosophy Supplements* 83(62): 65–80.

Nagel, Ernest. 1961. *The structure of science. Problems in the logic of scientific explanation*. London: Routledge.

Neander, Karen. 1991. Functions as selected effects: the conceptual analyst's defense. *Philosophy of Science* 58(2): 168–184.

O'Malley, Maureen A. 2014. *Philosophy of microbiology*. Cambridge: Cambridge University Press.

Rescher, Nicholas. 1996. *Process metaphysics: An introduction to process philosophy*. New York: SUNY Press.

Rescher, Nicholas. 2000. *Process philosophy: A survey of basic issues*. Pittsburgh: University of Pittsburgh Press.

Seibt, Johanna. 2003. *Process theories: Crossdisciplinary studies in dynamic categories*. Dordrecht: Kluwer Academic.

Seibt, Johanna. 2016. Process philosophy. In *The Stanford encyclopedia of philosophy* (winter 2016 edition), ed. Edward N. Zalta. http://plato.stanford.edu/archives/ win2016/entries/ process-philosophy/.

Simon, Herbert A. 1962. The architecture of complexity. *Proceedings of the American Philosophical Society* 106(6): 467–482.

Simons, Peter. 1987. *Parts. A study in ontology*. Oxford: Clarendon.

Smith, Barry, and Achille C. Varzi. 2000. Fiat and bona-fide boundaries. *Philosophy and Phenomenological Research* 60(2): 401–420.

Strawson, Peter. 1959. *Individuals. An essay in descriptive metaphysics*. London: Routledge.

van Inwagen, Peter. 1990. *Material beings*. New York: Cornell University Press.
Waters, C. Kenneth. 2014. Shifting attention from theory to practice in philosophy of biology. In *New directions in the philosophy of science. The philosophy of science in a European perspective*, eds. Maria Carla Galavotti, Dennis Dieks, Wenceslao J. Gonzalez, Stephan Hartmann, Thomas Uebel, and Marcel Weber, 121–139. Cham: Springer.
Whitehead, Alfred N. 1929. *Process and reality: An essay in cosmology*. New York: Macmillan.
Wilson, Robert A., and Matthew Barker. 2017. The biological notion of individual. In *The Stanford encyclopedia of philosophy* (spring 2017 edition), ed. Edward N. Zalta. https://plato.stanford.edu/archives/spr2017/entries/biology-individual/.
Wimsatt, William C. 1972. Complexity and organization. *PSA: Proceedings of the Biennial Meeting of the Philosophy of Science Association* 1972: 67–86.
Wimsatt, William C. 2007. *Re-engineering philosophy for limited beings. Piecewise approximations to reality*. Cambridge, MA: Harvard University Press.

{ PART II }

Experimental Practices of Individuation

{5}

Ask Not "What *Is* an Individual?"

C. Kenneth Waters

5.1 Introduction

Biologists and philosophers often ask "what is an X?" with an emphasis on the "is." As Arthur Fine (1996) might remark, it is as if one can hear the stamping of feet upon utterance of the word "is" when we ask, what *is* a species?, what *is* an organism?, or what *is* a gene? In the first part of this chapter, I motivate a more pragmatic approach to questions about individuals and individuation by presenting my account of conceptual practice in genetics. I maintain that instead of asking "what *is* a gene?," philosophers should ask: "how do biologists individuate genes?," "for what purposes?," and "do their practices of individuating genes serve these purposes?" In the second part of this chapter, I propose that we adopt this approach in our analyses of organisms and biological individuals. I illustrate how philosophical work on individuation typically focuses on questions about what it is to be an individual by discussing Peter Godfrey-Smith's analysis of Darwinian individuals (2009). According to his analysis, animals exemplify the paradigm of Darwinian individuality. I contend that we should not interpret this claim as a "philosophy of nature." Next, I turn to the idea that the real individuals are holobionts, not animals. I challenge the underlying presumption that one concept of individuals identifies the real individuals while others do not. But I reappropriate Scott Gilbert, Jan Sapp, and Alfred Tauber's argumentation (2012) for the holobiont view to advance pragmatism; instead of asking "what is an individual?" as if there is an essence to (or a paradigm of) individuality, philosophers should examine how biologists individuate animals, assemblages of animals, and commensals, and then ask what purposes these individuating practices can serve.

5.2 How Biologists Individuate Genes

The rise and practice of classical genetics involved individuating mutations and segments (i.e., genes) in which mutations occur. Individuating mutations and genes continues to be an important part of practice in contemporary genetics. In classical genetics, the individuation of genes was intimately connected to the individuation of mutations. Experimental practice involved identifying mutations and using mutations to individuate genes. Geneticists believed that genes were the fundamental units of heredity, that genes were located in cellular structures called 'chromosomes,' and that mutations were differences in genes. One of the central aims of classical genetics was to parse chromosomes into their basic functional units, genes (the "mapping project").[1]

Thomas Hunt Morgan and his collaborators established empirical strategies for this individuating practice. Working on fruitflies, they searched laboratory populations for the appearance of flies with distinctive outward differences (compared to other flies in the laboratory population). Upon finding such a difference, they conducted breeding experiments to determine whether the outward difference was caused by one or a few differences in chromosome(s). If they determined that the outward difference was indeed caused by one or a few chromosomal differences, they would call the outward difference a difference in 'phenotype' and the chromosomal difference a difference in 'genotype.' The term 'mutation' was used to refer to individual differences in genotype.

The Morgan group developed sophisticated strategies for breeding flies to determine whether a phenotypic difference was caused by one mutation or a few mutations. In cases where a phenotypic difference was caused by a few mutations, their breeding strategies enabled them to individuate the different mutations responsible for the phenotypic difference. By individuating mutations, they individuated genes in which the mutations occurred. Morgan's group also developed strategies for determining the relative locations of mutations, and hence the relative locations of the genes in which the mutations were located in a chromosome. The success of these strategies supported the idea that mutations and genes were located in linear fashion along the chromosomes.

Classical geneticists learned that different mutations could occur in one and the same gene, and, as the individuation practices of the science advanced through the first half of the twentieth century, geneticists established an experimental method, called the cis-trans test, for determining whether two mutations were located in the same gene or in different genes. This effectively provided them with a means for individuating functional units in the chromosome; that is, for individuating "genes." In time, geneticists learned how to determine the relative location of

---

[1] A more ambitious aim of classical genetics was to investigate a broad range of biological processes (see Waters 2004, 2008).

different mutations occurring within one and the same gene. Hence, they learned that genes are functional units positioned in linear fashion along chromosomes, that mutations are differences within genes, and that these differences are located in linear fashion within the genes.

The individuating practices of classical genetics did not depend on knowledge about what genes were made of, how genes functioned, what the function of any gene was, or even how differences in a gene could cause differences in phenotype. They did not need to answer the question, "what is a gene?" Of course, some theoretically inclined geneticists (especially Muller; see Carlson 1966) were carrying out investigations aimed at determining the material make-up of genes, the nature of gene action, and the individual functions of particular genes. One might view theoretical research into the question, "what is a gene?" to have been a metaphysics of sorts, a metaphysics that incubated later genetics.[2] But it was not central to the practice of the science. Most research in the science involved using genetics to investigate a wide range of basic biological processes, not just hereditary processes (Kohler 1994; Waters 2004). If one wishes to understand the knowledge of classical genetics and how the science worked, one needs to understand how classical geneticists individuated mutations and genes. This is required because geneticists' explanations of the transmission of phenotypic differences from one generation to the next (their major explanatory feat) depended on individuating mutations and genes. But one does not need to know how different geneticists might have speculated about the question "what *is* a gene?"

In sum, classical geneticists shared a minimal concept of the gene. Geneticists knew that differences in genes could cause uniform differences in phenotype in particular genetic and environmental contexts (Waters 2009) and that was an important part of their conception of genes. But ideas about the material make-up, function, or nature of gene action were not part of their basic concept. What about individuation practices in contemporary geneticists? What concepts do contemporary biologists employ?

Contemporary geneticists individuate numerous elements in genetic material: genes, introns, exons, regulatory regions, repetitive regions, transposable elements, and many more. For the purposes of this chapter, I focus on the individuation of genes at the molecular scale. According to textbook definitions, genes are segments of DNA that code for proteins. DNA consists of linear sequences of nucleotides. Proteins, roughly speaking, consist of polypeptides that are made up of linear sequences of amino acids. The textbook idea, as schoolchildren are taught, is that genes are segments of DNA that produce RNA molecules, and RNA molecules produce polypeptides. Linear sequences of nucleotides in DNA are said to "code for" linear sequences of amino acids in polypeptides.

---

[2] I thank an anonymous reviewer for pushing me on this point.

Philosophers have argued that this gene concept is ambiguous, hopelessly vague, and that it admits exceptions. Many contend that genes exist at the cytological scale, not at the molecular scale. I have addressed these arguments elsewhere (Waters 1994, 2001) and will not repeat the details here. Instead, I present my analysis of the molecular gene concept and show how it is used in contemporary genetics to individuate segments of DNA. I will then describe what purposes this practice of individuating serves across biological and biomedical sciences.

My analysis is cast in causal terms because the common use of informational language typically muddles important issues and easily leads to exaggerated claims on behalf of contemporary genetics. I therefore refrain from characterizing the process of RNA synthesis as "transcription" or the process of polypeptide synthesis as "translation." Biologists have investigated and explained these processes as biochemical syntheses, and it is helpful to conceive of them in these terms. But the question remains, "what is the relation between genes and polypeptides?"

Some biologists and philosophers say that the segments of nucleotide sequences within DNA called genes "code for" the amino acids comprising polypeptides. But this language is unclear, and, as I have already mentioned, gives rise to exaggerated claims about the significance of genes. Others say that the segments called genes "produce" polypeptides. But this is also misleading. The synthesis of RNA molecules depends on many molecules, not just DNA, and the synthesis of polypeptide molecules depends on many molecules, not just RNA molecules. So what, if anything, is the distinctive causal role of DNA in the synthesis of RNA molecules, and what, if anything, is the distinctive role of RNA molecules in the synthesis of polypeptides?

The distinctive role of genes is that they are makers of actual differences in the synthesis of RNA molecules (I have called them 'actual-difference makers'), and the distinctive role of RNA molecules is that they are makers of actual difference in the synthesis of different kinds of polypeptides. Cells contain many different kinds of polypeptides, each kind having a particular linear sequence of amino acids.[3] Although there are many causes of the synthesis of these molecules, most of the causes are uniform across the synthesis of different kinds of RNA molecules. RNA is a distinctive cause because it takes on different values in the synthesis of different kinds of polypeptides. That is, actual differences in the nucleotide sequences of different RNA molecules determine the actual differences in amino acid sequences in different kinds of polypeptides. Hence, they are actual-difference makers. Likewise, actual differences in nucleotide sequences in different genes determine the actual differences in nucleotide sequences in different RNA molecules. (In eukaryotes, differences in RNA molecules are often determined by

[3] Of course there are always exceptions to such general biological claims. In this case: some proteins, for example, prions, take on different conformations *in vitro*. These different conformations could be regarded as different kinds of proteins that have identical amino acid sequences. What this shows is that RNA is not an actual-difference maker of these different conformations.

differences in genes and different splicing agents—see later discussion.) This is to say that the DNA segments called genes are causes that determine differences in amino acid sequences that actually occur in a cell. Hence, DNA is a distinctive cause of these actual differences. It is an "actual-difference maker." In this sense, we can say that the sequence of nucleotides in a gene determines the sequence of nucleotides in an RNA molecule, which determines the sequence of amino acids in a polypeptide. This analysis treats the molecular gene concept as an indexical. DNA segments are not simply genes; they are genes for linear sequences. That is, the referent of an application of the term 'gene' depends on what sequence in an RNA molecule or polypeptide is set by investigative, explanatory, or technological context. DNA segments are not simply genes; they are genes for sequences at some stage of genetic expression in a particular cellular context. (See Waters 2007 for a fuller exposition of this account.)

The situation gets very complicated in eukaryotes because RNA molecules are often edited as they are being synthesized. That is, some regions of the RNA molecules, called 'introns,' are removed and some or all of the remaining regions, called 'exons,' are spliced together. The RNA molecule consisting of spliced exons moves from the nucleus to the cytoplasm, where it participates in polypeptide synthesis. Hence, one continuous segment of DNA determines the sequence of nucleotides in the primary RNA molecule, but a set of discontinuous segments in DNA (a set of exons) determines (or helps determine) the sequence of nucleotides in the mature RNA molecule and the sequence of amino acids in the polypeptide. What is individuated as the gene? Do biologists individuate the continuous segment that determines the linear sequence in the primary RNA molecules? Or do they individuate the discontinuous set of DNA segments that determines the linear sequences within the RNA exons and partly determines the sequence of nucleotides in the polypeptide?

It turns out that biologists sometimes individuate genes by including both exon and intron regions in DNA, but, at other times, they individuate genes by including only the exon regions. Many philosophers have taken this ambiguity to be a serious problem and a reason to doubt that genes exist at the molecular scale. But this is not a problem for individuating. When biologists are interested in primary RNA molecules, it serves their interests to individuate DNA segments containing both exons and introns. When they are interested in polypeptides, it serves their interest to individuate discontinuous DNA segments containing just exons (i.e., "split genes"). The molecular concept is an indexical concept, and its application involves individuating genes for particular linear sequences. This makes the practice of individuation both flexible and precise.

The flexibility of gene individuation practices is especially important in navigating the incredibly complicated processes of differential splicing. Biologists have learned that one and the same kind of primary RNA molecule is sometimes spliced in hundreds, perhaps even thousands of different ways. This makes it possible for one continuous segment of DNA and different splicing agents to determine the amino acid sequences in hundreds or even thousands of different polypeptides. In one tissue at a particular stage of development, one polypeptide is synthesized; in another tissue or at a different stage of development, a different polypeptide is synthesized. The molecular gene concept is indexed to these different contexts. It is used to individuate many different overlapping and discontinuous segments of a DNA molecule, as illustrated in Figure 5.1.

The molecular gene concept provides flexibility in individuating DNA segments in precise ways. But this individuation does not yield a unique partition of DNA. Classical geneticists might have speculated that there was a unique partition of the chromosome into its genes. To appropriate a commonly used metaphor, geneticists might have thought that DNA could be cut at its joints. But molecular biologists have learned that (in eukaryotes at least) this is impossible. There are simply too many joints, yielding overlapping genes. The division of DNA into molecular

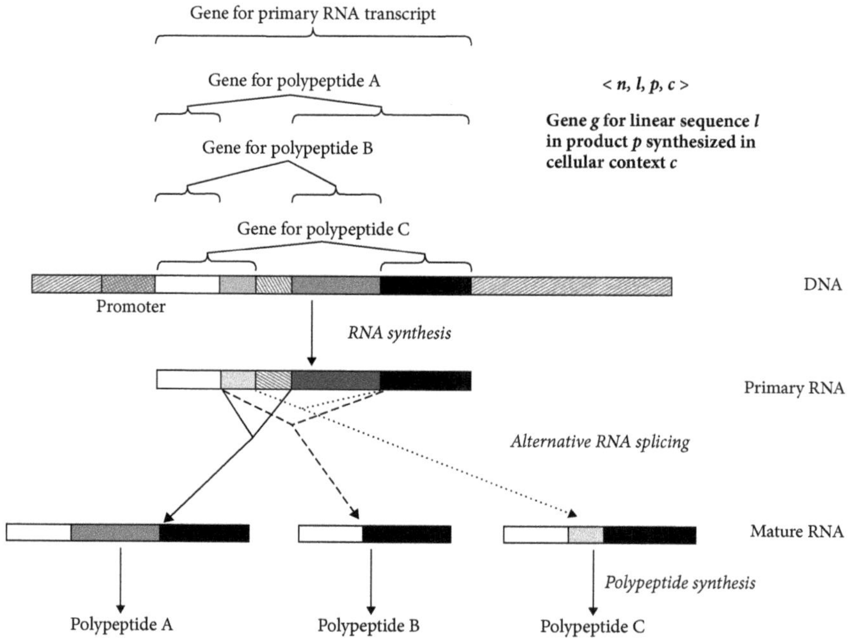

FIGURE 5.1 *The molecular gene concept enables biologists to partition DNA in multiple ways. With respect to this figure, if biologists are interested in the synthesis of the primary RNA transcript (see preceding discussion), then the DNA segment covered by the bracket directly under "Gene for primary RNA transcript" is the relevant gene (the continuous region of five segments in the DNA molecule starting with the solid white segment and ending with the solid black segment). If biologists are interested in the synthesis of polypeptide B occurring in some tissue at a particular stage of development, then the relevant gene is the discontinuous region marked by the brackets directly under "Gene for polypeptide B" (the white and gray segments in DNA and not the segments between).*

genes is a complicated mess. Nevertheless, the piecemeal practice of individuating genes for particular purposes is an important scientific activity across the biological and biomedical sciences.

5.3 Purposes Served by Individuating Genes

Individuating molecular genes is useful and serves a multiplicity of scientific aims: to explain, predict, manipulate, and, most importantly, investigate. Consider explanation. Individuating a molecular gene and identifying its linear sequence can provide a basis for explaining the linear sequence in an RNA and/or polypeptide molecule. Identifying a difference in a gene can explain the differences in linear sequences in RNA and/or polypeptide molecules. When the difference in linear sequence in RNAs and/or polypeptides is simply linked to an outward difference in phenotype, identifying the difference in the gene can also explain the difference in phenotype. But these explanations are partial, and their significance is typically modest.

Gene-based explanations are partial because genes do not control the timing of synthesis. At any one time, most genes are not participating in a synthesis. Complex regulatory processes determine when the synthesis of particular kinds of RNA and polypeptide molecules occur. In multicellular organisms, complex regulatory processes also determine where (in what tissue, for example) and when (for example, at what stages of development) the synthesis of particular kinds of RNA and polypeptide molecules will occur. The regulatory processes are incredibly complex, and it appears impossible to fully explain the development of a multicellular creature—or even the functioning of a single prokaryotic cell—in terms of genes. The significance of gene-based explanations is modest. Genes can be individuated to explain why particular molecules have the linear structure they do, but that by itself does not explain much. With respect to the countless processes going on within an organism, gene-based explanations leave a vast explanatory void.

Similar points can be made regarding prediction. Individuating genes is useful for making predictions, but most of these predictions are made in the context of experimentally manipulated results. This prompts the question: If genes do not explain or predict that much, why is the individuation of genes so central to biological and biomedical practice?

Individuating genes is central to practice because it provides a basis for manipulating many kinds of different processes within organisms. This is relevant to medicine, agriculture, and technology. It is also important for investigation. The *genetic approach* to investigation involves using genes to manipulate biological structures or processes in ways that will reveal the functions of those structures or important aspects of those processes (Waters 2004, 2008). The results of these investigations are often local and partial. In addition, it is worth emphasizing that the explanations yielded from intervening on genes are typically not gene-based.

Genes are like the scalpels of classical physiology. They are used as tools to investigate; they are not elements in the eventual explanations.

The foremost purpose of individuating genes is to forge epistemological entry points for explaining and predicting a very narrow range of biological phenomena and for investigating a much wider breadth of phenomena. Genes are not important categories of being (Waters 2017). Answering the question "what is it to be a gene?" does not provide important metaphysical insights into the functioning or development of organisms. The processes by which organisms function and develop are incomprehensibly complex. But, biologists have achieved partial and piecemeal accounts of aspects of these processes. They have also obtained a degree of control over parts of the processes. Given the complexities of the biological world, these are impressive achievements. A key to biologists' success, is their individuation of genes. So, ask not "what is a gene?" Ask "how do biologists individuate genes and for what purposes?"

5.4 Darwinian Individuality

Perhaps the stamping of feet is loudest when biologists and philosophers ask, "what *is* an organism?" The search has been for a universal analysis, one that gets at the essentials of what it is to be an organism. This question has recently drawn more attention as biologists have been investigating major evolutionary transitions, especially the transition from unicellular organisms to multicellular organisms. Theoretical evolutionary biologists need an explanatory target; they want to know what features make multicellular entities organisms rather than collections of cells so they can construct theoretical models of the transition to multicellularity (e.g., Buss 1987; Maynard-Smith and Szachmary 1995; Queller and Strassmann 2009). Experimental evolutionary biologists also need to have an experimental target (Ratcliff et al. 2012). They want to know what features would mark evolution toward becoming a multicellular organism so they can determine whether their experimental populations of unicellular organisms are evolving traits of multicellular individuality.

Meanwhile, philosophers interested in questions about individuality have moved the target of analysis. What began as the question "what is an organism?" has shifted to "what is a biological individual?" to "what is an evolutionary individual?" and, in some of the most recent work, "what is a Darwinian individual?" This shift helps make the inquiry more tractable. It is much easier to construct a universal answer to a question if it is reframed in terms of a theory. Trying to answer "what is an organism?" might force one to analyze the incredibly messy world of life, a world that resists essentialist analyses. Trying to answer the question, "what is a Darwinian individual?" leads one to analyze abstract principles, which are constructed in a tidy theoretical framework. One is much more likely to find an answer that appears universal, that seems to get at the essentials, if one moves from the question "what is an organism?" to the question "what is a Darwinian individual?"

Take, for example, Peter Godfrey-Smith's primary work on individuality (2009). Godfrey-Smith answers the question about Darwinian individuality by carrying out a careful examination of important elements of contemporary Darwinian theory. The details of his analysis do not matter for the purposes of this chapter, and I will not examine them here. I am not interested in joining the lively debate about what it is to be a Darwinian individual.[4] My interest is to advance a form of pragmatism by proposing that we shift attention away from seeking an analysis of individuality, as if it could be read off the best scientific theories, to seeking an understanding of individuation practices in science with respect to the purposes they serve in scientific inquiry.

Godfrey-Smith begins by arguing that reproduction is "at the heart of Darwinian evolution." His subsequent analysis of reproduction leads him to identify three parameters defining what it is to be a "paradigm" Darwinian individual:

B: Passing through a bottleneck between generations, such as through the production of a fertilized egg that consists of a single cell
G: Featuring reproductive specialization of parts, such as the distinction of a germ line from somatic cells
I: Exhibiting overall integration among parts

Godfrey-Smith points out that each of these parameters admits of degrees. For example, a bottleneck might be a narrowing to a single cell, as in the case of animals like us, or a narrowing to a relatively small number of cells as in the case of plants like aspens.

Godfrey-Smith claims that entities that exhibit the highest degree of all three features, what he calls *paradigm Darwinian individuals*, evolve most effectively through Darwinian processes. As he puts it, populations consisting of paradigm Darwinian individuals are "*the* kind of system that can produce novel and complex organisms, highly adapted to their circumstances" (p. 6, emphasis added). Populations made up of "marginal" Darwinian individuals (i.e., those that exhibit the three features to lesser degrees) can still evolve, but apparently not to the degree of novelty, complexity, and adaptedness of paradigm cases. He illustrates his analysis of Darwinian individuality with a cube. See Figure 5.2.

Godfrey-Smith applies his framework to major evolutionary transitions. He explains:

> The framework developed here is intended to be useful when thinking about transitions, especially their intermediate stages. We often find populations

[4] Although I am not interested in joining the debate, I believe it has led to fruitful philosophical inquiry about (what I would view as) conceptual tools in biology. For example, see James Griesemer's critique of Godfrey-Smith's analysis of reproduction and Griesemer's alternative (2014*a* and 2014*b*).

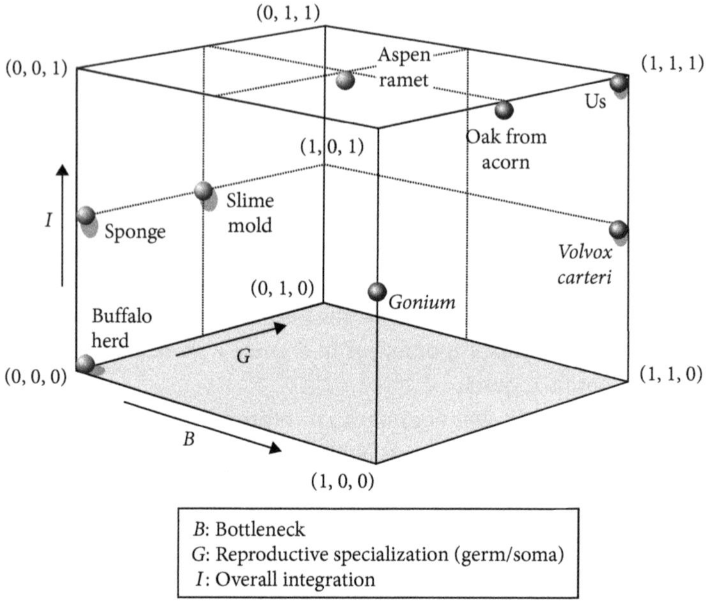

FIGURE 5.2

Godfrey-Smith's representation of his three dimensional analysis of Darwinian individuality. Each dimension comes in degrees, ranging from 0 to 1. Godfrey-Smith calls entities that meet the maxima of all three dimensions (1,1,1) "paradigm" Darwinian individuals. He says that entities that meet the dimensions to lesser degrees are "marginal" individuals from a Darwinian point of view. Image from Godfrey-Smith 2009.

> that have a marginal Darwinian status—for example, collective entities that are somewhat organism-like, but not all the way there. A great array of this can also be seen in aquatic life, in which there is a variety of forms of partial integration of cells and simple organisms into collective entities, including seaweeds, corals, and sponges. (Godfrey-Smith 2009: 8–9)

He posits that "marginal" Darwinian individuals can get "all the way there" by evolving into "paradigm" Darwinian individuals:

> As a transition of this kind occurs, a population may appear first as a marginal case, from a Darwinian point of view—an ensemble of collective entities who can only be said to reproduce in an extended, generous sense, who barely count as individuals at all. But there may be successive increases in integration [and presumably in B and G as well], until the entities display a well-defined mode of reproduction at the higher level, with heritable variation in the traits found at that level. The collective becomes a paradigm case. (Godfrey-Smith 2009: 9)

According to Godfrey-Smith's account, evolutionary processes are themselves evolving—a point John Beatty made forcefully to challenge the received view of theories (Beatty 1982). Beatty argued that there are no necessary—that is, no noncontingent—laws of evolution. He used the Hardy-Weinberg equilibrium from population genetics as an example. This principle depends on the individual alleles of a diploid pair being segregated in processes that give them equal probabilities of being transmitted to any given gamete. But, as Beatty argued, there are known exceptions to this kind of random segregation. In exceptional cases, some alleles have a greater than 50% chance of being transmitted. The Hardy-Weinberg principle, a central principle of the evolutionary theory called *population genetics*, does not apply to these alleles. Beatty argued that organisms could evolve such that the Hardy-Weinberg equilibrium no longer applied to any of them. The processes underlying evolution are evolving and so-called laws that describe today's processes might not describe tomorrow's.

Alison McConwell has recently extended Beatty's contingency thesis to argue that, since evolutionary individuality is a contingent result of evolution, we should expect it to take on a plurality of forms (McConwell 2017*a*, 2017*b*). Like Beatty's, McConwell's conception of biological diversity is not progressive. She does not assume that there is a directionality to second-order evolutionary processes. She also does not assume that there is an ideal of individuality, an essence of individuality, or a hierarchy of evolutionary processes associated with a hierarchy of forms of individuality. Instead, there is a contingent plurality.

Ellen Clarke has also argued for a plurality of individuality (2013), but Clarke's conception is quite different from McConwell's. Clarke's argument assumes that Darwinian processes are unchanging in the sense that the basic principles of Darwinian theory always apply. She argues that to evolve via these processes, organisms must have certain features and that these features define what it is to be an evolutionary individual. Plurality enters into her picture because these features can be instantiated in different ways. So, individuality is bounded by the stability of evolutionary dynamics, but, within those bounds, there can be abundant diversity. Whereas McConwell follows Beatty's idea that evolution itself evolves (what I am calling "second-order evolution") and hence imposes no fixed boundaries on individuality, Clarke's analysis assumes that evolution itself is stable.

Godfrey-Smith's conception of the plurality of individuality is different from both Clarke's and McConwell's conceptions. Like Beatty, he believes evolutionary processes are themselves evolving. Godfrey-Smith maintains that as collectives of entities in an evolving population take on the three features of being a paradigm Darwinian individual, the evolutionary process itself evolves. At an early stage of evolution, when the collectives are just marginal Darwinian individuals, the system can evolve these collectives. In some cases, as the collectives (or "marginal individuals") evolve, they take on the features of being a paradigm individual. In these cases, the evolutionary process itself changes and it "can produce novel and complex organisms, highly adapted to their circumstances" (p. 6). Godfrey-Smith's account of the transition to multicellularity proceeds from the evolution of collectives of unicellular organisms to the evolution of marginal Darwinian individuals, and then, from the evolution of marginal Darwinian individuals to the evolution of fully Darwinian individuals. The second-order evolutionary process is viewed as leading toward a more effectively evolving population with individuals that are more fully integrated; hence, in an important sense, evolution is viewed as being progressive.[5] The plurality of multicellular individuality evidently represents different stages of a second-order evolutionary process.

5.5 The Great Cube of Being?

What might readers take to be the metaphysical significance of Godfrey-Smith's analysis? On the one hand, Godfrey-Smith sometimes qualifies his statements with phrases including "from a Darwinian view," which suggests that his analysis is not intended to have metaphysical import. On the other hand, in the introduction of *Darwinian Populations* (2009), Godfrey-Smith explicitly states that his account does have such import. He distinguishes between science, philosophy of science, and philosophy of nature. The third kind of investigation, the philosophy of nature, is a form of metaphysics. He says that his book "moves constantly between these three kinds of investigation."

"The focus of *science*," Godfrey-Smith writes, "is the natural world." He envisions *philosophy of science* (as many philosophers do) as a theory-focused meta-discipline:

> The focus of philosophy of science is science itself. The aim is to understand how science works and what it achieves. Here we ask what kinds of contact with the world theories can have—how they function as representations, how they can yield understanding. We ask about the role of worthy but vexed goals such as truth, simplicity, and explanatory power and about the nature of evidence, testing and scientific change. (p. 2)

Philosophy of nature, Godfrey-Smith writes:

> is the project of taking science as developed by scientists, and working out what its real message is, especially for larger questions about our place in nature. (p. 3)

As mentioned earlier, Godfrey-Smith states that his book (2009) constantly moves between science, philosophy of science, and philosophy of nature. With this in mind, readers might ask about the philosophy of nature. What is the "real message" implied by his analysis about "our place in nature"? Godfrey-Smith doesn't say.[6] But readers searching for its "real message" might wonder whether his three-dimensional image of individuality represents the *Great Cube of Being*, with

[5] In later writings, Godfrey-Smith says that the transition can be reversed, that populations of paradigm individuals can evolve into populations of marginal individuals. This is consistent with the idea that the transition toward paradigm individuality is progressive in the sense that it leads toward populations that can evolve *more* effectively and individuals that are *more* integrated.

[6] In the rest of this section, my intention is to identify what readers (i.e., not Godfrey-Smith) might take to be "the real message . . . about our place in nature" of Godfrey-Smith's account of individuality. I am not claiming that what follows describes Godfrey-Smith's own views about the metaphysical

animals like us perched at the pinnacle, plants not too far away, and creatures like sponges and slime molds having a long ways to go. It is unclear where bacteria and archaea fit in. Nevertheless, viewed as a philosophy of nature, this is a grand view and possibly a seductive one. After all, it reinforces some traditional Western ideas about our place in nature.

I suggest we resist this interpretation by abandoning the project of philosophy of nature. Instead of viewing science as having some "real message" about our place in nature, or even about our place on a cube of Darwinian being, we can view science more modestly as a system for investigating the world that provides us means for describing, explaining, predicting, and manipulating aspects of parts of the world that are important to us.

5.6 Holobionts

A number of biologists and philosophers have recently argued that animals, as traditionally conceived, are not biological individuals at all. Rather, they are parts of individuals. The real individuals, they argue, are multispecies assemblages made up of the animal and the symbiotic microbes (bacteria, archeae, and viruses) living within it, for example within their guts or on their skins. More generally, they argue that holobionts, consisting of eukaryotic hosts and symbiotic microbes, are the real individuals.

Again, I am not interested in joining the debate about whether this or that kind of biological entity is a "real" individual. My interest is to advance pragmatism. I will not be defending or criticizing the idea that holobionts, rather than the eukaryotic hosts within holobionts, are the real individuals (just as I have not defended or criticized Godfrey-Smith's conception of Darwinian individuality). Instead, I will propose that the concepts of animal and of holobiont should be viewed as tools biologists use to individuate entities. On this view, philosophers should be analyzing how and why biologists use the holobiont concept (as well as other concepts) to individuate. The literature on holobionts is a rich source for motivating my pragmatic perspective on individuation because it connects to a broad range of scientific and medical interests. I will use a thought-provoking article "A Symbiotic View of Life: We Have Never Been Individuals" by Scott Gilbert, Jan Sapp, and Alfred Tauber (2012) to develop my position.

Before turning to Gilbert, Sapp, and Tauber's paper, it is useful to point out two features of holobionts that have figured into the debate about whether it is these assemblages, or alternatively whether it is the eukaryotic hosts within them, that are the real biological individuals. The first feature is that holobionts are

significance of his cubic analysis. Other than implying in the introduction that his account has metaphysical significance, he is coy with respect to the issue.

functionally integrated. Eukaryotic hosts could not survive or reproduce without their microbial collaborators. And many holobiont microbes critically depend on their hosts as well. Considerations about functional integration motivate the idea that holobionts are individuals. Another feature of holobionts, however, pulls in the other direction. Holobiont reproduction typically involves "horizontal" as well as "vertical" transmission. Parent–descendant relationships are usually conceived of in terms of vertical transmission. The cells in an offspring are all descendants of cells in the offspring's parent(s). This conception of transmission generally applies to plants and animals. But holobionts are different. The microbes in a holobiont are often recruited from the environment rather than inherited from the parent holobiont. That is, many of the microbes in a holobiont are not descendants of microbes in the parent(s); they are descendants of microbes that were transmitted from the environment. Considerations about horizontal transmission motivate the idea that holobionts are not individuals because evolution by natural selection presumably requires vertical transmission.[7]

Much of the philosophical literature in the debate about whether holobionts are real individuals concerns whether the means by which they reproduce are consistent with the mechanism of evolution by means of natural selection. But Gilbert, Sapp, and Tauber's paper takes a broader perspective to argue for two theses: (1) entities we usually consider individuals, such as multicellular plants and animals, are not genuine individuals; and (2) the real individuals are holobionts.

The central line of reasoning leading to the first thesis in Gilbert, Sapp, and Tauber (2012; hereafter GST) proceeds as follows:

1. from an *anatomical* conception of individuality, animals are not individuals
2. from a *developmental* conception of individuality, animals are not individuals
3. from a *physiological* conception of individuality, animals are not individuals
4. from a *genetic* conception of individuality, animals are not individuals
5. from an *immunological* conception of individuality, animals are not individuals

Although it is not necessary for my purposes to examine how they justify each of these claims, it will be instructive to briefly examine a couple of their arguments. I begin with their argument that, from a physiological conception, animals are not individuals.

The physiological view of individuality is traced by GST to the early and mid-nineteenth-century writings of Henri Milne-Edwards and Rudolf Leuckart. The basic

[7] For a different idea of what is required for evolution by natural selection, see Roughgarden et al. 2017.

idea was that the parts of a complex organism cooperate for the good of the whole. After Darwin, this developed into the view that evolution led to an "ever-increasing integration through the division of labor" (p. 329). This integration evolved within cells and also among the cells of multicellular organisms. The idea that cooperation of parts is an important feature of being an organism is indeed a central theme in the scientific literature on individuality. For example, it is one of the two axes on which Queller and Strassman set out to map the space of organismality (2009).

Biologists have learned that the division of labor favored by selection can be achieved through an integration of functioning among different species living together; for example, by an integration of activities carried out by an animal host and its microbial symbionts. GST list numerous findings about how microbial symbiosis contributes to the physiological function of animals ranging from insects to mammals. They cite research that shows that bacteria symbionts are required for lipid metabolism, detoxification, colonic pH regulation, vitamin synthesis, intestinal permeability, and suppression of dangerous immune responses. They conclude that examples such as these show that "on classic physiological grounds, animals are not individuals" (Gilbert et al. 2012: 330).

GST argue that animals are not individuals from a genetic conception of individuality on the grounds that allelic variations are not the basis for phenotypic variation in the host. Biologists have learned, they contend, that the genome of importance is what is called the *hologenome* (the combined genome of the host and microbial symbionts). They support this idea by describing how the synthesis of phenylalanine in the mealy bug *Planococcus* requires several enzymes. Only one of them is encoded in the genome of the host; the others are encoded in the genomes of two nested bacteria contained in the host. They argue:

> the genomes of *all three* organisms have been altered through this symbiosis. . . . genomes evolve in such a manner that they need their partners to achieve complex genetic integration. None of the three species . . . has a "complete" genome. It is the holobiont that does. (Gilbert et al. 2012: 329–330)

On the basis of examples such as this one, GST conclude, on classical physiological grounds, animals are not individuals.

Similar example-based arguments are offered for GST's claims that animals are not individuals from anatomical, developmental, and immunological conceptions (claims 1, 2, and 5 listed earlier). But arguing that animals are not individuals is not the same as arguing that holobionts are individuals. GST shift attention to their second main thesis, the thesis that holobionts are genuine individuals, when they consider individuality from an evolutionary perspective.

GST begin by noting that evolutionary individuality has been conceived in terms of "that which can be selected" in the process of evolution. They continue:

> Usually, these individuals are genes or monogenomic organisms. But, from the above discussion, it is evident that organisms are anatomically,

> physiologically, developmentally, genetically, and immunologically multigenomic and multispecies complexes. Can it be that organisms are selected as multigenomic associations? Is the fittest in life's struggle the multispecies group, and not an individual of a single species in that group? (Gilbert et al. 2012: 331)

They answer this question by describing an example of a holobiont, pea aphids and their endemic microbes, which they argue satisfies the conditions for group selection. Much of their discussion is aimed at showing that the holobiont immune system provides a mechanism against cheating, a potential constraint inhibiting group selection.

It is worth noting that the example chosen by GST to advance their case that holobionts are evolutionary individuals has a consequential feature: one of the bacterial symbionts is nested within the other bacterial symbiont and both bacterial symbionts are nested within the aphids. This means that reproduction of these assemblages is vertical. Hence, Godfrey-Smith classifies these holobionts as Darwinian individuals while he maintains that holobionts which exhibit horizontal transmission count as "organisms" but not as "Darwinian individuals" (2012). In contrast, GST do not seem concerned with modes of transmission and draw a general conclusion:

> Thus, animals can no longer be considered individuals in any sense of classical biology: anatomical, developmental, physiological, immunological, genetic, or evolutionary. . . . the holobiont, with its integrated community of species, becomes a unit of natural selection whose evolutionary mechanisms suggest complexity hitherto largely unexplored. (Gilbert et al. 2012: 334)

As mentioned, I will not take sides in debates about questions of the form "who are the real individuals?" I will, however, point out that just as the cube of Darwinian individuality resonates with the idea of the great chain of being, the idea that real individuals are assemblages of diverse organisms cooperating as functional wholes has its romantic appeal.

> For animals, as well as plants, there have never been individuals. This new paradigm for biology asks new questions and seeks new relationships among the different living entities on Earth. We are all lichens. (Gilbert et al. 2012: 336)

But microbiologists and biomedical scientists would note that many of the microbes within us and on us are not cooperating, and many that cooperate at certain times or in certain contexts do not cooperate at other times or in other contexts. Gilbert and other proponents of the holobiont view explicitly acknowledge this. For example, Roughgarden et al. (2017) include pathogens as integral parts of the holobiont assemblage. Relationships between hosts and their microbes are dynamic, potentially adversarial, and often fragile.

Readers should resist interpreting the viewpoint offered by GST as a philosophy of nature, as a metaphysics that identifies the "real message" of the science about our place in nature. On the other hand, the idea they develop in the concluding section of their paper—namely, that conceiving holobionts as individuals opens up new opportunities for numerous fields ranging from medicine to conservation biology—is an idea worth philosophical attention.

5.7 Purposes Served by Individuating Organisms and the Like

Maureen O'Malley suggests that questions about individuality should be framed in terms of the explanatory usefulness of different concepts (forthcoming). She begins by pointing out that distinguishing between biological individuals (or "metabolic individuals") and evolutionary individuals (or "Darwinian individuals") is becoming standard practice among philosophers writing on questions about individuality. In this paper, she challenges the idea that one of these concepts has explanatory priority. Instead, she maintains

> [t]here are explanatory and descriptive aims that can only be achieved when the constituent entities of multispecies biological individuals are separated, and other aims that rule out such separation. These explanatory differences rest on the distinction between biological and evolutionary (i.e., Darwinian) individuals, and an acknowledgement that neither has out-and-out priority free of explanatory context. (O'Malley forthcoming: 4)

Although O'Malley's focus is on explanatory utility and (in the paper quoted here) on the explanation of major evolutionary transitions in individuality, her idea that different concepts of individuality serve different aims can be generalized and localized. First, it can be generalized with respect to more than just two distinctive sets of explanatory aims (to explain biological functioning or evolutionary change). The idea that biologists tune their individuations of entities to different explanatory interests is a powerful one. We can use this idea as a basis for reappropriating the argumentation of GST.

The examples discussed in GST connect to a broad range of explanatory interests: anatomical, developmental, physiological, genetic, and immunological. Their mode of argumentation identifies numerous examples where a property or process depends not just on the animal host, but also on its symbiotic microbes. They infer from these dependencies that the animal host is not a genuine individual. Recall their example of the synthesis of phenylalanine in the mealy bug *Planococcus*. The synthesis, they point out, depends on an enzyme from the animal host and on several enzymes from two different symbiotic bacteria. The individual, they reason, could not be the animal host, and presumably, it must be the assemblage of host plus bacterial symbionts. We can reframe this reasoning in terms of the usefulness of individuating entities. If one's aim is to explain (or for that matter, to manipulate) the synthesis of phenylalanine in these creatures, then it is useful to individuate the holobiont. Individuating the animal within might obscure explanatorily relevant enzymes and DNA.

I propose that we interpret the arguments of GST as arguments about the usefulness of individuating assemblages for various purposes. We can drop O'Malley's tacit assumption that there must be one correct way to individuate biological individuals. And we can think of the purposes in a more fine-grained, localized manner. One could entertain the possibility that for the purpose of explaining the synthesis of phenylalanine in *Planococcus* it is best to individuate assemblages of host and two microbes, whereas for the purposes of explaining other biochemical syntheses, it might be more useful to individuate just the animal host.

The idea that biologists can individuate in a multiplicity of ways to serve a variety of explanatory purposes can be generalized in a second way. The purposes of individuation extend beyond the descriptive and explanatory. They also include purposes of investigation, manipulation, and prediction. Broadening philosophical attention to these purposes could move inquiry well beyond the areas of biological science that currently receive the most philosophical attention. It could also reveal that the conceptual division between biological individuals and evolutionary individual is too blunt; biologists fine-tune their individuation practices to serve specific interests. Consider the concept of "evolutionary individual."

Philosophers writing on the concept of "evolutionary individual" generally assume that evolutionary individuals must be the units of Darwinian evolution, which is generally taken to mean the units of selection. This is why Godfrey-Smith grounds his analysis of Darwinian individual in terms of selection theory. It is also why, in seeking to establish that holobionts are individuals from an evolutionary perspective, GST (and Roughgarden et al. 2017) argue that holobionts meet the conditions for group selection. But Darwinian theory cannot be reduced to the theory of transmutation by natural selection. Darwin's *Origin of Species* was based on two main ideas: (1) transmutation by means of natural selection and (2) common ancestry. The idea that species differentiate into varieties and varieties further differentiate by means of natural selection was critical to the explanatory power of Darwin's theory (Waters 2003).

In contemporary Darwinian (or neo-Darwinian) theory, the processes responsible for common ancestry (for the origin and differentiation of varieties) are conceived as speciation. Speciation has been and continues to be a central investigative topic in evolutionary biology. Yet it has largely escaped the attention of philosophers writing about evolutionary individuality. Starting with Hull (1978), philosophers have written a lot about whether species are individuals, but they haven't paid much attention to the question of how entities (organisms) making up species should be conceived for the purposes of explaining speciation processes.

An integral part of speciation involves populations becoming reproductively isolated from one another. A prominent idea is that differential mating preferences play an important role in the evolutionary development of reproductive isolation. Laboratory biologists have investigated how this development might proceed. Consider the following line of experimentation. In laboratory experiments with *Drosophila* flies, Diane Dodd randomly divided a homogeneous population of flies into two groups. She propagated one group on a molasses medium and the other group on a starch medium. The populations developed mating preferences. Flies propagated on one medium developed a preference to mate with flies propagated on the same medium. That is, molasses flies (flies evolved on the molasses medium) preferred to mate with other molasses flies rather than starch flies (Dodd 1989). Recently, experimenters demonstrated that this diet-induced mating preference depends on microbial symbionts (Sharon et al. 2010, 2011). When the two populations were treated with antibiotics, thus eliminating the bacteria, the flies no longer exhibited mating preferences. These experiments indicate that it can be useful to individuate holobionts, rather than just animals, when investigating processes of speciation. In some scientific contexts (e.g., the context of theorizing about transmutation by means of natural selection), individuating evolutionary entities at the level of animal hosts might be most useful. But in other scientific contexts (e.g., the context of investigating speciation), individuating evolutionary entities at an assemblage level might be best. The brute conceptual division between biological individual and evolutionary individual does not do justice to what should be nuanced practices of individuation.

This case reveals three things. First, biologists have a multiplicity of purposes for individuating organisms. Second, purposes for individuating, even evolutionary biologists' purposes for individuating, are best served by fine-grained practices of individuation that cross-individuate entities. Third, biologists' individuation of entities in the context of experimentation may reveal scientific aims, such as the aim of investigating speciation, that can be lost in the context of abstract theorizing. Philosophers' preoccupation with analyzing abstract theories obscures the diversity of purposes in science and leads us to oversimplify our understanding of biological practices of individuation.

5.8 Three Lessons

Following philosophical pragmatism, we should view practices of individuation in terms of three-place relations: between the world, ideas, and human purposes and actions. My three lessons can be viewed in terms of Charles Morris's account of semiosis (1938). Although I draw these lessons with respect to philosophy of biology, I believe they apply to philosophy of science more generally and, in fact, to philosophy at large.

The first lesson concerns the world—that is, ontology. Much of the general philosophical literature on individuality is based on the assumption, or perhaps I should say the faith, that the world is neatly divided into individuals. This is also evident in philosophy of biology. With respect to genes, when philosophers learned that DNA is not neatly divided into genes, they decided that genes must not exist at the molecular scale. The real entities of heredity, they inferred, must be found elsewhere. Indeed the findings of genetics indicate that DNA is not neatly divided into elements that line up with a uniquely correct and comprehensive partition of developmental processes. But the ontological conclusion we should draw is that the processes by which organisms function and develop are, *as a whole*, incomprehensibly complex. They are a mess; a mess with lots of structure, but no overall structure.

With respect to the ontology of organisms, philosophers have become increasingly open to the idea that the world might not be neatly divided into just one kind of individual, that there might be multiple kinds of individuals, perhaps even cross-identifying kinds of individuals (e.g., organisms vs. evolutionary individuals). But almost all recent philosophical attention has been devoted to analyzing what it is to be an evolutionary individual, not what it is to be an organism. It is as if the processes of evolution must define what it is to be an organism because, after all, organisms are products of evolution. In addition, the distribution of philosophical attention suggests there is an assumption among many philosophers that what it is to be an evolutionary individual can be established by analyzing the best scientific theory, which is taken to be the theory of natural selection as developed in the framework of population genetics. But scientific findings discussed in this chapter indicate that the empirical world (as contrasted with the abstract world of Darwinian theory) is not neatly divided into evolutionary individuals. The processes of evolution are extremely complex, and the results of these complex processes have not yielded a neat parsing of life into individuals. The processes of evolution are a mess; a mess with lots of structure, but no overall structure. That's the way the world is (Waters 2017).

The second lesson concerns epistemology. Concepts should be conceived as tools, used to help achieve particular purposes. Biologists use concepts to individuate different entities and processes. These tools serve a number of scientific and technological interests, but they do not provide the basis for a philosophy of nature, or for grand views about the living world and our place within it. To be clear, my targets are not particular analyses of individuality such as Godfrey-Smith's or Gilbert, Sapp, and Tauber's. My target is the metaphysical presumption that scientific theories have a "real message" about "our place in nature." The molecular gene concept is used to individuate segments of DNA in ways that serve a number of purposes, including purposes to investigate, explain, predict, and manipulate biological entities and processes. But this tool does not provide the means to draw a univocal parsing of DNA that structures the development or the functioning of organisms. Likewise, concepts of organismality, including those based on metabolic, immunological, symbiotic, and evolutionary viewpoints, are used to individuate biological life in a multiplicity of ways that serve different purposes. Philosophers have emphasized explanatory purposes and particularly purposes related to explaining the evolution of creatures like us. But description and explanation do not exhaust the aims of science. Scientists also aim to manipulate, control, investigate, and predict. This brings us to the third lesson.

The meta-philosophical lesson is future looking and pertains to what questions should guide philosophical inquiry. I conclude that instead of asking, "what is a biological individual?" philosophers should ask:

- what ways of conceiving biological individuality could be useful?
- in what contexts and for what purposes would they be useful?
- could one concept be useful for situations in which biologists tend to use several concepts?
- in situations where biologists use one concept, could a different concept or a multiplicity of concepts better serve biologists' purposes?
- might there be purposes in medicine, public health, or environmental science that would be better achieved by using new concepts of individuality?

Pursuing these kinds of questions, couched in terms of purposes, could move philosophers beyond analyzing science solely in terms of description, explanation, and prediction. It could decrease our obsession with the philosophical methodology of grounding analysis in abstract theories. Pursuing these questions could also lead to deeper and more inclusive understandings of science and a more useful philosophy of science.

Acknowledgments

I thank audiences at Taiwan National Chung Cheng University, the University of Pittsburgh, and the Final Conference of the ERC Project "A Process Ontology for Contemporary Biology" for helpful feedback. I am grateful that a number of individuals offered useful critiques and discussion of earlier drafts of this chapter: Ananya Chattoraj, Antoine Dussault, Scott Gilbert, Brian Hanley, Ka Ho Lam, Oliver Lean, Celso Neto, Maureen O'Malley, Isobel Ronai, Ruth Shaw, Bill Wimsatt, Marcel Weber, Helen Zhao, and especially Marc Ereshefsky, Alan Love, and Dan Molter. I am indebted to Javier Suarez Diaz for sharing an enlightening manuscript of his with me. Anonymous referees provided helpful critiques. I thank my patient editors, Otávio Bueno, Ruey-Lin Chen, and Melinda Bonnie Fagan. The research and writing of this chapter was supported in part by a grant from the John Templeton Foundation: #50191; From Biological Practice to Scientific Metaphysics; http://biological-practice-to-metaphysics.org and Canada's generous CRC program.

References

Beatty, John. 1982. What's wrong with the received view of evolutionary biology? In PSA 1980, *Proceedings of the 1980 Biennial Meetings of the Philosophy of Science Association, Volume 2*, eds. P. D. Asquith and R. N. Giere. East Lansing: Philosophy of Science Association.

Buss, Leon W. 1987. *The evolution of individuality*. Princeton: Princeton University Press.

Carlson, Elof A. 1966. *The gene: A critical history*. W. B. Saunders, republished in 1989. Ames: Iowa State University Press.

Clarke, Ellen. 2013. The multiple realizability of biological individuals. *Journal of Philosophy* 110(8): 413–435.

Dodd, Diane M. B. 1989. Reproductive isolation as a consequence of adaptive divergence in Drosophila pseudoobscura. *Evolution* 43: 1308–1311.

Fine, Arthur. 1996. *The shaky game: Einstein, realism, and the quantum theory*. Chicago: University of Chicago Press.

Gilbert, Scott F., Jan Sapp, and Alfred I. Tauber. 2012. A symbiotic view of life: We have never been individuals. *Quarterly Review of Biology* 87(4): 325–341.

Godfrey-Smith, Peter. 2009. *Darwinian populations and natural selection*. Oxford: Oxford University Press.

Godfrey-Smith, Peter 2012. Darwinian individuals. In *From groups to individuals: Evolution and emerging individuality*, eds. Frédéric Bouchard and Philippe Huneman, 17–36. Cambridge, MA: MIT Press.

Griesemer, James. 2014*a*. Reproduction and the scaffolded development of hybrids. In *Developing scaffolds in evolution, culture, and cognition*, eds. Linnda R. Caporael, James R. Griesemer, and William C. Wimsatt, 23–55. Cambridge, MA: MIT Press.

Griesemer, James. 2014*b*. Reproduction and scaffolded developmental processes: An integrated evolutionary perspective. In *Towards a theory of development*, eds. Alessandro Minelli and Thomas Pradeu, 183–202. Oxford: Oxford University Press.

Hull, David. 1978. A matter of individuality. *Philosophy of Science* 45: 335–360.

Kohler, Robert. 1994. *Lords of the fly: Drosophila genetics and the experimental life*. Chicago: Chicago University Press.

Maynard-Smith, John, and Eörs Szathmáry. 1995. *The major transitions in evolution*. Oxford: Oxford University Press.

McConwell, Alison. 2017*a*. *Individuality, the major transitions and the evolutionary contingency thesis*, PhD dissertation. Calgary: University of Calgary.

McConwell, Alison. 2017*b*. Contingency and individuality: A plurality of individuality types. *Philosophy of Science* 84: 1104–1116.

Morris, Charles. 1938. Foundations of the theory of signs. In *International encyclopedia of unified science, volume 1*, ed. Otto Neurath, 79–137. Chicago: University of Chicago Press.

O'Malley, Maureen A. Forthcoming. Is evolution fundamental when it comes to defining biological ontology? No. In *Current controversies in philosophy of science*, eds. Shamik Dasgupta and Brad Westlake. London: Routledge.

Queller, D., and J. Strassmann. 2009. Beyond society: The evolution of organismality. *Philosophical Transactions of the Royal Society B* 364: 3143–3155.

Ratcliff, William C., R. Ford Denison, Mark Borrello, and Michael Travisano. 2012. Experimental evolution of multicellularity. *Proceedings of the National Academy of Sciences* 109(5): 1595–1600.

Roughgarden, Joan, Scott F. Gilbert, Eugene Rosenberg, Ilana Zilber-Rosenberg, and Elisabeth A. Lloyd. 2017. Holobionts as units of selection and a model of their population dynamics and evolution. *Biological Theory*. https://doi.org/10.1007/s13752-017-0287-1.

Sharon G., D. Segal, J. M. Ringo, A. Hefetz, I. Zilber-Rosenberg, and E. Rosenberg. 2010. Commensal bacteria play a role in mating preference of Drosophila melanogaster. *Proceedings of the National Academy of Sciences USA* 107: 20051–20056. http://dx.doi.org/10.1073/pnas.1009906107.

Sharon G., D. Segal, I. Zilber-Rosenberg, and E. Rosenberg. 2011. Symbiotic bacteria are responsible for diet-induced mating preference in Drosophila melanogaster, providing support for the hologenome concept of evolution. *Gut Microbes* 2: 190–192. http://dx.doi.org/10.4161/gmic.2.3.16103.

Waters, C. Kenneth. 1994. Genes made molecular. *Philosophy of Science* 61: 163–185.

Waters, C. Kenneth. 2001. Molecules made biological. *Revue Internationale de Philosophie*, 4/2000-n 214-(2000): 539–564.

Waters, C. Kenneth. 2003. The arguments in Darwin's *Origin of species*. In *The Cambridge companion to Darwin*, eds. Jonathan Hodge and Gregory Radick, 116–139. Cambridge: Cambridge University Press.

Waters, C. Kenneth. 2004. What was classical genetics? *Studies in History and Philosophy of Science* 35: 783–809.

Waters, C. Kenneth. 2007. Causes that make a difference. *Journal of Philosophy* CIV(11): 551–579. [pre-print on PhilSci-Archive.pitt.edu]

Waters, C. Kenneth. 2008. Beyond theoretical reduction and layer-cake antireduction: How DNA retooled genetics and transformed biological practice. In *Oxford handbook to the philosophy of biology*, ed. Michael Ruse, 238–262. New York: Oxford University Press, 2008. [pre-print on PhilSci-Archive.pitt.edu]

Waters, C. Kenneth. 2017. No general structure. In *Metaphysics and the philosophy of science: New essays*, eds. Matthew Slater and Zanja Yudell, 81–108. Oxford: Oxford University Press. [pre-print on PhilSci-Archive.pitt.edu]

{6}

Individuality, Organisms, and Cell Differentiation

Melinda Bonnie Fagan

6.1 Introduction

This chapter explores the implications of stem cell experiments and the artificial entities they produce for ideas about biological individuality and organisms. There is no consensus among philosophers of biology (or biologists) concerning criteria for biological individuality or organismality or even the proper way to construe the relation between these two concepts. This chapter does not aspire to settle these debates, but rather to add to them a perspective that has received little attention as yet from philosophers of biology: experimental studies of cell and embryonic development. I take as my starting point the experimental practices and concepts of stem cell biology, and, from consideration of these, I draw conclusions that bear on questions of biological individuality, the cell–organism relation, and philosophical accounts of organisms. The main claim I will defend is that we should understand stem cells as including a developmental process—a cell lineage—and key features of organismal organization. This in turn supports a claim I have made elsewhere, that at least one variety of stem cell can be considered a model organism for studying development. It also comports well with a process view of stem cells and of organisms more generally, although I do not defend such a view here.

The phenomenon of development (ontogeny) is central to questions of biological individuality and organismality (Buss 1987). Every organism undergoes processes of change in the passage from birth to death. Those processes that are directional and robust across members of a species comprise the phenomenon of development.[1] In multicellular organisms, development includes the transformation of a single cell (the zygote) to a multicellular embryo and then to a mature organism, which eventually dies. The developing entity, while remaining in some

[1] Waddington (1970) parses development into three subprocesses: growth (increase in size, mass, and cell number), differentiation (appearance of diverse structures/functions among parts), and morphogenesis (generation of complex structures, e.g., organs, out of simpler ones, e.g., cells). This chapter

sense a single individual, transitions from the cellular to multicellular level of biological organization. Stem cells are crucial to this transition. Therefore, clarifying their nature and role is important for understanding the development of multicellular organisms, in turn a central issue for theories of biological individuality and organismality.

Stem cells are defined functionally as undifferentiated cells that can both self-renew and differentiate (e.g., Ramalho-Santos and Willenbring 2007: 35; Melton and Cowan 2009: xxiv). Self-renewal for a cell is defined as producing one or two copies of the original cell, while to differentiate is to give rise to more specialized cells. Stem cells are thus conceptualized in terms of developmental and reproductive processes. This general definition applies to all varieties of stem cell, of which there are many: embryonic, pluripotent, multipotent, induced, mesenchymal, neural, hematopoietic (blood-forming), embryonal carcinoma, and more. The model organism thesis that I proposed in earlier work (2013*a*, 2016*a*) concerns one variety in particular: cultured pluripotent stem cells.[2] Cultured pluripotent stem cells are produced by extracting cells from a multicellular organism and placing them in an in vitro environment that maintains or induces stem cell phenotype and behavior. In this artificial setting, cultured stem cells continue to divide ("self-renew") and grow in clumps, which are "thinned" on a biweekly basis ("passaging"). The experimental procedures of in vitro cell culture and passaging produce a continuous stem cell *line*. Artificially produced lines of this sort furnish pluripotent stem cells for an enormous range of experiments: drug development, producing more specialized varieties of stem cell (e.g., neural, hematopoietic [blood-making], muscle) for further experimentation, genome editing, disease modeling, and studying core mechanisms of mammalian development, among others.

In previous work, I have argued that there is reason to consider cultured pluripotent stem cells as biological individuals by analogy with multicellular organisms (Fagan 2013*a*, 2016*a*). This argument hinges on the thesis that cultured pluripotent stem cells are model organisms designed to render features of early mammalian development accessible to observation and experiment. These stem cells (or rather, stem cell lines) are constructed to exhibit features of organismal individuality in a highly simplified, accessible way. So these experimental entities can be considered *model organisms* for studying mammalian development. I review this argument in Section 6.3. Section 6.2 provides background for the model organism argument, as well as the lineage view of stem cells which figures prominently in the argument of this chapter. Having sketched the model organism thesis and its background, I then

---

focuses primarily on differentiation at the cellular level, although other aspects of development are discussed as well.

[2] Some of the claims in this chapter extend to other kinds of stem cell. However, as much of the argument concerns experimental methods for producing and identifying stem cells, extrapolation of my conclusions to unmanipulated stem cells in the body is a matter for further consideration, which must await another paper.

consider an objection to that thesis: cultured pluripotent stem cells do not exhibit the internal complexity, integration, or autonomy requisite for organisms according to many theories. These stem cells therefore cannot be model organisms (although they might be models). I synthesize several prominent accounts of organismal organization to formulate a strong version of this requirement and present the argument that cultured pluripotent stem cells fail to satisfy it (Section 6.4).

Section 6.5 rebuts this objection to the model organism thesis. The key idea of this rebuttal is that pluripotency, and developmental potential more generally, is partly constitutive of what it is to be a stem cell. Experiments that establish this potential for cultured stem cell lines produce whole multicellular organisms and other entities exhibiting some but not all features of organismal organization. If our understanding of the concept "stem cell" is expanded to include these experiments, as well as those that demonstrate self-renewal, then cultured pluripotent stem cells satisfy the strong organismal organization requirement (and, *a fortiori*, weaker versions of that requirement). The organismal organization objection is thus dispelled. Section 6.6 discusses the broader significance of this result, clarifying the process view of stem cells that fits well with this study of experimental practices and concepts.

6.2 Preliminaries: Cell Individuality

Cells have long been understood as the simplest "units of life," with standard parts and strikingly clear individuation criteria. Indeed, a cell is arguably our clearest example of a biological individual according to common-sense criteria: a living entity that can be counted, picked out from its environment, and distinguished from others of its kind. The criterion for such common-sense individuation of cells is unambiguous: an enclosing membrane, such that the cell consists of the membrane and what it contains. Of course, signals from the environment pass through the membrane continuously and vice versa—the membrane is not a barrier. It does, however, provide a clear criterion for determining whether there is one cell, two cells, no cells, or any number, in a given situation. When cells reproduce by division, the process is finished when the membranes "pinch off" to separate. This topological criterion for cell individuation is one of the clearest, least ambiguous boundary conditions in all of biology. So there is a strong *prima facie* case for considering cells to be biological individuals according to common-sense criteria.

The core tenets of cell theory offer a more systematic statement of cell individuation conditions:[3]

(CT1) Every organism begins as a single cell, which, in multicellular organisms, gives rise to all the body's cells.

---

[3] Historically, the principles of cell theory have been expressed in many different ways (Virchow 1859/1958; Reynolds 2007; Nicholson 2010). The preceding statements are very widely accepted in

(CT2) Cells reproduce by binary division.[4]

(CT3) The life of a cell begins with a division event and ends with either a second division event yielding two offspring or cell death (and no offspring).

(CT4) Generations of cells linked by reproductive division form a lineage.

These four tenets are background assumptions shared by many biological fields today, including stem cell research. CT1 relates cells to organisms. Every organism begins as a single cell, and most continue that way. For single-cell organisms such as bacteria and yeast, questions of cell and organismal individuality coincide. As will be seen later, stem cells are implicated in the developmental transition between these two "levels" of individuality. CT2 specifies how cells engage in a significant biological process: reproduction. CT3 indicates a single cell's spatial and temporal boundaries. And CT4 introduces the notion of a lineage, a biological entity composed of successive cell generations, organized by reproductive relations.

CT1–4 apply alike to single-cell organisms and to cells of multicellular organisms. Following on the listed points, both categories of cell satisfy some common-sense criteria for individuality: they are physically contiguous, exhibit clear spatial and temporal boundaries, bear properties, and function as agents within an environment (Wilson and Barker 2013). Single-cell organisms also straightforwardly satisfy many widely accepted criteria for *biological* individuality: participation in biologically significant causal processes (e.g., reproduction and evolution), physiological integration, and functional autonomy.[5] For cells of multicellular organisms, biological individuality is less clear-cut. Different theoretical perspectives yield different answers, based on different criteria (for a selection, see Guay and Pradeu 2016). From the perspective of cell biology, one consideration in favor of taking cells of multicellular organisms to be biological individuals is that they are reliably classified into types. The properties of cells that compose multicellular organisms do not vary continuously but cluster into discrete groups, each of which shares a suite of integrated structural and functional traits. These groupings are referred to as "cell types" (e.g., neurons, muscle, B lymphocytes). Cell types exhibit considerable robustness across organismal taxa and can be

---

life sciences today but should not be confused with tenets of cell theory across all relevant historical contexts.

[4] There are two modes of cell division: mitosis and meiosis. In mitosis, the genome replicates once before the cell divides. In meiosis, the genome replicates once, but two rounds of cell division follow, yielding four offspring cells with half the complement of DNA. Because stem cell and somatic cell phenomena involve mitosis, the term "cell division" is used throughout this chapter to refer to that mode only.

[5] One often-cited criterion of biological individuality does not apply to single-cell organisms alongside the rest: genetic homogeneity (Wilson 1999). However, this criterion has recently come under criticism, not least because multicellular organisms also fail to satisfy traditional formulations of it (see Wilson and Barker [2013] and references therein).

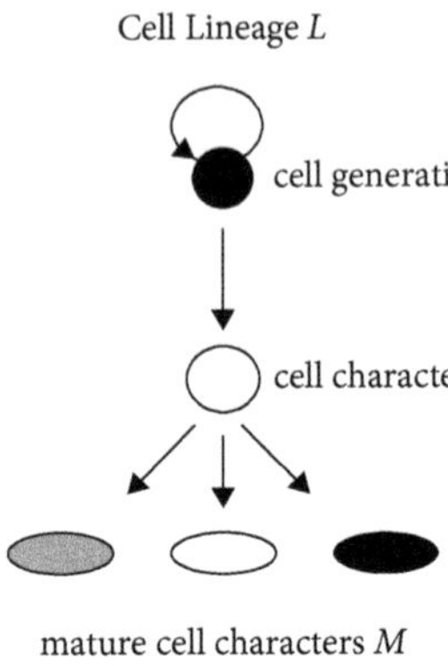

FIGURE 6.1 *Minimal stem cell model.*

clearly distinguished from one another by their particular combinations of structural and functional traits. The features of cell types are typically stable across cell generations—liver cells divide to produce more liver cells, fibroblasts beget more fibroblasts, and so on. Furthermore, cells of multicellular organisms also satisfy physiological and evolutionary criteria for individuality, each having a metabolism, heritable structural and functional traits that vary in response to mutation, and clear criteria for reproductive success. So neurons, muscle cells, blood cells, and so on satisfy at least some important criteria for biological individuality. They therefore qualify as biological individuals according to some significant theoretical perspectives within biology.

Stem cells, however, are not a cell type of this sort.[6] As noted in Section 6.1, they are defined as cells capable of generating progeny like themselves and more specialized, differentiated cells. It follows from this definition that stem cells are not lone individual cells but starting points of cell *lineages*. Self-renewal is a reproductive process spanning two or more cell generations; stem cells are defined in terms of this process, which generates a cell lineage. Differentiation is a process of change in cell phenotype, relative to a presumed order of developmental stages—it is a directed process of change. In previous work (2013*a*), I have explicated the general definition of "stem cell" as an abstract model consisting of four variables specifying the lineage (or source), time interval, traits of candidate stem cells, and traits of relevant differentiated cells. This model defines a stem cell by its position as the unique stem of cell lineage L, with maximal self-renewal and differentiation potential relative to sets of cell characters C and M, values of which vary within L over time interval n (Figure 6.1). Variable C denotes a set of characters for individual cells, which can be attributed to candidate stem cells (e.g., size,

[6] Readers who find the preceding considerations about individuality for cells of multicellular organisms unpersuasive may thus concur with the main argument of this chapter; my claims about stem cells rely on those points for contrast only.

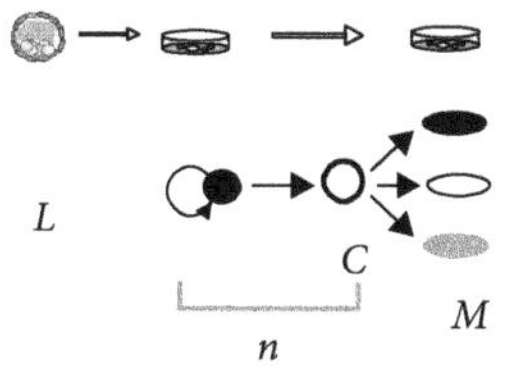

FIGURE 6.2 *Applying the model via stem cell experiments.*

shape, concentration of a particular protein). Variable *M* refers to a set of distinctive characters of a mature cell type(s) of interest; these characters mark the endpoint(s) of cell development. This explication of the general definition makes evident the lineage aspect of stem cells: the very concept presupposes a lineage produced by balanced processes of cellular self-renewal and differentiation, spanning multiple cell generations and developmental stages. To put the point in metaphysical terms, a stem cell is not an individual cell of a particular type that can be conceptualized in isolation. Rather, a stem cell as such is inseparable from a cell lineage in which processes of self-renewal and differentiation occur.

Furthermore, stem cells cannot be experimentally shown to correspond to single cells in biological systems.[7] That is to say, they cannot be experimentally individuated at the single-cell level, as cells of the familiar types can. This is because stem cell experiments involve two sets of measurements, while no single cell persists through both (Figure 6.2). To experimentally establish a candidate stem cell's differentiation potential, that cell must be placed in an environment conducive to differentiation and character-values of progeny that indicate differentiation measured later. To establish self-renewal, a candidate stem cell must be maintained in an environment that blocks differentiation and its progeny measured for character-values that indicate similarity to the original. It is not possible to perform both experiments on a single cell.[8] So stem cells cannot be experimentally individuated at the single-cell level.

These two claims, which I have defended in detail in previous work, are significant for the biological individuality of stem cells. Stem cells as currently conceptualized and experimentally identified do not correspond to individual cells. Therefore, while specialized mature cells of a multicellular organism can be defined and experimentally characterized as individual cells, stem cells cannot. The contrasts block arguments for biological individuality of stem cells that parallel those for single cells, whether isolated or of multicellular organisms.

---

[7] The term "correspond" is here used loosely to refer to a mapping relation from components of the model to real biological systems. My argument is independent of any technical account of the correspondence relation.

[8] The evidential problem, which I elsewhere term "the stem cell uncertainty principle," is more complex and severe than this brief treatment can convey. For details, see Fagan (2013*a*: 3) and Fagan (2013*b*); a more compressed discussion appears in Fagan (2016*a*).

However, another line of argument supports a claim of biological individuality for at least some stem cells. Cultured pluripotent stem cells, I propose, are biological individuals in a way analogous to multicellular organisms. Central to this proposal is the notion of a *model organism*. The next section briefly introduces this notion and summarizes the argument that some stem cells can reasonably be considered model organisms.

6.3 Model Organism Thesis

Since the early twentieth century, biologists have concentrated research efforts on a few select organisms, including fruitfly (*Drosophila melanogaster*), nematode worm (*Caenorhabditis elegans*), the laboratory mouse *Mus musculis, Escherichia coli* bacteria, and a few others.[9] These "classic" model organisms are members of species taken from the wild and modified to suit researchers' purposes. Model organism research provides much of the experimental basis for twentieth- and twenty-first-century biology, including its enormous advances in genetics, molecular biology, and medical research. Classic model organisms are constructed to be tractable for laboratory use and to exhibit phenomena of interest in a way accessible to controlled manipulation. Model organisms for studying development, in particular, share several characteristic features: small body size, large offspring number, rapid growth rate, robust processes of reproduction and development, tractability, simplicity, and accessibility to observation and measurement (Bolker 1995).[10]

Cultured pluripotent stem cells exhibit all the features of model organisms for studying development.[11] As described in Section 6.1, stem cells of this kind are experimental entities, produced by removing cells from an organismal source and placing them in cell culture with chemical factors added to prevent differentiation. Artificial cell cultures consist of a transparent dish or flask of liquid media enriched with nutrients, kept at a constant warm temperature (Landecker 2007). Cells that rapidly divide under these conditions form colonies, which are selected and put into new cultures. This process of selection and transfer is repeated every

---

[9] The literature on model organisms is extensive and spans multiple disciplines. Influential studies of particular model organisms include Kohler (1994) and Rader (2004). See Fagan (2013*a*: chapt. 7) for additional references.

[10] Fagan (2016*b*) discusses discontinuities between interdisciplinary studies of model organisms and recent work on models and modeling in philosophy of science. Briefly, the modeling role of model organisms overlaps that of abstract scientific models, but has several distinguishing features. Studies of classic model organisms are, for the most part, concerned with their epistemic modeling role rather than with their ontological status as organisms; the latter quality is taken for granted in case studies of, for example, laboratory mice and fruitflies.

[11] A more detailed version of this argument appears in Fagan (2016*a*), an earlier version in chapter 7 of Fagan (2013*a*).

few weeks, to maintain a continuously growing lineage of undifferentiated cells. Self-renewal is therefore imposed by the experimental design. Differentiation potential is demonstrated by moving small "clumps" of cultured self-renewing cells to a new environment including chemical factors that induce differentiation into a certain cell type: cardiac muscle, neurons, bone marrow, and the like. Continuously growing lines of undifferentiated cells that can differentiate under appropriate culture conditions qualify as stem cell lines. Cultured *pluripotent* stem cells are capable of differentiating into all the cell types that make up a multicellular organism.[12]

These experimentally constructed entities exhibit features associated with individuality in several theoretical perspectives. In common-sense terms, they have clear spatial boundaries in three dimensions and they endure over time—having, indeed, apparently unlimited lifespans. Cultured pluripotent stem cells also have molecular, cellular, morphological, and physiological properties and participate in causal processes of self-renewal, experimental manipulation, and—under suitable conditions—cell development (see later discussion). They are, in addition, evolutionary individuals, subject to artificial as well as natural selection. Cycles of transfer to fresh cultures, which select for cells "with a uniform undifferentiated morphology," even impose a version of the germ–soma distinction analogous to that of multicellular organisms (Thomson et al. 1998: 1147, n6). Moreover, as stated earlier, cultured pluripotent stem cells exhibit the characteristic features of model organisms. Their body size, though flexible in principle, is in practice constrained by the dimensions of cell culture dishes and flasks. Rapid, continuous growth is ensured by selection of cultured cells for frequent division. Transfer to new cultures occurs about once a fortnight, ensuring a rapid reproductive rate. And cultured pluripotent stem cells make cellular developmental processes visible and accessible—in striking contrast with normal mammalian development. Tractability is at present more goal than reality, as researchers work to streamline and standardize stem cell culture conditions. However, when any new stem cell line is established, the next tasks are to reduce variation among cells and manipulate culture conditions to yield reproducible experimental results. These efforts resemble earlier standardization practices for classic model organisms such as inbred mice and fruitfly strains.

Cultured pluripotent stem cells also play the dual epistemic role of a model organism, being objects of study in their own right as well as tools for representing target phenomena. More precisely, model organisms are designed to exhibit a phenomenon of scientific interest that is shared by a wide range of other (nonmodel) organisms, in a simple, accessible, exaggerated way. The phenomenon of interest in this case is multicellular development itself. Cultured pluripotent stem cells are

---

[12] Stem cells with more restricted developmental potential are termed "multipotent," "oligopotent," or "unipotent" (though the latter two monikers are rarely used).

designed to display paradigmatic features of mammalian (specifically, human) development in a way that is accessible to observation and experimental manipulation. To sum up: in their epistemic role and methodologically significant features, cultured pluripotent stem cells resemble the classic model organisms of experimental biology. On the strength of this analogy, it is reasonable to consider these stem cells to be model organisms for studying early mammalian development. To put the point in metaphysical terms, cultured pluripotent stem cells are rudimentary multicellular organisms that, in virtue of shared features, provide insight into the developmental processes occurring in more complex multicellular organisms at early embryonic stages.

6.4 Organization Objection

An obvious objection to this model organism thesis is that cultured pluripotent stem cells might be *models* without being *organisms*. The features of classic model organisms that are shared by cultured pluripotent stem cells have primarily to do with their epistemic modeling role, not the entities' ontological status as organisms. So the case that cultured pluripotent stem cells should be considered organisms must be made independently of the case that they are models. Can cultured pluripotent stem cells reasonably be considered organisms?

There is a dialectical difficulty in answering this question, as there is no consensus among philosophers of biology (or among biologists) concerning the criteria for being an organism or concerning the relation of organismality to biological individuality. The nature of an organism is a perennially contested issue. My strategy for navigating this dialectical difficulty is to turn the question around, directing my argument to the strongest reason *not* to consider cultured pluripotent stem cells as organisms. To the question "Why not consider cultured pluripotent stem cells to be organisms?" the most intuitive answer is: "Because they lack sufficient internal organization." Pluripotent stem cells growing in culture are maintained as undifferentiated clumps, lacking the internal functional organization exhibited by each component cell or by paradigmatic biological organisms like hawks and horses. This objection presupposes that some degree of internal organization is required for an entity to qualify as an organism: an *organization* requirement for organismality. As stated, of course, this requirement is very vague. Most accounts of organismality include some version of this requirement, but differ widely in their details. Rather than survey and address all the alternatives on offer, I will direct my argument to the strongest version of the organization requirement that can be extracted from the philosophical literature.

By so doing, I do not mean to endorse this strong version of the requirement. My purpose in this section is dialectical rather than ontological: if I can show that cultured pluripotent stem cells do (despite appearances) satisfy this strong version of the requirement, then it is safe to assume that they also satisfy weaker versions.[13] This approach allows me to address what I take to be the main objection to the model organism thesis, despite wide disagreement about the nature of organisms. However, rebuttal of one key objection does not thereby establish that cultured pluripotent stem cells are organisms. There may be other criteria for organismality which give rise to other objections—the organization requirement is not itself a full account of biological organisms. I concentrate here on this requirement, in its strong version, not because I think it is the only or most important criterion for organismality, but because this requirement seems to be one that cultured pluripotent stem cells fail to satisfy. In the next section, I will show that, contra appearances, they do satisfy it. If the requirement itself turns out to be wrongheaded, then the objection to my model organism thesis evaporates in any case. Any other objections, based on different criteria for organismality, must be dealt with in another paper.

In the remainder of this section, I explicate the strong version of the organismal organization requirement. I proceed by surveying several prominent accounts of organismal organization from the philosophical literature. This survey is not intended to be comprehensive but, again, to articulate the strongest version of this requirement that can be extracted from the philosophical literature. Internal organization and functional integration have long been considered essential features of organisms. Indeed, the term "organism," coined in the early eighteenth century, originally referred to the "organizing of a body" by an immanent principle of order (Cheung 2006). The conceptual link between "organism" and "organization" has been influential in discussions of biological individuality, because (multicellular) organisms are often taken to be paradigmatic biological individuals (although, as noted earlier, this is a contested issue). The starting point for philosophical discussion of organisms as organized beings is Kant's account of living things as self-organizing systems, in which the parts reciprocally produce one another and are continually replenished under the influence of the organized whole. Recent accounts update and develop this view in light of twentieth-century advances in biology, dynamical systems theory, and conceptions of autonomy.

One influential account defines organisms as "autopoietic systems" exhibiting a kind of double circularity: differentiated parts bring forth and are brought forth by one another, and the same reciprocal relation holds between parts and the whole (Maturana and Varela 1980; Weber and Varela 2002).[14] The circular, nonlinear relations connecting parts with one another and (collectively) with the overall whole have a dual aspect, being both causal and means-end. The latter ensures that "organisms are subjects having purposes," each relating to its environment as a "unitary locus capable of experience" (Weber and Varela 2002: 101–102). An organism's unity and "autonomy" emerge from the non-linear causality

[13] Readers who object to the strong version of this requirement can substitute the version they prefer. If weaker than that rebutted here, the overall argument of this chapter is unaffected.

[14] See also the discussion in Wolfe (2010).

that constructs the boundary between organism and environment. Autopoiesis theorists focus on the minimal case of an organism: a single cell. In this case, the boundary is the cell membrane, produced and maintained by intracellular metabolism, which in turn is defined and constituted by the boundary between cell and environment. Intracellular metabolism also reproduces itself, in that its operation produces the components involved in ongoing metabolic processes. More generally, "an autopoietic system . . . is one that continuously produces the components that specify it, while at the same time realizing it (the system) as a concrete unity in space and time, which makes the network of production of components possible" (Weber and Varela 2002: 115). So the autopoietic system continually regenerates itself, as well as the boundaries that establish it as "a distinguishable unity" separate from the environment with which it is inextricably engaged.

Other accounts of organismal organization also emphasize nonlinear causality and the internal–external boundary.[15] For example, Mossio and Moreno (2010) characterize organisms as a special case of self-maintaining systems; i.e., systems that causally affect the material boundary conditions essential for their continued existence. On their view, organisms are distinguished from other self-maintaining systems by "organisational closure": dynamic stability arising from a network of internal structures that operate to mutually constrain one another's boundary conditions. The overall system is thus maintained as a functioning whole distinct from its environment. Organizational closure requires structural and functional diversity internal to the self-maintaining organism: "*various* localizable structures make a specific and *distinguishable* contribution to the maintenance of the whole organisation" (p. 281; emphasis added).[16] Similar ideas are explicated in Collier's (2004) account of "cohesion" for self-organizing systems. Collier's "cohesion" is a logical closure relation unifying the internal causal relations among components of a system, such that the system as a whole exhibits ongoing self-maintenance.[17] More recently, Arnellos, Moreno, and Ruiz-Mirazo (2014) propose organizational requirements for autonomous multicellular organisms, with the key condition being regulation by an intercellular network controlling mechanisms of cell differentiation and division.

McShea (2001) takes a minimalist approach, identifying degrees of individuation for different kinds of biological entity. Lowest is a homogeneous aggregate of

---

[15] Means–end relations and the notion of organisms as subjects of experience do not feature in these other accounts, however.

[16] See also: "*various different* structures contribute by playing *different* mutually dependent causal roles to the global organisation, which collectively achieves self maintenance" (Mossio and Moreno 2010: 276; emphasis added). Mossio and Moreno do not consider organisational closure a sufficient condition for "organismality." Additional requirements include "modular templates" and control mechanisms involving "dynamical decoupling between constraints."

[17] Collier (2004) proposes three conditions for self-organization: an external energy source, internal interactions among components, and at least one critical point between different system attractors. Organization is generated from the latter two (internal) requirements.

interacting parts, highest is an aggregate of parts showing structural organization in their own right, and intermediate is a differentiated aggregate of interacting heterogeneous parts. These idealized points mark a continuum of individuality within each level of compositional hierarchy (e.g., prokaryotic cells, eukaryotic cells, multicellular organisms), indicating "the degree to which [a biological entity] constitutes a unified whole" (2001: 409).[18] Though McShea's account is purely structural, his three degrees of individuation highlight several shared themes in the various accounts of organismal organization: interactions among components, diversity of components, and "nesting" of parts and wholes. To these can be added two "dynamic" features: self-reproduction of the system of interacting components and maintenance of the boundary between system and environment.

This survey, although necessarily brief, points us toward a strong organization requirement for organismality, consisting of multiple conditions. An organism is composed of structurally and functionally differentiated parts, integrated together to form a unified whole. Integration is understood as mutual causal interdependence (e.g., Wilson 2000), which can be characterized on at least two dimensions: degree of coordination (how tightly orchestrated or constrained are the structure and behavior of parts) and degree of differentiation (heterogeneity of parts' structures and behaviors). The mutual interrelations of parts constitute the organized whole, the unified organism. Over time, and constrained by the overall system organization, the interacting parts reproduce themselves. Also reproduced is the physical boundary with the external environment. Organisms' autonomy consists in robustness to external perturbation that results from maintenance of this boundary and interaction of internal differentiated components. These ideas can be expressed as five conditions, which collectively comprise the strong organization requirement for organismality:

Autonomy: The boundary between organism and environment is internally generated and maintained.

Heterogeneity: An organism's components are structurally and functionally diverse.

Interaction: An organism's components form a network of mutual dependence relations.

Generation: The interdependent network of an organism's components is self-maintaining/reproducing.

Unity: The self-reproducing network and boundary constitute an organism's unity over time.

---

[18] McShea's criteria are designed to provide an operational scale for determining whether there is a trend toward increased "nesting" of hierarchical levels of organization in the history of life. So they are intended to be minimal and abstract, for inclusive application.

Cultured pluripotent stem cells growing in artificial culture do not appear to satisfy these conditions. Though each cell is tidily enclosed in its self-maintaining membrane, there is no established boundary for a colony of these cells. As a group, the self-renewing cells simply divide and take up more space, bounded only by the walls of their containers. So there is no internally generated boundary for cultured pluripotent stem cells. The autonomy condition is not satisfied.

The heterogeneity condition is similarly flouted. Although each cell in a cultured stem cell colony is made up of structurally and functionally diverse components, the cells as a group are homogeneous in many respects. Self-renewing cells transferred to new cultures are deliberately selected for "a uniform undifferentiated morphology" (Thomson et al. 1998: 1147, n6)—that is, they lack both variation and specification. Furthermore, cultured pluripotent stem cells are selected to be homogeneous with respect to certain morphological, molecular, and developmental traits: a flat, round morphology; large nuclei surrounded by thin layers of cytoplasm; prominent nucleoli; normal chromosome number and appearance; high telomerase activity; high or low expression of specific molecules; and, most recently, certain global gene expression patterns. Developmentally, of course, pluripotent stem cells share common features as well: "(i) derivation from the pre- or peri-implantation embryo, (ii) prolonged undifferentiated proliferation, and (iii) stable developmental potential to form derivatives of all three embryonic germ layers even after prolonged culture" (Thomson et al. 1998: 1145).[19] Similarities across the various kinds of pluripotent stem cell (e.g., embryonic stem cells, induced pluripotent stem cells, germline stem cells) are no accident. Construction of these model systems proceeded in stepwise fashion, with new kinds of pluripotent stem cells designed to resemble earlier ones (see Fagan 2013*a*). Each such lineage is created by selecting *against* structural and functional diversity among constituent cells.

The sources from which cultured pluripotent stem cells are derived, in contrast, do exhibit such diversity. For embryonic stem cells (ESC), the source is an early embryo in tissue culture. At this stage, embryos consist of two layers of cells: an outer layer that can give rise to extraembryonic tissues and an inner cell mass (ICM) that can give rise to embryonic tissues. To create pluripotent stem cells, part of the ICM is removed to a new artificial culture. Some ICM cells divide rapidly under these conditions, producing clumps of "outgrowth" (Thomson et al. 1998). Colonies that exhibit the morphological and molecular features noted earlier are selected and transferred to new cultures. The other main variety of pluripotent stem cell is "induced" by adding a specific set of genes to nuclei of mature differentiated cells (Takahashi and Yamanaka 2006). After several weeks, a very few cells treated in this way transform to resemble

[19] One of the features that distinguish embryonic stem cells (ESC) from their predecessor stem cell lines is that they are reliably pluripotent, rather than exhibiting an unpredictable range of developmental potential.

ESC. For example, skin cells in culture form a flat, evenly distributed sheet of fibers. But after the genes are added, a few cells show a different morphology—round and clumped, like ESC—along with ESC-like molecular traits. Induced pluripotent stem cells (iPSC) can be made from a variety of differentiated cells and tissues. The key point here is that successful construction of pluripotent stem cells involves *loss* of differentiation in the original source, highlighting the contrast with multicellular organisms.

For interaction among cells the situation is not as clear-cut, because the molecular and cellular pathways underlying stem cell capacities are not well understood. It is clear that pluripotent stem cells are extremely sensitive to their external environment—which includes other cells. A "cocktail" of different chemical and molecular genetic substances is required to maintain cultured pluripotent stem cells in their undifferentiated, self-renewing state. The original human ESC protocol included a layer of cultured "feeder cells" that supported the ICM-derived cells, both physically and by producing growth factors (Thomson et al. 1998). Such feeder layers can be replaced by chemical supplements to media—however, their role is proof of principle that pluripotent stem cells are responsive to signals from other cells. Interactions with the extracellular environment are crucial for "stemness." But it is not at all clear that individual cells comprising a pluripotent stem cell colony are responsive to *one another*, such that the colony forms a cohesive system analogous to a multicellular organism.

The generation condition, in contrast, is satisfied. Pluripotent stem cells are self-renewing by design and definition, selected for continuous cell division under culture conditions. So pluripotent stem cells are constitutively self-renewing. In addition, some of the molecular characters used to identify them (chromosome number, chromosome appearance, and telomerase activity) are directly implicated in self-renewal. However, there are two caveats to consider. The first is that, as noted earlier, the interdependent network comprising a pluripotent stem cell colony may be minimal; no more than cells in physical contact with one another so as to form undifferentiated colonies. Second, the self-renewing character of pluripotent stem cells at the colony level is partly due to scientists' intervention. Transferring cells through successive cultures creates a lineage of pluripotent stem cell colonies. Without this intervention, the cells would simply grow and expand across available space. It is not clear whether, left to themselves, pluripotent stem cells would reproduce as distinguishable colonies.

Finally, the unity condition for organisms combines the requirements of self-reproducing network and boundary. The latter is not satisfied in the stem cell case, while the former is questionable at best. Overall then, cultured pluripotent stem cells do not meet all (or even most) of the conditions for the strong organismal organization requirement. Therefore (so runs the objection), cultured pluripotent stem cells are not organisms, and so it follows that they are not model organisms.

6.5 Development and Individuality

In this section, I rebut this objection to the model organism thesis. The key idea of this rebuttal is that the characterization of cultured pluripotent stem cells in the previous section is inadequate, because it omits *pluripotency*. The conception of cultured pluripotent stem cells that fails to satisfy autonomy, heterogeneity, interaction, and unity conditions for organismal organization takes account only of their self-renewal. This amounts to conceptualizing cultured pluripotent stem cells as a stable cell type analogous to neurons, red blood cells, skin cells, and the like (Section 6.2). The stable cell type "cultured pluripotent stem cell" is undifferentiated, reproduces rapidly in artificial culture conditions, and exhibits a species-specific morphology and characteristic gene expression profile (a few key genes highly expressed relative to more differentiated cell types, including other stem cells). But this characterization is incomplete, because self-renewal is only one of two defining abilities of a stem cell. The other is a capacity for differentiation: change of phenotype into cells that belong to more specialized, mature cell types.

Self-renewal and differentiation are complementary processes that involve comparison across cell generations. Self-renewal is cell reproduction in which parent and offspring resemble one another. Differentiation is the process by which parts of a developing organism acquire diverse, specialized traits over time. At the cellular level, differentiation is the transformation of cell phenotype in accordance with a linear developmental process. The more differentiated a cell, the further along it is in that developmental process. Both abilities are equally essential to stem cells as those entities are currently defined. Their definition in terms of cellular processes of reproduction and development supports a *process view* of stem cells; I discuss this idea further in Section 6.6. The main point for now is that the organization objection presupposes an incomplete characterization of the stem cells of interest. If their differentiation potential is considered, the objection fails. To see this, we need to look more closely at the experimental procedures that establish pluripotency for cultured cell lines.

In general, pluripotency is established by placing samples of a cell line in environments that encourage differentiation. In practice, these environments are of two main types: artificial cell culture and organismal bodies (Figure 6.2). The "gold standard" test for pluripotency is, for obvious reasons, development of a viable whole organism. This procedure generates a continuous cell lineage from stem cells to an adult organism. Briefly, stem cells are transplanted from artificial culture into an early-stage embryo. The embryo is then implanted in a uterus and develops in normal mammalian fashion into a fetus, then a neonate, and finally a mature organism. If the transplanted stem cells are pluripotent, then all the resulting organism's tissues should include transplant-derived cells, including its germline. This way of establishing pluripotency identifies stem cells as a source of organismal organization; they can literally generate all the diverse cell types of a whole organism.[20] However, this method is not ethically permissible for human cells. Pluripotency in our species must be established in other ways.

Two main methods are used to demonstrate pluripotency in human stem cells. The simplest is to suspend the stem cells in fluid, rather than growing colonies on a solid surface (usually of other cells; see earlier discussion). Under these conditions, cultured stem cells arrange themselves into spheres about 0.1–0.2 mm in diameter, consisting of an inner core of undifferentiated cells wrapped in a layer of more differentiated cells resembling embryonic endoderm. These structures are termed "embryoid bodies," due to their resemblance to early embryos. Embryos at the gastrula stage[21] are organized into three germ layers with characteristic developmental fates: an outer layer of ectoderm, which gives rise to nervous tissue; an inner layer, endoderm, which gives rise to major organs; and mesoderm in between, which gives rise to muscle, bone, and fat. Cells from embryoid bodies show patterns of developmental potential similar to germ layers. So the embryoid body serves as a proxy for the whole organism.

Although far simpler than most multicellular organisms, embryoid bodies do satisfy most conditions for organismal organization. An embryoid body is autonomous, in that it forms and maintains a clear boundary with the environment, reflected in its distinctive spheroid shape. It exhibits internal heterogeneity, being comprised of inner and outer layers. That its cellular parts interact is indicated by their generative properties: embryoid bodies reprise, in simplified form, the developmental potential of an early embryo with three germ layers. Embryoid bodies are unified, as stable bounded entities maintained over time. The only condition that is evidently not satisfied is generation: embryoid bodies do not reproduce themselves, but rather (under appropriate environmental conditions) give rise to more differentiated cell types. Recall, however, that the generation condition is satisfied by cultured pluripotent stem cells, due to their self-renewal. It is also notable that very little experimental intervention is needed for cultured pluripotent stem cells to form embryoid bodies—a change in surface contacts between cells suffices. This suggests that cell–cell interaction may be more significant for colonies of pluripotent stem cells than is first apparent. Along the same lines, it is known that cell–cell interactions are required for human ESC to differentiate, although the details of these interactions remain unknown (Thomson et al. 1998). Furthermore, cultured pluripotent stem cells tend to differentiate spontaneously if left to themselves, often in uncontrolled and unanticipated ways.[22] This occurs even in artificial culture that encourages self-renewal, particularly when the dividing cells contact or

[20] For brevity, this summary combines two different whole-organism tests for pluripotency: production of all the differentiated cell types within a whole organism and production of the organismal body itself (tetraploid complementation).

[21] This is the stage at which "twinning" is no longer possible; i.e., "individuation" of the whole organism.

[22] Wernig (personal communication, October 16, 2009).

"overgrow" one another. Seen in this light, experimenters' manipulations actually *block* stem cells innate tendency to differentiate, rather than pushing them toward differentiation. So cultured pluripotent stem cells may satisfy heterogeneity and interaction conditions in their capacity to transform into embryoid bodies. And the latter, as we have seen, satisfy all but the generation condition for organismal organization.

The second standard test for pluripotency is the oldest—the original test for the first pluripotent stem cells, which were actually a form of cancer. Teratocarcinoma is a cancer that manifests as tumors containing diverse cell types: fat, muscle, skin, nerves—even teeth. Experiments in the 1950s showed that teratocarcinomas arising in a particular strain of inbred mouse contained undifferentiated cells that gave rise to more specialized cells. Teratocarcinoma development was accordingly conceived as a simplified version of embryonic development.[23] The main contrast between the two involves spatiotemporal organization. Teratocarcinomas "appear to recapitulate many of the events that occur during early embryonic development but in a disorganized manner" (Andrews 2002: 406). Like early embryos, undifferentiated teratocarcinoma cells give rise to a wide range of cell types, including products from each of the three germ layers (ectoderm, mesoderm, and endoderm). What is lacking is the orchestrated integration of diverse cell types, to produce a whole organism. Teratocarcinomas were originally propagated in mice, with transplants giving rise to new tumors. Cell culture eventually replaced mice as the environment for undifferentiated teratocarcinoma cells—the first cultured pluripotent stem cells. ESC were later constructed to resemble these peculiar cancer cells, morphologically and developmentally. These experimental origins are inversely recapitulated in the most rigorous pluripotency test for human stem cells: cultured cells are transplanted into inbred mice, where they develop into a nonmalignant version of teratocarcinoma. If the resulting tumors contain cells from all three germ layers, albeit disorganized and jumbled, then scientists conclude the transplanted stem cells are pluripotent. Three germ layers in a tumor are considered a proxy for a whole organism.

Teratomas that experimentally demonstrate pluripotency for a line of cultured stem cells satisfy some conditions for organismal organization. Like embryoid bodies, teratomas are autonomous in the sense that they form a clear boundary from their environment. As structurally distinct growths within the bodies of host mice, they exhibit unity of a sort. By definition, teratomas' cellular components are diverse; the heterogeneity condition is satisfied. But it is not clear that these cellular components interact in any systematic way; although some interactions are presumably required to maintain a teratoma as a discretely bounded entity within an organism's body, these interactions are not well-understood and appear unlikely to be robust across tumors. So the interaction condition may not be

---

[23] See Fagan (2013*a*: chapt. 7) for details.

satisfied. The generation condition is also not satisfied—but this may be an experimental artifact, as the original form, teratocarcinoma, is tumorigenic. Overall, then, teratomas satisfy most of the conditions for strong organismal organization.

Bringing together the different experimental entities implicated in pluripotency experiments, we can distinguish several modes of differentiation (Table 6.1). Development of a multicellular organism from an early embryo is the "normal," unmanipulated case. Teratocarcinoma development presents a pathological contrast: specialized cells and tissues appear, but in a "disorderly" fashion. The disorder is of two kinds: specialized cells and tissues are not arranged into functional organs, nor are they arranged so as to constitute an autonomous organism. The life of a teratocarcinoma (which makes it malignant, a threat to any containing organism) is sustained by proliferation of its undifferentiated cells; "cancer stem cells" (cf. Laplane 2016). Development of specialized cells and tissues from an embryoid body offers yet another contrast. In this case, development is simplified and minimal rather than unpredictably disordered. The range of specialized cell types deriving from an embryoid body is limited; nothing like a whole organism results. Instead, cells from an embryoid body can be distributed to different culture environments, each designed to encourage differentiation in a particular "direction"; toward a different kind of specialized cell type: skin, muscle, blood, neurons, and the like. The result is a more thorough disarticulation of organismal organization, "reduced" to homogeneous populations of specialized cell types. If one considers the full set of "directed development" experiments conducted on a single cultured pluripotent stem cell line, the result is a fragmentary, spatiotemporally distributed approximation of organismal development. But this result does not, intuitively, constitute an organism.

Pluripotent stem cells, uniquely, encompass all the preceding modes of multicellular organismal development: normal, pathological, simplified, and dispersed. Which is instantiated in any particular case depends on the particulars of the environment. So cultured pluripotent stem cells are capable not only of differentiating to produce the full range of cell types, but of realizing different processes or modes of differentiation. Most experiments with cultured pluripotent stem cells aim to produce specialized cells of one particular type: neuron, muscle, blood, and the like. But tests for pluripotency reveal greater developmental versatility. In their undifferentiated state, cultured pluripotent stem cells are at best a "limiting case" of a multicellular organism: drastically simplified, lacking all or most

TABLE 6.1 Modes of differentiation based on pluripotency tests for cultured stem cells

Source	Process	Product
Early embryo	Normal	Multicellular organism
Cancer cells	Pathological	Disordered organs, tissues
Embryoid body	Simplified	Dispersed specialized cells
PSC colony	Inclusive	All of the above

features of organismal organization. But, in a suitable environment, this minimal state gives way to more complex multicellular arrangements, ranging from a simple two-layered embryoid body to a normal mature organism. These developmental products, taken together, satisfy the strong organismal organization requirement. Cultured pluripotent stem cells' capacity for differentiation affords continuity between undifferentiated cell colonies and intricately structured multicellular organisms, as well as intermediates like embryoid bodies and teratomas. So, in addition to pluripotency, these stem cells' differentiation potential consists in producing (or being able to produce) key aspects of organismal organization. When this broader developmental potential is taken into account, cultured pluripotent stem cells satisfy conditions of the strong requirement for organismal organization. This rebuts the organization objection presented in Section 6.4. If the preceding argument is successful, then what I take to be the primary objection to the model organism thesis is avoided. The view of stem cells invoked in this argument also suggests that we should think of stem cells as processes rather than as substantive structural entities. I conclude by discussing this idea, although full analysis of the implications is a task for another paper.[24]

6.6 Process View of Stem Cells

On the view of stem cells that I have presented here (see Section 6.2) and defended in earlier work (2013*a*, 2016*a*, 2016*b*), a stem cell is not a single cell bearing the marks of biological individuality, but rather a lineage consisting of two or more cell generations and a set of variable characters relative to which comparisons are made, such that parent and offspring cells are determined to be the same or different. The proposal that we conceptualize a stem cell as a process is just the proposal to think of stem cells as encompassing all the components of the model that explicates their general definition—a continuous generative process that includes both self-renewal and differentiation in a cell lineage.

This process view of stem cells can be clarified by contrast with a "substantivist" alternative. On the latter view, stem cells are substantive cellular entities that causally produce another kind of entity—organisms or simplified proxies for organisms.[25] This way of conceptualizing stem cells identifies them with the undifferentiated entities maintained in artificial cell culture; self-renewing stem cells conceived as a stable cell type. The aspects of organismal organization that satisfy the strong requirement are, on this view, effects of which stem cells are causes. Stem cells so conceived do not themselves exhibit those organized features but causally contribute to the production of entities that do. To claim that cultured

---

[24] Thanks to Ruey-Lin Chen for proposing this interpretation.

[25] This alternative view is based on comments by Thomas Pradeu and an anonymous reviewer.

pluripotent stem cells satisfy the strong organismal organization requirement in virtue of their being causal sources of such organization seems to presume an implausibly preformationist view of development—amounting to the claim that cultured pluripotent stem cells somehow "contain" the organized features of an organism that could develop from those cells in future. This is not the view of stem cells that is proposed here. The substantivist view severs stem cells from one of their defining capacities: differentiation.

In contrast, I have argued that we should not conceptualize stem cells as lacking differentiation potential. If we conceptualize them as, for example, pluripotent, then a developmental process is implicated in the concept of a stem cell. To put the point in metaphysical terms, a stem cell *is* a coupled developmental-reproductive process: differentiation and self-renewal. This process view of stem cells is supported by the general definition explicated by the model presented earlier: developing lineages of cells related by reproduction and a presumed developmental order that serve as the basis of comparison across cell types. The diverse developmental products that show cultured stem cells to be pluripotent are later stages of the process; undifferentiated self-renewing cells are earlier stages. But the concept of a stem cell encompasses both. This is, I suggest, the most perspicuous way to understand "the stem cell": as a cell that initiates a lineage of cells engaged in both reproduction (self-renewal) and development (differentiation). If we take the stem cell concept to have ontological significance (as a framework for understanding the worldly phenomena of interest for stem cell biology and other related fields), then we should say that stem cells *are* generative processes (lineages) which include organismal organization in their later (or future) stages. This then rebuts the organization objection.

The process view of stem cells, which undercuts an objection to my thesis that cultured pluripotent stem cells can be considered model organisms, is indirectly supported by arguments for a process ontology for organisms (e.g., Dupré, Chapter 2, this volume; Pemberton, Chapter 3, this volume). The features of such an ontology comport well with the view of stem cells proposed in this chapter: that we conceptualize cultured pluripotent stem cells as processes extending from the extraction of cells from an organismal source into cell culture, to the organized entities that demonstrate pluripotency. The core idea is that cultured pluripotent stem cells are artificial, simplified organism-processes, which include all the stages needed to experimentally identify them as being those entities. For some purposes, we will be more interested in the early, undifferentiated stages, which persist through self-renewal. In other contexts, later stages, exhibiting some or all features of organismal organization, will be the focus. The modeling role of stem cells, which comprises half the model organism thesis, encompasses the full range of stages.

Clearly there is more to be done, to fully examine the implications of the process view of stem cells. But the argument of this chapter has reached a stopping point. I have previously argued that stem cells are not a cell type like those that structurally compose the bodies multicellular organisms (Fagan 2016*a*). However, I have also argued, pluripotent stem cells maintained in artificial culture can be reasonably considered model organisms.[26] This chapter addresses an objection to that proposal: cultured pluripotent stem cells cannot be model organisms, because they lack the requisite integration and cohesive unity to be organisms. I synthesize several prominent accounts of organismal organization to articulate a strong version of this requirement, and thereby present the objection in a strong, dialectically useful form. I then show that this objection depends on an incomplete characterization of stem cells, ignoring their defining capacity for differentiation. A proper understanding of the stem cell concept, including both defining capacities, rebuts the organization objection to my thesis that stem cells can be considered model organisms.

Acknowledgments

Many thanks to Ruey-Lin Chen for arranging the conference where I presented the first version of this chapter (Cheng Chung University, Chiayi, Taiwan, December 8–9, 2014). This version has benefited from discussions with Louiseneh Arakelian, Julia Bursten, Ruey-Lin Chen, Sara Green, Matt Haber, Lucie Laplane, Anne Peterson, Thomas Pradeu, and Joe Rouse, and from comments and questions by audiences at Cheng Chung University (December 2014), the University of Utah (April 2016), IHPST (Paris, April 2016), and the University of Bordeaux (May 2016). Thomas Pradeu, my co-editors, and two anonymous reviewers provided helpful comments on earlier drafts. Funding was provided by the Humanities Research Center at Rice University, the Mosle Research Foundation, the University of Utah College of Humanities, and a generous donation from the family of Sterling M. McMurrin.

References

Andrews, P. 2002. From teratocarcinomas to embryonic stem cells. *Philosophical Transactions of the Royal Society of London, Series B* 357: 405–417.

Arnellos, A., A. Moreno, and K. Ruiz-Mirazo. 2014. Organizational requirements for multicellular autonomy: Insights from a comparative case study. *Biology and Philosophy* 29: 851–884.

Bolker, J. 1995. Model organisms in developmental biology. *BioEssays* 17: 451–455.

[26] It is an open question whether other kinds of stem cell can be considered model organisms. I am inclined to think that only stem cells that are pluri- or totipotent can be considered so, while stem cells of more restricted potential can be considered model organs. Careful consideration of this question is beyond the scope of this chapter.

Buss, L. W. 1987. *The evolution of individuality*. Princeton: Princeton University Press.
Cheung, T. 2006. From the organism of a body to the body of an organism: Occurrence and meaning of the word "organism" from the seventeenth to the nineteenth centuries. *British Journal for the History of Science* 39: 319–339.
Collier, J. 2004. Self-organization, individuation and identity. *Revue Internationale de Philosophie* 59: 151–172.
Fagan, M. B. 2013*a*. *Philosophy of stem cell biology*. London: Palgrave Macmillan.
Fagan, M. B. 2013*b*. The stem cell uncertainty principle. *Philosophy of Science* 80: 945–957.
Fagan, M. B. 2016*a*. Cell and body: Individuals in stem cell biology. In *Individuals across the sciences*, eds. A. Guay and T. Pradeu, 122–143. New York: Oxford University Press.
Fagan, M. B. 2016*b*. Generative models: Human embryonic stem cells and multiple modeling relations. *Studies in History and Philosophy of Science* 56: 122–134.
Guay, A., and T. Pradeu (eds.). 2016. *Individuals across the sciences*. New York: Oxford University Press.
Kohler, R. 1994. *The lords of the fly*. Chicago: University of Chicago Press.
Landecker, H. 2007. *Culturing life*. Cambridge: Harvard University Press.
Laplane L. 2016. *Cancer stem cells: Ontology and therapies*. Cambridge, MA: Harvard University Press.
Maturana, H., and F. Varela. 1980. *Autopoiesis and cognition*. Boston: D. Reidel.
McShea, D. W. 2001. The hierarchical structure of organisms. *Paleobiology* 27: 405–423.
Melton, D., and C. Cowan. 2009. Stemness: Definitions, criteria, and standards. In *Essentials of stem biology*, 2nd ed., eds. R. Lanza et al., xxii–xxix. San Diego, CA: Academic Press.
Mossio, M., and A. Moreno. 2010. Organisational closure in biological organisms. *History and Philosophy of the Life Sciences* 32: 269–288.
Nicholson, D. 2010. Biological atomism and cell theory. *Studies in History and Philosophy of Biological and Biomedical Sciences* 41: 202–211.
Rader, K. 2004. *Making mice*. Princeton: Princeton University Press.
Ramalho-Santos, M., and H. Willenbring. 2007. On the origin of the term "stem cell." *Cell Stem Cell* 1: 35–38.
Reynolds, A. 2007. The theory of the cell state and the question of cell autonomy in nineteenth and early twentieth-century biology. *Science in Context* 20: 71–95.
Takahashi, K., and S. Yamanaka. 2006. Induction of pluripotent stem cells from mouse embryonic and adult fibroblast cultures by defined factors. *Cell* 126: 663–676.
Thomson, J., J. Itskovitz-Eldor, S. Shapiro, M. Waknitz, J. Swiergel, V. Marshall, and J. Jones. 1998. Embryonic stem cell lines derived from human blastocysts. *Science* 282: 1145–1147.
Virchow, R. 1859/1958. Atoms and individuals. In *Disease, life, and man: Selected essays by Rudolf Virchow*, translated and with an introduction by L. J. Rather, 120–141. Stanford: Stanford University Press. Original published 1859.
Waddington, C. H. 1970. Concepts and theories of growth, development, differentiation and morphogenesis. *Towards Theoretical Biology* 1: 1–41.
Weber, A., and F. J. Varela. 2002. Life after Kant: Natural purposes and the autopoietic foundations of biological individuality. *Phenomenology and the Cognitive Sciences* 1: 97–125.
Wilson, J. 1999. *Biological individuality*. New York: Cambridge University Press.
Wilson, J. 2000. Ontological butchery: Organism concepts and biological generalizations. *Philosophy of Science* 67(supplement): S301–S311.

Wilson, R., and M. Barker. 2013. The biological notion of individual. In *The Stanford encyclopedia of philosophy* (spring 2013 edition), ed. Edward N. Zalta. http://plato.stanford.edu/archives/spr2013/entries/biology-individual/.

Wolfe, C.T. 2010. Do organisms have an ontological status? *History and Philosophy of the Life Sciences* 32: 195–231.

{7}

Individuation of Developmental Systems

A REPRODUCER PERSPECTIVE

James Griesemer

7.1 Process-Relativity of Individuality Concepts

In my view, the concept of "individuality," especially as it is used in the sciences, is a scientific-theoretical concept rather than a logical or metaphysical concept to which scientific usage is somehow obliged to conform. While philosophers have had much to say about individuality (e.g., Goodman and Leonard 1940; Strawson 1959; Wiggins 1980), I think we learn more about the concepts, claims, and consequences of individuality in the sciences by supposing it is a theoretical term that can mean different, even conflicting, things in different theoretical contexts, such as in different scientific specialties and disciplines with different research questions, projects, and phenomena to consider, rather than a logical or metaphysical term that should have a single, best, all-purpose fixed meaning. In this chapter, I limit my argument to concepts of biological individuality, focusing on concepts used in some parts of evolutionary biology. Aspects of what I shall argue may generalize. It is beyond my scope to speculate about other sciences.

There is great variety in usage of the term "individual" among biological specialties, from a ramet, genet, or a module in botany; to a zooid, person, or sorus in zoology; to a cell in microbiology; to a behaving subject in ethology; to an organism, species, or unit of selection in evolutionary biology. Most biologists use the term "individual" interchangeably with "organism" except when they are discussing questions of units and levels of evolution, though the problem of individuality as philosophers understand it is then merely transferred to the question of what is an organism.

My thesis is that, despite all the variety of usage, concepts of biological individuality suitable for evolutionary biology involve two senses of *process-relativity*. The concept depends on (1) some *empirical process* we take to be operating in the world that we humans talk about, try to understand, predict, explain, or control; and (2) a *tracking process* that we humans perform, through which we attempt to do the things described in (1). In other words, concepts of individuality depend on the empirical process of concern to the humans who investigate, use, or otherwise rely on it to understand and interpret biological entities and phenomena and also on what tracking processes they perform to investigate them with. Individuality concepts thus have a dual character: they are partly about processes in the world beyond and around us and partly about us, in so far as our concepts are linked to the mode and manner of our tracking engagements with the world.

We need concepts of *individuality* in science as well as in other practices of everyday life in order to tell stories about what goes on in the world, do science, and make attributions of properties, relations, responsibility (causal or moral), and standing (e.g., epistemic, moral, legal). We also need concepts (plural) of *individuation* in order to count, distinguish, and relate what we find when we engage the empirical world of processes. Differently put, we need to individuate—distinguish one individual from another—in our varied tracking practices to track the varied processes we discover and order them into countable and accountable "things": lineages, groups, kinds, and so forth. Scientists have need of, as nominalists like Nelson Goodman might say, a logic of individuals whatever their metaphysical commitments or lack thereof.

My goals for the chapter are to (1) illustrate process-relativity of biological individuality by discussing just two different kinds of attempts to articulate concepts of evolutionary individuality, one based in natural selection theory, the other based in a theory of biological reproduction; and (2) display how the tracking activities of scientific practices are entwined with the empirical processes on which both individuation and individuality depend. I will not attempt to survey the vast and growing literature on individuality in biology (e.g., see Clarke 2010; Bouchard and Huneman 2013; Guay and Pradeu 2016). In the final section, I will make some remarks intended to motivate a search for extrapolations and applications beyond the narrow scope I have set for my discussion.

7.2 Individuality, Individuation, and the Practice of Evolutionary Theory

Much of the discussion of biological individuality in philosophy of biology has been driven by the units of selection problem. To what "units" does Darwin's theory of evolution by natural selection apply? Is there one (kind of) unit of selection or many? Organisms? Groups? Species? Genes? What evolutionary "interests" lead potentially self-interested entities to function as parts serving the collective interest of a whole, possibly even in conflict with their own self-interest? Can Darwin's theory of the evolution of populations and species by means of natural selection among organisms be generalized by abstracting from a characterization of individual *as organism* (Griesemer 2005), across the compositional hierarchy from molecules to man or from genes to species (Lewontin 1970), or perhaps across a functional hierarchy of replicators and interactors (Dawkins 1976; Hull 1988)? Differently put: Is "organism" just one form or manifestation of biological individuality among many? What is an organism anyway? Are colonial animals individual organisms? Are modular plants lacking (early) germline segregation individual organisms (Jackson, Buss, and Cook 1985; Buss 1987)?

I need not rehearse the history of this problem. It is sufficiently well-known to be reviewed in encyclopedia articles (Lloyd 2012). My problem starts, instead, with two prior questions. Is the generalization of Darwin's theory the right starting point for inquiring into the nature and characterization of evolutionary individuality, even for the sake of answering questions about the units of selection? Many philosophers of biology assume so, or at least became aware of the problem through pondering the nature and structure of evolutionary theory (e.g., Sober 1984; Sober and Wilson 1998; Godfrey-Smith 2009). Many began thinking about the problem at a time (1960s–1980s) when the status of biology and theories in biology were in question by philosophers of science of an earlier era, trained to valorize physics as exemplary of science in general, who judged biology to be lacking important qualities of "mature" sciences. At stake was not only how to understand units of selection, but whether Darwin's Principles filled the role of laws of nature in a (neo-)Darwinian theory.

On the other hand, many biologists discovered the problem of units of selection because they were dealing with empirical phenomena that seemed puzzling against the backdrop of the old catechism that genes are the units of heredity, organisms are the units of selection, and populations are the units of evolution (e.g., Wade 1976, 1977, 2016). Many plants, animals, fungi, and bacteria have complex life cycles or colonial social organization, suggesting that looser organization than that of vertebrate animals is sufficient for individuality. A fair few of these behave in ways strikingly contrary to familiar individual human behavior; for example, the cellular slime molds whose physically attached "parts" disaggregate and aggregate (Bonner 1967). Organisms may have tissues that are not at all germ-like yet which can initiate new reproductive lineages (buds, branches, cuttings). Groups of organisms may themselves reproduce, develop, behave, or evolve in organism-like ways (eusocial colonial animals).

Groups of organisms can be made to respond *as groups* to selection in the laboratory (e.g., flour beetles; see Wade 1976). Their evolutionary responsiveness seems to suggest selection may operate at more "levels" than only that of self-interested organisms competing for resources or mates (e.g., house mice carrying t-alleles in some gametes and some groups but not others; Lewontin 1970; discussed in Wimsatt 1980; Lloyd 1988). Not to mention that there are entities which seem to evolve, yet are distinctly un–organism-like, such as retroviruses and maybe prions and some transposons (see Godfrey-Smith 2009) as well as symbiotic associations of organisms, such as the Hawaiian bobtail squid and its light-emitting bacterial symbiont, *Vibrio* (McFall-Ngai et al. 2013).

In those early decades of the units of selection problem, only a few philosophers, in collaboration with scientists, approached the problem of units of selection from the perspective of *experimental* rather than theoretical investigation of group phenomena (e.g., Antonovics et al. 1988; Griesemer and Wade 1988). In experiments, the problem of *individuation* raises immediate and practical concerns for experimental design, management, and data collection, whatever one's theoretical or metaphysical attitude toward concepts of *individuality*. Empirical entities such as vials of flies or beetles, Petri dishes of bacteria, or hives of bees must be labeled and tracked, regardless of whether they are treated as groups or individuals. I suggest the question of biological individuation is practically, empirically, logically *prior* to the theoretical units of selection individuality question, even if the "ultimate" process of evolution is historically prior to the "proximate" processes structuring individuality. Empirical interest in some particular (or particular kind of) biological "system" whose physiology, ecology, development, reproduction, life cycle, or behavior has some organism-like and some group-like (or at least non–organism-like) aspects or activities demands attention be paid to the tracking problem—that is, the problem of individuation—regardless of theoretical commitments. Moreover, theory must eventually face the practical, empirical, and logical problems with regard to the problem of adducing evidence anyway.

Mayr's (1961) labeling of biological problems and processes as proximate or ultimate and Dobzhansky's declaration (1973) that nothing in biology makes sense except in the light of evolution notwithstanding, it is equally clear that evolution does not make sense except in the light of the rest of biology (Griesemer 1994). So, it is a toss-up which *concepts* must "come first" in order to understand "the rest," never mind which *processes* must have come first in order to *evolve* the rest. After all, those biologists who arrive at the units of selection question starting with puzzles about the nature and scope of Darwinian theory must have in mind some biological "content," some empirical phenomena, through which heritable variation in fitness can be generated, else the theory is merely an empty formalism. What salience do these biologically, epistemically, logically prior concerns have for the problem of evolutionary individuality? And, if the prior concerns grow out of empirical practices of inquiry into the behavior of biological systems, what relevance does that grounding in non- or pre-evolutionary practice, rather than in evolutionary theory, have for understanding biological individuality? What salience, for example, does the fact that John Tyler Bonner (1952, 1967) sought to understand slime mold *morphogenesis* and to use it as a model system for understanding *developmental regulation* have for the role his "system" now plays in arguments about the *evolution* of individuality (Buss 1987; Godfrey-Smith 2009; Queller and Strathmann 2009)?

My own philosophical interest in biological individuality grew out of the units of selection problem—I was taught by Bill Wimsatt while units of selection was *the* problem in philosophy of biology; I was also taught by biologists who came to the problem through experimental and field studies of evolution (Michael Wade, Stevan Arnold), as well as by theoreticians (Russ Lande). I have become convinced that units problems in evolutionary biology must be addressed first and foremost, *prior* to questions about natural selection, drift, and other "forces" of evolution, in terms of developmental, reproductive, and ecological processes. And I have become convinced that the way to interpret how biologists understand developmental and reproductive processes is to study how they individuate processes in biological practice; that is, how they track reproduction and development empirically (Griesemer 2007), as parts, Ken Waters would say, of their "investigative strategies" (Waters 2008). Even if the "individuality" problem is logically (or metaphysically) prior to the "individuation" problem, individuation in practice is practically prior to the "individuality" problem in the sense that we can discover new ways to think about individuality through practical commitments to individuate in particular ways, respects, and degrees, but the reverse is an armchair problem for metaphysicians alone.

One way to grasp the priority of the problem of individuating reproduction processes to a problem of biological individuality founded on Darwin's principles is to consider the place of Malthus's "principle of population" in relation to Darwin's principles. Recall Malthus's principle: population, unfettered, tends to grow geometrically while food supply tends to grow, at best, arithmetically, so population will inevitably exceed food supply and individuals will suffer unless population is regulated (Malthus 1826). It remained unclear, after Lewontin's (1970) work to regiment and generalize Darwin's principles, whether or not Malthus's principle should be considered prior to (as presupposition or axiom), as one of, or rolled into (as conceptually embedded in) Darwin's three principles of variation, fitness difference, and heritability. Lewontin (1970) presented Malthus's principle as a presupposition of Darwin's principles. Maynard Smith (1987) made the connection between Malthus and Darwin in terms of a principle of "multiplication," which he included *within* his formulation of "Darwin's Principles"—multiplication, variation, and heritability—demoting fitness differences to an add-on in the special cases where evolution occurs *by means of* natural selection, rather than, say, by means of drift alone. Most formalizers after Lewontin seem to embed Malthus's principle within Darwin's rather than stating it as a separate principle or presupposition, but are never entirely clear how the embedding works. I favor the clarity of Maynard Smith's approach, but Maynard Smith didn't go very far toward characterizing his principle of multiplication, nor in distinguishing a *biological* concept of multiplication (i.e., reproduction) from other, nonbiological processes of duplication, multiplication, or copying (see Griesemer 2000*a*, 2000*c*, 2005). My view is that the units of selection question must begin with an answer to the question of what sorts of "things" can *potentially* be units of selection, and only then can we move on to the question under what circumstances and conditions does one of these potential units *actually* become a unit with an "opportunity for selection" (Moorad and Wade 2013). Therein lies the question of how and when something *becomes* an evolutionary individual. Emphasis on becoming shifts the problem to individuation of processes from individuality of things.

One kind of answer to what can potentially be a unit of selection, and therefore an evolutionary individual, is "biological individuals." Richard Lewontin (1970) and David Hull (e.g., 1980, 1988) explored this kind of answer in depth. Lewontin appealed to traditional notions of entities that reside at "levels" of organization other than that of organisms—genes, cells, groups—to get his abstract notion of individuality going. Hull (and Dawkins 1976, 1982) appealed to functions of replication and interaction, though these concepts depended on received views of DNA replication and molecular, cellular, and organism interaction, together with a metaphysical interpretation of Kripke's theory of proper names as rigid designators (see Hull 1980, 1981) to get his notion of individuality going.[1]

Both of these approaches to what can be a unit of selection have been challenged. My own challenge was based on the argument that each ends up question-begging because they have to assume that these highly evolved entities at levels of composition or function can nevertheless exist prior to the evolution of genes and cells (Griesemer 2000*a*, 2000*b*, 2000*c*, 2005). I think these approaches carry a germ of truth, but it is tricky to avoid the question-begging circularity of appeal to already evolved entities to characterize evolutionary individuality. Theories of evolutionary transition (Buss 1987; Maynard Smith and Szathmáry 1995; Michod 1999) put pressure on these philosophical approaches to units of selection by recognizing that the compositional levels of evolution and the functioning of entities at those levels *themselves* evolved. Dawkins (1976, 1982), and many origins of life researchers who favor a "replicator first" approach (see Griesemer 2008), attempt to cut this Gordian knot by proposing that life began with a spontaneously formed entity which was at once a first replicator and an interactor.[2] As is well-known to philosophers, "unmoved mover" arguments are mainly trained on metaphysical rather than scientific problems.

More recently, some authors have widened these approaches to regard biological or evolutionary "individuality" as a degree property (e.g., Queller and Strathmann 2009; Godfrey-Smith 2009, 2013). Queller and Strathmann (2009: 3143), for example, take a hallmark of organismality to be "high cooperation among the parts . . . nearly free of conflicts." We might generalize that idea directly from "organismality" to "biological individuality." Their paper implies that many traits usually associated with biological individuality ("physical contiguity,

[1] For a disentangling of Hull and Dawkins, see Griesemer (2005).

[2] "Spontaneous" here means "by chemistry alone," but as this is often characterized as historically prior to the origin of systems capable of evolution, it also becomes question-begging to appeal to chemistry as prior to biology in these evolutionary terms. This form of "tautology problem" can be stated thus: "What is alive? Those chemical systems which can evolve by means of natural selection. What can evolve by means of natural selection? Those chemical systems which are alive."

indivisibility, clonality or high relatedness, development from a single cell, short-term and long-term genetic cotransmission, germ–soma separation and membership in the same species") really are "non-essential." Queller and Strathmann (2009: 3143) suggest treating organismality as a degree property relating degree of cooperation to degree of conflict among parts implies that "we accept some unconventional organisms, including some social insect colonies, some microbial groups and viruses, a few sexual partnerships and a number of mutualistic associations" (2009: 3143).

While this degree-property characterization of biological individuality opens a path to the *evolution* of multiple levels of evolutionary units (evolutionary transition) or evolution of novel functional properties of replicators and interactors (evolution of genetic coding, evolution of genotype/phenotype distinction) by adaptive changes in degree, it still does not address what kinds of entities are candidates—have the potential—to evolve these properties of individuality from a low to a high degree and thus what biological individuality means. Moreover, it does not distinguish a scientific-theoretical view of "part-hood" as involving something like "processes that hang together which interested scientists may track" from a metaphysically permissive view admitting as individuals whatever collections of entities satisfy a logic of mereology. When is an entity a part, and when is it not? The answer cannot be that an entity is a part so long as it, along with other parts, "shows a high degree of cooperation and is nearly free of conflict with the others" on pain of the same kind of circularity (part-hood delimited by individuality; individuality delimited by part-hood) that threatens earlier views on biological individuality.

So long as the characterization, criterion, or definition is "instrumental" or "pragmatic"—part of an empirical tracking practice rather than embedded in a theoretical concept of a formalized theory of evolution—I think such circularity is relatively harmless because *pragmatically* the limits of investigator interest or ability to measure empirical quantities of interest—such as fitness, cooperation, or conflict—halts circular reasoning with a generic, doubt-stopping solution: "it's what we could afford to do, so that's the best we can do under the circumstances of our investigation."

A different kind of answer to the question of what can *potentially* be a unit of selection, looks to the processes (rather than properties or qualities) which must operate for there to be biological individuals of the above kinds in the first place, let alone individuals subject to natural selection. Must there already be, in other words, some sort of biological individual to be a candidate Darwinian (or evolutionary) individual? And what sort of biological properties underpin biological individuality, if not status as replicators or interactors or cooperators? What processes are necessary for replication, interaction, cooperation, or selection to operate but which do not thereby constitute replication, interaction, cooperation, or selection processes (i.e., which are not in themselves instances of those processes)?

This second, "prior process" kind of approach takes biological individuality to be a problem prior to questions of units of selection and evolutionary individuality, though, in a way, so did the direct approach through appeal to biological individuals since they ended up appealing to some more fundamental property of replication, interaction, cooperation, or selection to interpret concepts of part, organism, and biological individuality.[3] But those previous "direct" approaches all embedded the individuality concept in solutions to the units of selection question, which is to court circularity if those properties (replication, interaction, cooperation) are themselves to be understood as *outcomes* of selection processes. If to be a biological individual is to be (potentially) a unit of selection, then appeal to biological individuality cannot answer our questions about what can be a unit of selection.

My own prior process approach is to take reproduction to be a process conceptually prior to selection/evolution and to take "reproducers" as units which can potentially be units of evolution and selection (Griesemer 2000*a*, 2000*b*, 2000*c*, 2002, 2006*a*, 2014*a*, 2014*b*, 2016). I characterize reproduction in a way that does not depend on properties which themselves require evolutionary explanation. In pursuing that approach, the abstract conceptual considerations are paired with practical, empirical considerations. To *characterize* a biological process of reproduction is also to characterize (or presuppose or imply) practices for empirically tracking reproduction, development, and heredity. For evolutionary purposes (typically) a reproducer approach *foregrounds* hereditary rather than developmental aspects (see Griesemer 2007). That foregrounding, however, is an artifact, I believe, of the way the history of the units of selection and Darwin's Principles problems unfolded. The foreground–background balance between heredity and development is shifting with the rise of evo-devo (e.g., Wagner 2014), eco-devo (e.g., Gilbert and Epel 2009; Sultan 2015), and calls for extended or alternative synthesis to the "modern synthesis" of genetics, systematics, and evolution (e.g., Pigliucci and Müller 2010). In my view, formal characterization in a theory, or "formalization," of a concept specifies not only its meaning in a web of theoretical meanings, but also its empirical role in process tracking because, to have empirical content, the theory must also specify a domain over which the theory delimits and guides empirical inquiry. Formalization includes this latter empirical guidance aspect and not only production of a theoretical "formula" (Griesemer 2013).

In the remaining sections, I consider two kinds of approaches to the question of evolutionary individuality which consider prior processes and empirical tracking together. In Section 7.3, I consider an approach which seems to me to

[3] Examples of prior process approaches include focus on properties like autonomy (e.g., Ruiz-Mirazo and Moreno 2012; Moreno and Mossio 2015), autopoiesis (Maturana and Varela 1980), and self-organization (e.g., Kauffman 1993). A synthetic approach combining many of these can be found in Gánti (e.g., 2003). A critique of these many diverse kinds of definitions of individuality can be found in Clarke (2010).

avoid circularity despite appealing to evolutionary concepts in order to articulate definitions of individuality. It avoids circularity by focusing on the *practice* of evolutionary theorizing and evaluating evidence rather than on designing concepts and principles of a generalized evolutionary theory per se. Charles Goodnight (2013) considers the practice of *assigning* fitness and considers, as one of three definitions, *that to which fitness is assigned in an evolutionary explanation* as a concept of evolutionary individuality. While depending heavily on concepts central to Darwinian and neo-Darwinian theory, Goodnight's approach does not define "individual" in terms of fitness (or selection), but rather appeals to certain kinds of practices (fitness *assignment*) which are necessary to *individuate* entities subject to evolutionary explanation (i.e., necessary to detect and track things that *have* [their own] fitness values). Before turning to Goodnight's approach, I need to consider tracking practices more generally.

A hallmark of tracking is the practice of introducing a "mark" into a process by causally intersecting with it and collecting traces from the marking interaction for use in representations of the process as a sequence of traces. That's a complicated, fancy way of saying that if you want to track something, you either have to keep your eyes on it, mentally noting its location continuously, or mark it in some way so that you can track the marks, noting its location episodically by collecting the traces the mark leaves (Griesemer 2007). I track my carry-on bag through airport screening by watching it go down the conveyor belt, into the scanner, out of the scanner, and down the belt on the far side. I cannot track a package in a similar way after I turn it over to FedEx, nor can any single FedEx employee, so they affix a tracking label with a barcode to my package, give me the tracking number associated with the barcode, and every employee who subsequently handles my package (I assume) scans the barcode with a scanner, reporting its whereabouts to a central database that I can consult online in order to track my package through the sequence of location "traces" collected by the database. Employees must not forget to scan my package, the database must work, and the Internet must be available or my tracking activity fails. This empirical tracking amounts to what sociologist Bruno Latour (1999) called establishing a "chain of reference."

Scientists interact with a process by observing ("mental" marking by taking note) or experimenting ("physical" marking by perturbation or alteration) and thereby create "traces": physical remnants that can be used as representations of the state of a marked process at multiple times. A sequence of these traces represents the process as it was tracked. This concept of tracking draws on Salmon's (1984) concept of "mark transmission" as a way of characterizing what it is to be a causal (as opposed to noncausal) process. The idea here is not to rely on Salmon's account as an analysis of causal process, but rather to use it as a characterization of scientific practices of following causal processes by causally interacting with them and using causal interactions as means of detecting, tracking, and individuating processes (Griesemer 2007). Processes that are subject to identity-changing fusions (merging of multiple processes into one) or fissions (splits of one process into several) call for renaming processes by remarking interactions or by relabeling traces so as to track fused and fissioned processes *as* individuals (see Griesemer [2014*a*] on relabeling entities in HIV replication). Differently put, traces only represent processes in so far as they reflect the numerical identity of the process(es) tracked. Reflecting numerical identity in representational, referential use is another way to describe individuation.

Returning to evolutionary individuality and selection processes, Goodnight uses fitness assignment as a way of describing what selection processes are tracked in empirical inquiries into evolution. In operational terms, to receive a fitness assignment is to be given a unique fitness label, w_i (a subscript value i = 1, or 2, or . . . added to a representation of relative fitness, w), or to be subjected to a measurement of fitness value, to which a fitness label is given, such that w_i = w. The things that take fitness assignments are the individuals corresponding to selection processes tracked in an empirical study. Evolutionary individuality, in other words, lies in two aspects: (1) the causal intersection of a fitness-assigning process (the scientist) and some biological process (the biological subject/phenomenon) and, prior to that, (2) the individuation of biological processes in a tracking practice performed by the scientist.

I emphasize *assignment* in Goodnight's approach as the lesson to be learned from what went wrong in previous efforts to "solve" the units of selection problem with a definitive *analysis* of that concept or the concept of biological (or evolutionary) individuality. Those attempts had to *presuppose* that to which the properties in play *could* be assigned (replicator, interactor, etc.). Goodnight focuses on the assignment practice itself. He shows that fitness assignment can work as an approach to characterizing evolutionary individuality not because it provides a definitive conceptual analysis, but because it serves a tracking process and practice of individuation. Using Goodnight's definition, one picks out, as the individuals in a contextual analysis (see later discussion), those processes to which fitness had been successfully assigned in a tracking and empirical measurement process. If fitness assignments do not track selection processes—for example, if fitness is assigned to entities that do not vary, or that do not reproduce differentially, or which do not fuse or fission to form fitness-bearing entities in the process(es) tracked—then the empirical results of fitness measurement (in either observation or experiment) may be statistically nonsignificant, so that the fitness assignment failed to track a selection process after all.

In Section 7.4, I consider another approach to the question of evolutionary individuality that is different from, but compatible with, Goodnight's story of fitness assignment. It involves my account of reproducers and how to track them. However, it characterizes processes of reproduction, *with or without* fitness differences or consequences, in a way that can be understood in terms that are far more general than is usual in biology, prior to the problem of units of selection and prior to the tracking problems of fitness assignment and measurement that Goodnight addresses.

Reproduction is a process through which fitness and variation in fitness can be *generated*. "Reproducers" are units with the potential (given the right circumstances) to be potential units of selection (i.e., fitness *bearers* and thus takers of fitness *assignment*), hence they are conceptually and biologically prior to units of selection. By itself, my reproducer approach is abstract and "theoretical," but it is specified together with a view of how to detect and track reproduction processes beyond the "ordinary" or familiar cases that guide so much of our thinking about candidate units of selection. That is, the concept of a process of reproduction provides insight into the character of biological *individuality* because it is complemented by a tracking practice of *individuation* by means of a "material overlap" criterion for reproductive processes. Material overlap picks out reproduction processes by specifying tracking conditions in ways that help explain why exemplary cases of biological individuality are exemplary and problem cases are problematic. In this way, the concept is heuristic and tied to tracking practice rather than definitional and tied to a conceptual analysis (Griesemer 2014*a*).

Neither Goodnight's nor my approach need be correct, empirically or theoretically adequate, nor the only ones of their kinds to do the modest illustrative work I aim to do with them in this chapter. The points I want to make are that both approaches, from very different starting points, tie concepts of individuality to particular kinds of empirical processes and both link understanding and interpretation of concepts of individuality to empirical practices of process tracking, either fitness assignment or material overlap. Moreover, with certain additional assumptions, one can view fitness assignment as part of a reproduction process tracking practice or, conversely, view a reproduction process tracking practice as part of a fitness assignment practice, though this further demonstration is beyond the scope of this chapter.

7.3 Goodnight's Definitions of the (Evolutionary) Individual

Goodnight (2013) entertains the possibility that selection may occur at multiple levels rather than just at the level of organisms and that there might be evolutionary individuals in addition to organisms that are potentially subject to selection at those other levels. His discussion and analysis tacitly assume the primary candidate levels are gene, cell, organism, group (of organisms), species, and multispecies communities. My interest is not in evaluating the argument or evidence for multilevel selection or whether those are the right levels to consider, but rather with Goodnight's strategy for defining "individual." His approach involves an interesting blend of theory and practice.

He observes that intuitive concepts of individuality fail to reflect all relevant aspects needed by the science of evolutionary biology, even when considering such canonical organisms as humans. This is because humans only function as organisms because "we" are organized as parts of complex communities or "holobionts" that include mites, skin and gut bacteria, and other, intracellular, symbionts such as mitochondria. Are humans "individual" organisms or communities (Dupré and O'Malley 2009; Gilbert, Sapp, and Tauber 2012; Chiu and Gilbert 2015)? Are intracellular symbionts individuals in their own right or parts? Can individual *organisms* be comprised of other individual *organisms*? In doubting intuitive concepts, Goodnight's approach to individuality is philosophical business as usual: look for problematic cases that challenge orthodoxy and demand conceptual clarification to fit complex reality. But Goodnight also introduces his preferred statistical *method* for analyzing multilevel selection, "contextual analysis," which changes the tenor of the argument for a concept of individuality. Again, my interest is not to defend this method, which is admittedly controversial, nor assess its merits relative to other methods (cf. Lloyd 1988; Okasha 2006). My interest is in how the method suggests an individuation principle for evolutionary individuals and how it is implicated in Goodnight's definitions of individuality.

The background operative idea since the emergence of population genetics in the first half of the twentieth century (Provine 1971) and quantitative evolutionary genetics in the second half (Lande 1976, 1979; Lande and Arnold 1983) is that evolution by natural selection can be interpreted as the operation of a causal process that changes the frequency distribution of genes, genotypes, or phenotypes within a population within a generation and that an evolutionary "response" to selection within generations is due to the transmission of genes or traits between generations (measured by allele frequency change or phenotype "heritability"). One measure of the operation of selection is Price's covariance between relative fitness and phenotype (Price 1970, 1972; Frank 1995; Okasha 2006). Another is the regression of relative fitness on phenotype (Lande and Arnold 1983), discussed by Goodnight (2013).

Contextual analysis extends standard regression analysis to represent effects of "context" on variables measured at a focal level (Heisler and Damuth 1987). For example, if phenotypic traits of organisms in a population are measured along with some measure of organism fitness (e.g., number of progeny), a standard way to assess selection is to calculate the statistical regression of relative fitness on phenotype. When regression coefficients in this sort of analysis are statistically significantly different than 0 (i.e., the fitness surface has non-zero slope), that is taken as evidence that selection is operating. But operating on what? In the case as described, it is operating on organisms, the things to which fitness and phenotype values are assigned and thus have been (or could be) measured.

If the population is a "meta"-population subdivided into multiple "sub"-groups, however, then there might be an effect on organism phenotype of membership in a particular *group* or an effect on organism phenotype due to selection operating on groups themselves. Contextual analysis adds to the calculation group phenotypes, as described in a meta-population story, allowing representation of these "contextual effects" by computing and comparing partial regression of fitness on both individual phenotype and on group mean phenotype. If group phenotype is a statistical function of individual phenotypes (e.g., average individual phenotype within each group), that is called a "composite variable" (Goodnight 2013: 38), which can be used to identify a group effect on individual phenotype. If a group phenotype that is *not* a function (or not the right sort of function) of individual phenotype is also measured—that is, a "contextual variable" such as population size or emigration rate (which cannot be assigned to individual organisms but can be assigned to groups as wholes)—then a contextual analysis can show that group selection is operating by detecting a significant regression of fitness on *group-level trait* (Goodnight 2013: 39, cf. box 2.1). For my purposes, further details of the method are not important. I draw attention to some key implications of the method for individuation and individuality.

First of all, contextual analysis admits of only one level for fitness assignment. The other level considered in the analysis is a context for the focal level. In the examples Goodnight (2013) considers, these levels resolve to organism (individual) and group (context). It does not follow that contextual analysis can only discover, display, or represent fitness at the level of assignment, however. As Goodnight points out, the discovery that some measured variables are "contextual" rather than "composite" provides evidence that selection operates at the group level and thus that, although investigators may have assigned (and tracked) fitness at the organism level in a particular study, they could have assigned, modeled, and tracked fitness at what in this study is group level because the analysis revealed that there is a significant causal process of selection operating at that level. Thus, fitness *assignment* is a function of investigator interest, ability, and opportunity and a matter of investigator choice. It is part of a practice of empirical measurement, which contextual analysis represents in particular respects. In some of the examples Goodnight describes (e.g., paleontological studies of fossil species or higher taxa), there is no doubt that paleontologists believe organisms in populations within species vary in fitness, but there is seldom opportunity for paleontologists to measure those fitness differences in fossils, so paleontologists do not assign fitness below the level of species because (typically) they cannot measure it: they cannot get the data; they cannot track the process operating at that level.

Second, because contextual analysis describes and represents selection at a particular focal level of measurement chosen by the investigator and admits discovery of composite and contextual effects at higher levels, processes operating *below* the focal level are not *described* as selection processes because the analysis does not track or evaluate *fitness* below the focal level. Because fitness is not *assigned* at any lower level than the focal level, processes operating below the focal level must be described as something else. Goodnight suggests that these processes are typically described as developmental rather than selective:

> [C]hanges at or above the level at which fitness is assigned can be described in terms of evolution and natural selection. This is because there is variation among units for the level at which fitness is assigned, and changes at

> or above this level can be described in terms of fitness differences. Below the level at which we assign fitness, such distinctions are not possible, since all entities that make up one of the units for which we define fitness by definition have the same fitness. Therefore, we have to use another term for changes taking place below the level at which fitness is assigned. Thus, for example, it is customary to assign fitness at the level of the organism, and changes in phenotype at the organismic level or higher can be described in terms of evolution. Below the level of the organism, we cannot describe changes in terms of evolution, and instead use the term *development* to describe changes within the organism. (Goodnight 2013: 39)

If we can fairly characterize changes in context above the focal level at which fitness is assigned as "ecological," then we get the following picture: contextual analysis describes selection operating at a focal level at which fitness is assigned. Above that level, ecological processes operate through "composite" variables that may have effects on the focal level (so-called multilevel selection 1 or MLS1) and selection processes have effects on the group level that can be detected at the focal level through "contextual" variables (so-called multilevel selection 2 or MLS2).[4] Below the focal level, developmental processes operate that play a constitutive role in the generation of units at the focal level. It may be that, below focal level, selection processes do operate, but the choice to assign (and measure) fitness at the focal level forces description of those processes as developmental. It may be that above focal level, selection processes operate, but the choice to assign (and measure) fitness at the focal level forces description of those processes as ecological. Yet the discovery that some of those ecological processes operate through contextual (rather than composite) variables indicates selection may be occurring at the ecological level, and a full understanding of selection would thus require further work to represent, assign, and measure fitness at that contextual level by taking population structure explicitly into account (see also Lloyd 2001).

If evolutionary individuality is understood as that to which fitness is assigned, Goodnight's first definition of the individual (p. 42), then "the concept of 'individuality' is an artificial construct imposed by the observer" (Goodnight 2013: 48). By that, I take it, Goodnight means that "the individual" is imposed in the analysis by the choice of fitness assignment, not that the reality of what causal *processes* operate in the world are constructed by the investigator. His is not a *pernicious* relativism, but rather a process relativity of concept to what it is of interest to the investigator to track by assigning and measuring fitness in a particular way. To give it a philosophical label, we might call this a "process realism with entity (or level) relativism" or simply "processualism." Or, eschewing metaphysics, we could say that Goodnight identifies the "logical individuals" or "units of analysis" in a

[4] Note that Goodnight considers the MLS1/MLS2 distinction misleading. See Heisler and Damuth (1987), Damuth and Heisler (1988), for the origins of this distinction. See Okasha (2006) for a review.

contextual analysis as those units to which fitness is assigned and measured. As Goodnight goes on to say:

> [T]here is no one level that can clearly be called "the individual" to the exclusion of other levels. One level may be the individual with respect to one selective force, and in the same system a different level may be the individual with respect to a different selective force, and finally, the investigator may choose yet a third level of organization to define as the individual. (Goodnight 2013: 49)

In Goodnight's picture, processes within organisms might be selection processes (e.g., cancers might be selection processes operating at cellular level) but, due to the organization of those organisms in which there is a cellular bottleneck in the production of germ cells, somatic cell cancers only rarely exhibit evolutionary responses to selection beyond the reproductive generation of the organisms that suffer them. Thus, for practical reasons, it rarely is of value or interest to assign and measure fitness at cell rather than organism level and thus more effective to interpret cancer as a developmental than a selective process. That does not preclude studying cancer as a within-body selection process, but it does explain why it is impractical to do so for studies that focus on other levels.

In the interest of full understanding of selection in a particular system, we would perhaps, Goodnight argues, want to consider a definition of the individual as the unit of analysis (that which takes fitness assignment) at the *lowest* level at which selection is *actually* acting (Goodnight's definition 2, p. 44), rather than at a *focal* level of investigator interest. This would facilitate discovery, through contextual analysis, of all levels above the lowest at which selection is acting through the detection of contextual effects operating through contextual variables. This definition still relies on the notion of fitness assignment to units at a level as the core concept of individuality, or rather, principle of individuation. However, for empirical reasons (evolutionary response is presumed rare below organism level) and practical reasons (fitness and selection response measurement below organism level is hard), the value repaid in research results is rarely worth the effort required. It is cheaper, faster, better to pick a promising focal level than to start at "the bottom" of a presupposed hierarchy of levels and assign/measure fitness all the way up, "just in case." A theoretician's reply to this pragmatic strategy is: "How do you know evolutionary response below organism level is rare unless you test for it?" The same critique applies to measurement above organism level in deciding whether to test for population structure.

Goodnight's third definition—the individual is (a unit at) the lowest level at which a *response* to selection can occur (p. 47)—does not add to the points I want to make about individuation in practice, so I shall set consideration of it aside for present purposes.

The methodological commitment to track individuality through fitness assignment at a focal level, as required by contextual analysis, relegates to "development" those processes occurring below the focal level. This is a concept of development needed for *evolutionary* explanation. What about development as needed for *developmental* explanation "at" a focal level? Even though organisms are typically "the individuals" assigned fitnesses in evaluations of evidence of the operation of selection, whether through contextual, regression, or covariance analysis, it is also the case that organisms—the entities at the conventional focal level—develop (and reproduce, metabolize, behave, and interact ecologically). Organisms are not *mere* fitness bearers engaged only in selection processes "at their own level." They may be individuated with respect to other processes in addition to selection. Fitness assignment and methodological commitment to use contextual analysis may support ontological commitment to certain sorts of *evolutionary* individuals (assuming the empirical evidence is favorable), but that does not preclude other practices and methodological commitments for tracking other sorts of processes at organism level or any other level. Goodnight acknowledges that selection may operate at multiple levels, which leads to a process-relative, pluralistic perspective, but we should also acknowledge that other kinds of processes might operate "at" a level and simultaneously at multiple levels unless there are special reasons differentiating kinds of processes in this particular respect regarding levels of organization.

Moreover, if it is the case that to be a fitness bearer, to take a fitness assignment at a level, the individual must also be an individual with respect to processes other than selection, it may be that to have the potential to be an evolutionary individual, an entity or process may need to have the potential to be an individual with respect to those other processes. Thus, the potential to take a fitness assignment might be understood as the potential (in virtue of those other processes) to have the potential to take a fitness assignment. If that is the way things are, then it would seem that evolutionary individuality rests, conceptually, on other forms of individuality.

In the next section, I sketch an account of reproduction supporting a concept of "developmental individuality." I also suggest a tracking criterion for developmental individuality—material overlap—that demarcates developmental and reproductive "generations" and thus serves as an individuation criterion.

7.4 Reproducers and Reproductive Individuality

My purpose in this section is not to defend my account of reproducers (Griesemer 2000*a*, 2000*b*, 2000*c*, 2002, 2005, 2006*a*, 2014*a*, 2014*b*, 2016), but to use it, as in the previous section, as an example of an account of concepts supporting a notion of evolutionary individuality linked to a concept of individuation. As in the previous section, a focus on biological processes together with a tracking procedure and criterion of individuation leads to a "process realism with entity relativism" or "processualism." Here, the process is reproduction (and a recursively entwined process, development) rather than selection.[5] The tracking procedure is to look for "material overlap" relations and to relabel what is tracked each time a certain kind of fission or fusion event among processes, called a "scaffolding interaction," occurs, rather than to track whatever receives a fitness assignment, as in Goodnight's account. This relabeling practice delimits or individuates processes as biological individuals (reproducer or developer individuals) that have the potential to be potential units of selection. Thus, an evolutionary individual is a trackable reproducer process with the potential to be a potential unit of evolution (or selection).

However, the account also blurs lines between what conventionally count as reproduction and development processes, revealing a new problem of individuation *of generations* that is fundamental for a generalized population biology and thus for an extended theory of evolution. Like Goodnight's fitness assignment and contextual analysis, what lies below (above) a focal level of reproduction is development (ecology), but that lower level of development can be focal-level reproduction in its own right with a developmental (ecological) level below (above) that. Unlike Goodnight's fitness assignment approach, however, the reproducer account's individuation criterion—material overlap through scaffolding interactions—does not have the inherent asymmetry of fitness assignment at focal levels because it does not depend on a concept of statistical variation (at a level) that by definition disregards variation at lower levels.

The motivation is threefold, to (1) emphasize the materiality of reproduction rather than the abstract, informational properties of genes and replication; (2) reconnect problems of heredity and development after they were sundered in twentieth-century conceptual dichotomies like auto-/hetero- catalytic functions of DNA, gene transmission/expression, and genotype/phenotype; and (3) characterize the process of reproduction in terms that do not themselves require evolutionary analysis to avoid the circularity problems discussed earlier.

Maynard Smith (1987), in considering where to place Malthus's principle in relation to Darwin's, noted that the entities subject to evolution (and its special case: evolution by means of natural selection) must *multiply*. Whether Malthus's principle is a law of nature or merely a likely condition in which population growth tends to outstrip food supply is less important than the constraint on Darwin's principles to apply to things which multiply. I interpret the multiplication condition in Maynard Smith's principles of evolutionary units as a principle of reproduction (Griesemer 2000*c*). That is, evolution operates on "reproducers," or, more precisely, evolutionary processes are a special class of reproduction processes.

Evolution can be understood as descent with modification of a population of "reproducers." Reproduction is multiplication (of processes or entities) with material overlap conveying developmental capacities (i.e., material propagule

---

[5] On process entwinement, see Griesemer (2007).

generation or "pro-generation"). Material overlap means that at least some material parts of the "offspring" were formerly material parts of the "parents." Thus, reproduction involves a bond of material continuity, not merely one of resemblance or formal transmission of information. If material can be suitably marked through causal interaction, it can be tracked. So, the material overlap condition in the characterization of reproduction also yields a tracking criterion. Materials which "track together" can be counted an individual for a purpose of study.

To count as reproduction, rather than mere duplication or copying, at least some materially overlapping parts must convey developmental capacities to offspring (Griesemer 2000*a*, 2014*a*, 2014*b*, 2016). Development is the recursive acquisition (or refinement) of a capacity to reproduce. Recursion bottoms out in "null development"—progenerated entities originate with a capacity to reproduce, rather than having to acquire it by exercise of developmental capacities. Development in the special case of multicellular organisms, for example, requires reproduction of cells, reproduction of cells requires development of cellular capacities, and cell development requires chemical autocatalysis (null development) of cellular constituents. There is no fixed, asymmetrical relation between reproduction and development via material overlap as there is between selection and development (or ecology) in Goodnight's fitness assignment criterion of individuality. Although contextual analysis can shift focal levels, it must do so in order to treat a given level as selection rather than development. The reproducer account allows a plurality of interpretations of processes at each level as either development or reproduction, *depending on the choice of what material overlap relations to track at that level* because reproduction and development are mutually embedding processes at each level.

If one looks at the details of development, at any level of organization, one finds processes that look materially like reproduction: material processes fuse and fission; in the hybrid, fused state transformations take place that constitute acquisition of new developmental capacities; exercise of those capacities facilitates transition to a new developmental state, perhaps involving further fusions or fissions. Clearly, development of multicellular organisms involves reproduction by fissioning of cells as well as aggregation and differentiation (fusion in a different sense than fertilization) of cells interacting in weak to tight cell junctions to form organized tissues and organs. If we were to track these fusions, fissions, and transformations with cells, or even molecules, rather than organisms as the entities filling the roles of parents and offspring, it would be tempting to treat the fusion and fission events as *reproductive* processes in complex life cycles delimiting reproducers at a lower level than organisms, with many "developmental generations" separating reproductive generations (Griesemer 2016), just as cell generations mark developmental stages of the reproductive generations of multicellular organisms.

Griesemer (2014*a*) argues that (complex) molecules, such as HIV retrovirus genomes, *develop* and that hybrid complexes of them with various proteins *reproduce*, all within the span of a single host cell's life cycle. Griesemer (2016) argues that this and other cases indicate an important sense in which all life cycles are complex and that what is often called "development" can be evaluated as reproduction by tracking material overlap relations among the fusing and fissioning entities. The complexity of life cycles can be interpreted as due to transitions in the "ecological" contexts of development, marked by certain sorts of developmental interactions with elements of a developmental environment. Complex life cycles are not solely the province of parasites with multiple hosts.

One further concept completes the picture of reproduction: developmental scaffolding. Most developmental processes are facilitated rather than autonomous in the sense that the aid of an "external" process interacting with the developing system makes easier, or more likely, or with lower fitness cost, the acquisition, refinement, or exercise of a new developmental capacity (Caporael, Griesemer, and Wimsatt 2014). I say "external" because, on traditional accounts of compositional levels of organization, the scaffold is one thing, the thing scaffolded is another thing, and the pair in interaction may or may not be treated as a thing at a higher level. A human parent may scaffold a child to learn to cross the street safely—a capacity that can only be acquired after a certain state of cognitive development has been attained by the child and a certain state of cognitive maturity has been attained by the parent. To say that the child is our focal unit of development and the parent is part of the environment or ecological context of child development, albeit a very special one, is to track the material relations within the child over time and place without regard for the material relations between parent and child. In a sense, we commit to the child as an individual before the interaction, as an individual after the interaction, and as the same individual spanning the interaction, though perhaps changed in properties and capacities. From the point of view of focal interest in the child, little or no commitment is made to the ontological status of the parent. If attention does shift to her, then similar considerations apply.

A different view in keeping with the reproducer perspective (Griesemer 2014*a*, 2014*b*, 2016) is that the parent and child form a hybrid process by their temporary fusion (e.g., by holding hands), *intra*-action (tugging, guiding, pulling, and also talking, pointing out, exclaiming) as parts of a whole in a street-crossing phase, and fission (letting go hands) on the other side of the street.[6] In other words, the child and parent form a new individual due to their developmental interaction that "fuses" them in a psychosocial sense, a social developmental individual that persists for a short while, and then fissions into two more autonomous individuals after their interaction. This hybrid, dyadic individual is itself composed of individuals—the child and the parent; they do not lose their individuality (in that they can be tracked through the interaction), though they each cede a measure of autonomy for the duration of the interaction. Hybridization of this kind is the

[6] I take the concept of "intra-action" from Barad (2007).

formation of a new entity—a new individual—by fusion. Separation marks the formation of two entities from one.

For a variety of reasons, we are typically inclined to individuate and track with proper names or labels the parent as numerically the same parent and the child as numerically the same child who entered the street-crossing episode and exited it and to think of the pair as a mere temporary association of separate, distinct, more-or-less autonomous individuals and to discount claims that the dyad is itself an individual. Chief among these reasons is that when we track material relations, we tend not to regard material relations as an aspect of social relations nor to think of the temporal relations as well as the spatial relations when we note that the parent and child did not "exchange" any significant material parts with one another; there is no material overlap between *them* in the street-crossing episode. But there is material overlap between the child pre–street-crossing experience and the parent–child pair, and between the parent–child pair and the child post–street-crossing experience. The same holds for the parent. The fact that each emerges with more or less the same overall structure and properties as they entered the hybrid makes this social learning interaction *more* like semi-conservative DNA replication rather than less (Griesemer 2014*a*).

In general, the reproducer perspective suggests that a developmental individual be understood as a reproductive process individuated and marked, from start to finish, by developmental scaffolding interactions. Each time new developmental capacities are gained through hybrid-forming scaffolding interactions (in a process-relative sense specific to the kinds of scaffolds and developers causally interacting) or exercised in ways that result in new scaffolding interactions, we can recognize fusions and fissions in a material sense that mark off distinct "developmental generations" within a life cycle delimited by progeneration events demarcating "reproductive generations" (Figure 7.1A). In a similar vein, Salmon (1984) recognized that each pair of causal intersections, no matter how brief or seemingly insignificant, delimited distinct causal processes. However, although this means that a hit baseball traveling from the bat that strikes it to the window it breaks is "really" a sequence of many processes (as the ball also interacts with many molecules in the air on the way to the window), for many practical purposes, we discount these interactions and concatenate the many elemental component processes into a single struck-baseball process.

While we need not regard distinct developmental generations as delimiting numerically distinct *organisms* or reproducers (any more than metamorphosis marks the tadpole as one organism and the frog as another), we can consider that scaffolding events give us another way to interpret the organization of developmental processes into temporal sequences comprising higher level individuals in a way analogous to cell generations marking developmental stages of a multicellular, higher level individual organism. Since scaffolding in development creates new hybrid material entities through which new developmental capacities can be acquired, and conveyance of developmental capacities is the characteristic of

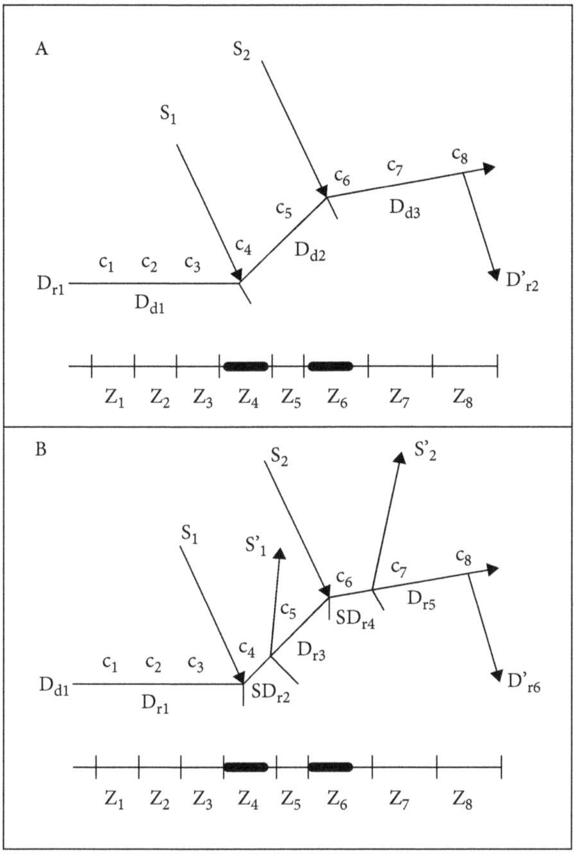

FIGURE 7.1 *Scaffolded development and reproduction.* **A**: *Traditional interpretation of development in which reproduction results in numerically distinct, spatially individuated reproducers. The transitions between reproductive generations (r1, r2) are marked by developmental stages (d1, d2, d3). Z is a Vygotskian zone of proximal development (see Caporael, Griesemer, and Wimsatt 2014)—a window in which the developer is capable of acquiring a new skill or capacity. D is a developmental process (indicated by the bent arrow running from left to right), subscripted r1 for the "parent" reproductive generation. D' is its reproductive offspring in generation r2. In between, in the development of D, new capacities* c_i *(i = 1 to 8) are acquired and exercised. S is a scaffold that interacts (causal arrow intersecting the causal arrow of D) at a particular time in development. Here,* S_1 *scaffolds D in a scaffolding interaction with onset in zone* Z_4 *and continuing to the end of* Z_4*, at which time D has acquired* c_4 *which is necessary for D to acquire* c_5*, which facilitates interaction with scaffold* S_2*. Thick bars on the developmental time line show the duration of scaffolding interactions in* Z_4 *and* Z_6*.* **B** *is similar to* **A** *except each developmental "stage" in* **A** *called a "developmental generation" (d1, d2, d3) is here called a "reproductive generation" to note the material overlap relations between* D_{r1}*,* SD_{r2}*,* D_{r3}*,* SD_{r4}*,* D_{r5}*, and* D'_{r6}*.* SD_{r2} *and* SD_{r4} *denote hybrid entities formed by the fusion of a scaffold, S, and developer, D. (Development of the scaffolds not shown.) D and S in generation r1 are "parents" of the hybrid offspring SD in generation r2. SD is the "parent" of offspring S1 and D in r3. And so forth.*

material reproduction processes, how we choose to mark, track, and measure material overlap relations amounts to the constructed aspect of reproductive and developmental individuality on par with Goodnight's constructed aspect of evolutionary individuality via fitness assignment. We can, in other words, interpret scaffolding interactions as delimiting developmental individuals and the markable material that passes through them as assignable to distinct developmental or reproductive generations (compare Figures 7.1A and B). In Figure 7.1A, three "developmental generations" separate two reproductive generations. In Figure 7.1B, six reproductive generations are separated by scaffolding interactions. The latter concept is analogous, but more fine-grained, to saying that human parents are really grandparents of their children: they form gametes as their immediate offspring; gametes are the parents that fuse to form a zygote, or grand-offspring (Griesemer 2014*a*, 2016).

On this kind of processualism, an *organism* is a construct depending on investigator interest in and tracking ability for certain kinds of reproducers, together with the reproducer and developer capacities of the processes tracked. These developmental individuals and generations can be tracked and delimited by marking developing processes that hybridize with scaffolds. They can be remarked or relabeled when they form new hybrid individuals constituting a new developmental or reproductive generation (depending on tracking interest) and relabeled again when constituting a third generation after scaffolds and developers separate (Figure 7.1B). Marking, tracking, and relabeling material that enters or exits a reproduction process delimited in a number of developmental or reproductive generations does not depend on tracking statistical variation in a population of processes, so tracking material overlap relations does not carry the same sort of asymmetry between levels that Goodnight's fitness assignment tracking does. There are still focal choices, but here they are not dictated by a relation between levels of organization but rather by a choice between what to track and what not to track.

7.5 Individuality as a Partial Ordering Rather than Degree Property

One further implication of this view suggests an alternative to treating individuality as a degree property tuned by the value of several parameters in a cooperation/conflict model, as in Queller and Strathmann's approach and as examined in depth by Godfrey-Smith (2009) in his Darwinian spaces model that maps the menagerie of reproductive modes in the three dimensions of cellular bottlenecks, germ–soma specialization, and integration. I have argued that individuality can be understood in terms of tracking material overlap relations—the analog of tracking fitness assignment in Goodnight's contextual analysis of statistical variation in populations. I also suggested that this notion of individuality is process-relative: individuality depends on individuation that happens in marking and tracking choices relative to a process of interest. Because of this process-relativity and the evident multitude of processes that can interact, we can think of the "degree property" of individuality rather as a partial ordering relation of "orders of individuality."

An individual is "stronger," more robust, when more processes track together than when fewer track together. If a marking interaction can be used to track material overlap relations with respect to several processes, it individuates a stronger individual than if it tracks only a subset of those processes. Vertebrate organisms strike us as very coherent, cohesive, "strongly" individual in part because the cellular parts of vertebrates seldom go their separate ways such that a variety of biological processes of interest (from physiology to behavior, development to reproduction, to ecological interaction) have to be tracked separately. Not so with a cellular slime mold, whose cellular parts can go their separate spatial ways and still be "one organism." In the case of sponges, the separate parts can come back together after being disaggregated, so we think of the period of separation (perhaps at the hands of an experimenter) as merely a stage in the peculiar developmental history of "the" sponge. Sponges and slime molds are weaker individuals than we are because their looser need for spatial contiguity of parts means we cannot track the "whole" organism by, say, marking the genome of a single stem cell. That would do for a vertebrate.[7] The holobiont perspective raises new questions not only about the individuality of the human organism, but about the salient tracking criteria for such entities. If humans plus their microbiomes are holobionts with a similarly distributed "hologenome" (Gilbert et al. 2018), then tracking humans through reproductive generations by tagging genes of the *Homo sapiens* component of human holobionts is inadequate unless microbiome transmission is strictly "vertical."

The notion that individuality can be partially ordered relative to a set of trackable processes of interest suggests a way of extending this processualism to other sciences. I have focused on reproduction and development because I believe these are fundamental to biological problems of evolution and, indeed, the evolution of biological individuality. But the view may be more general. What we take to be individuals rests in part on our ability to mark and track processes of interest, and, to the extent that many trackable processes "track together," we find individuals.

[7] Modulo the skin cells we tend to slough off and thus not *count* as part of the body any longer. In that case, our strong-individual marking/tracking criterion would fail when we happened to mark the genome of that skin cell. Thanks to Melinda Fagan for stressing that, even for vertebrates, marking some body cells will not do to track vertebrates throughout their life trajectories unless it is a stem cell that is marked.

Acknowledgments

I thank the organizers, Ruey-Lin Chen and Alexandre Guay, of the Taiwan Conference on Scientific Individuation in the Physical, Biological, and Experimental Sciences for putting together an exciting and productive meeting, and National Chung Cheng University, Taiwan for providing an excellent meeting venue and hosting our visit. I am deeply grateful to the volume editors, Melinda Bonnie Fagan, Otávio Bueno, and especially Ruey-Lin Chen for their helpful comments on my chapter and patience while I made extensive revisions. I extend special thanks and gratitude to the discussant of my talk at the conference, Bai-Chang Yang. His challenges and questions about my presentation were instrumental in helping me redirect my project and reach a final chapter much different, and much better, than it otherwise would have been. I thank Charles Goodnight for input that helped shape the argument. I also thank the people of the State of California and UC Davis for financial support.

References

Antonovics, Janis, Norman Ellstrand, and Robert Brandon. 1988. Genetic variation and environmental variation: expectations and experiments. In *Plant evolutionary biology*, eds. Leslie Gottlieb and Subodh Jain, 275–303. New York: Chapman and Hall.

Barad, Karen 2007. *Meeting the universe half way: Quantum physics and the entanglement of matter and meaning*. Durham, NC: Duke University Press.

Bonner, John T. 1952. *Morphogenesis: An essay on development*. Princeton: Princeton University Press.

Bonner, John T. 1967. *The cellular slime molds*, 2nd ed., revised and augmented. Princeton: Princeton University Press.

Bouchard, Frédéric, and Philippe Huneman (eds.). 2013. *From groups to individuals: Evolution and emerging individuality*. Cambridge, MA: MIT Press.

Buss, Leo W. 1987. *The evolution of individuality*. Princeton: Princeton University Press.

Caporael, Linnda, James Griesemer, and Willam Wimsatt (eds.). 2014. *Developing scaffolds in evolution, culture, and cognition*. Cambridge, MA: MIT Press.

Chiu, Linn, and Scott Gilbert. 2015. The birth of the Holobiont: Multi-species birthing through mutual scaffolding and niche construction. *Biosemiotics* 8: 191–210.

Clarke, Ellen 2010. The problem of biological individuality. *Biological Theory* 5(4): 312–325.

Damuth, John, and Ida L. Heisler. 1988. Alternative formulations of multilevel selection. *Biology and Philosophy* 3: 407–430.

Dawkins, Richard. 1976. *The selfish gene*. New York: Oxford University Press.

Dawkins, Richard. 1982. *The extended phenotype: The gene as the unit of selection*. New York: Oxford University Press.

Dobzhansky, Theodosius. 1973. Nothing in biology makes sense except in the light of evolution. *American Biology Teacher* 35(3): 125–129.

Dupré, John, and Maureen O'Malley. 2009. Varieties of living things: Life at the intersection of lineage and metabolism. *Philosophy and Theory in Biology* 1: e003.
Frank, Steven A. 1995. George Price's contributions to evolutionary genetics. *Journal of Theoretical Biology* 175: 373–388.
Gánti, Tibor. 2003. *The principles of life, with a commentary by James Griesemer and Eörs Szathmáry*. Oxford: Oxford University Press.
Gilbert, Scott, and David Epel. 2009. *Ecological developmental biology*. Sunderland: Sinauer Associates.
Gilbert, Scott, Eugene Rosenberg, and Ilana Zilber-Rosenberg. 2018. The Holobiont with its hologenome is a level of selection in evolution. In *Landscapes of collectivity in the life sciences*, eds. Snait Gissis, Ehud Lamm, and Ayelet Shavit, 305–324. Cambridge, MA: MIT Press.
Gilbert, Scott, Jan Sapp, and Alfred Tauber. 2012. A symbiotic view of life: We have never been individuals. *Quarterly Review of Biology* 87: 325–341.
Godfrey-Smith, Peter. 2009. *Darwinian populations and natural selection*. New York: Oxford University Press.
Godfrey-Smith, Peter. 2013. Defining individuals. In *From groups to individuals: Evolution and emerging individuality*, eds. Frédéric Bouchard and Philippe Huneman, 17–36. Cambridge, MA: MIT Press.
Goodman, Nelson, and Henry S. Leonard. 1940. The calculus of individuals and its uses. *Journal of Symbolic Logic* 5: 45–55.
Goodnight, Charles J. 2013. Defining the individual. In *From groups to individuals: Evolution and emerging individuality*, eds. Frédéric Bouchard and Philippe Huneman, 37–53. Cambridge, MA: MIT Press.
Griesemer, James. 1994. Tools for talking: Human nature, Weismannism and the interpretation of genetic information. In *Are genes us? The social consequences of the new genetics*, ed. Carl Cranor, 69–88. New Brunswick: Rutgers University Press.
Griesemer, James. 2000*a*. Development, culture and the units of inheritance. *Philosophy of Science* 67 (Proceedings): S348–S368.
Griesemer, James. 2000*b*. Reproduction and the reduction of genetics. In *The concept of the gene in development and evolution: Historical and epistemological perspectives*, eds. Peter Beurton, Raphael Falk, and Hans-Jörg Rheinberger, 240–285. New York: Cambridge University Press.
Griesemer, James. 2000*c*. The units of evolutionary transition. *Selection* 1: 67–80.
Griesemer, James. 2002. Limits of reproduction: A reductionistic research strategy in evolutionary biology. In *Promises and limits of reductionism in the biomedical sciences*, eds. Marc H. V. Van Regenmortel and David Hull, 211–231. Chichester: John Wiley and Sons.
Griesemer, James. 2005. The informational gene and the substantial body: On the generalization of evolutionary theory by abstraction. In *Idealization XII: Correcting the model, idealization and abstraction in the sciences (Poznan studies in the philosophy of the sciences and the humanities, volume 86)*, eds. Martin Jones and Nancy Cartwright, 59–115. Amsterdam: Rodopi.
Griesemer, James. 2006*a*. Genetics from an evolutionary process perspective. In *Genes in development*, eds. Eva M. Neumann-Held and Christoph Rehmann-Sutter, 199–237. Durham, NC: Duke University Press.

Griesemer, James. 2006*b*. Theoretical integration, cooperation, and theories as tracking devices. *Biological Theory* 1(1): 4–7.

Griesemer, James. 2007. Tracking organic processes: Representations and research styles in classical embryology and genetics. In *From embryology to evo-devo: A history of developmental evolution*, eds. Manfred Laubichler and Jane Maienschein, 375–433. Cambridge, MA: MIT Press.

Griesemer, James. 2008. Origins of life studies. In *The Oxford handbook of philosophy of biology*, ed. Michael Ruse, 263–290. New York: Oxford University Press.

Griesemer, James. 2013. Formalization and the meaning of "theory" in the inexact biological sciences. *Biological Theory* 7: 298–310.

Griesemer, James. 2014*a*. Reproduction and the scaffolded development of hybrids. In *Developing scaffolds in evolution, culture, and cognition*, eds. Linnda Caporael, James Griesemer, and William Wimsatt, 23–55. Cambridge, MA: MIT Press.

Griesemer, James. 2014*b*. Reproduction and scaffolded developmental processes: An integrated evolutionary perspective. In *Towards a theory of development*, eds. Alessandro Minelli and Thomas Pradeu, 183–202. New York: Oxford University Press.

Griesemer, James. 2016. Reproduction in complex life cycles: A developmental reaction norms perspective. *Philosophy of Science* 83: 803–815.

Griesemer, James, and Michael J. Wade. 1988. Laboratory models, causal explanation and group selection. *Biology and Philosophy* 3: 67–96.

Guay, Alexandre, and Thomas Pradeu (eds.). 2016. *Individuals across the sciences*. Oxford: Oxford University Press.

Heisler, Ida L., and John Damuth. 1987. A method for analyzing selection in hierarchically structured populations. *American Naturalist* 130: 582–602.

Hull, David. 1980. Individuality and selection. *Annual Reviews of Ecology and Systematics* 11: 311–332.

Hull, David. 1981. The units of evolution: A metaphysical essay. In *The philosophy of evolution*, eds. Uffe Jensen and Rom Harré, 3–44. Brighton: Harvester Press.

Hull, David. 1988. *Science as a process*. Chicago: University of Chicago Press.

Jackson, Jeremy, Leo Buss, and Robert Cook (eds.). 1985. *Population biology and evolution of clonal organisms*. New Haven: Yale University Press.

Kauffman, Stuart. 1993. *The origins of order: Self-organization and selection in evolution*. New York: Oxford University Press.

Lande, Russell 1976. Natural selection and random genetic drift in phenotypic evolution. *Evolution* 30: 314–334.

Lande, Russell. 1979. Quantitative genetic analysis of multivariate evolution, applied to brain-body size allometry. *Evolution* 33: 402–416.

Lande, Russell, and Stevan J. Arnold. 1983. The measurement of selection on correlated characters. *Evolution* 37: 1210–1226.

Latour, Bruno. 1999. *Pandora's hope: Essays on the reality of science studies*. Cambridge, MA: Harvard University Press.

Lewontin, Richard. 1970. The units of selection. *Annual Review of Ecology and Systematics* 1: 1–17.

Lloyd, Elisabeth A. 1988. *The structure and confirmation of evolutionary theory*. New York: Greenwood Press.

Lloyd, Elisabeth A. 2001. Units and levels of selection: An anatomy of the units of selection debates. In *Thinking about evolution: Historical, philosophical, and political perspectives*, vol. 2, eds. Rama S. Singh, Costas B. Krimbas, Diane B. Paul, and John Beatty, 267–291. New York: Cambridge University Press.

Lloyd, Elisabeth A. 2012. Units and levels of selection. *The Stanford encyclopedia of philosophy* (winter 2012 edition), ed. Edward N. Zalta. Accessed October 29, 2016. http://plato.stanford.edu/archives/win2012/entries/selection-units/.

Malthus, Thomas R. 1826. *An essay on the principle of population*. London: John Murray.

Maturana, Humberto, and Francisco Varela. 1980. *Autopoiesis and cognition: The realization of the living*. Boston studies in the philosophy of science volume 42. Boston: D. Reidel.

Maynard Smith, John. 1987. How to model evolution. In *The latest on the best: Essays on evolution and optimality*, ed. John Dupré, 119–131. Cambridge, MA: MIT Press.

Maynard Smith, John, and Eörs Szathmáry. 1995. *The major transitions in evolution*. Oxford: W. H. Freeman Spektrum.

Mayr, Ernst. 1961. Cause and effect in biology. *Science* 134(3489): 1501–1506.

McFall-Ngai, Margaret, Michael G. Hadfield, Thomas C. G. Bosch, Hannah V. Carey, Tomislav Domazet-Lošo, Angela E. Douglas, et al. 2013. Animals in a bacterial world: a new imperative for the life sciences. *Proceedings of the National Academy of Sciences USA* 110(9): 3229–3236. doi: 10.1073/pnas.1218525110. Epub Feb. 7, 2013.

Michod, Richard. 1999. *Darwinian dynamics: Evolutionary transitions in fitness and individuality*. Princeton: Princeton University Press.

Moorad, Jacob A., and Michael J. Wade. 2013. Selection gradients, the opportunity for selection, and the coefficient of determination. *American Naturalist* 181(3): 291–300.

Moreno, Alvaro, and Matteo Mossio. 2015. *Biological autonomy: A philosophical and theoretical enquiry*. History, philosophy and theory of the life sciences, volume 12. New York: Springer.

Okasha, Samir. 2006. *Evolution and the levels of selection*. Oxford: Oxford University Press.

Pigliucci, Massimo, and Gerd Müller (eds.). 2010. *Evolution: The extended synthesis*. Cambridge, MA: MIT Press.

Price, George R. 1970. Selection and covariance. *Nature* 227: 520–521.

Price, George R. 1972. Extension of covariance selection mathematics. *American Journal of Human Genetics* 35: 485–490.

Provine, William. 1971. *The origins of theoretical population genetics*. Chicago: University of Chicago Press.

Queller, David, and Joan Strathmann. 2009. Beyond society: The evolution of organismality. *Philosophical Transactions of the Royal Society B* 364: 3143–3155.

Ruiz-Mirazo, Kepa, and Alvaro Moreno. 2012. Autonomy in evolution: From minimal to complex life. *Synthese* 185: 21–52.

Salmon, Wesley. 1984. *Scientific explanation and the causal structure of the world*. Princeton: Princeton University Press.

Sober, Elliott. 1984. *The nature of selection*. Cambridge, MA: MIT Press.

Sober, Elliott, and David S. Wilson. 1998. *Unto others: The evolution and psychology of unselfish behavior*. Cambridge, MA: Harvard University Press.

Strawson, Peter F. 1959. *Individuals: An essay in descriptive metaphysics*. London: Methuen.

Sultan, Sonia. 2015. *Organism and environment: Ecological development, niche construction, and adaptation*. New York: Oxford University Press.

Wade, Michael J. 1976. Group selection among laboratory populations of *Tribolium*. *Proceedings of the National Academy of Sciences* 73(12): 4604–4607.

Wade, Michael J. 1977. An experimental study of group selection. *Evolution* 31: 134–153.

Wade, Michael J. 2016. *Adaptation in metapopulations: How interaction changes evolution*. Chicago: University of Chicago Press.

Wagner, Günter. 2014. *Homology, genes, and evolutionary innovation*. Princeton: Princeton University Press.

Waters, C. Kenneth. 2008. How practical know-how contextualizes theoretical knowledge: Exporting causal knowledge from laboratory to nature. *Philosophy of Science* 75: 707–719.

Wiggins, David. 1980. *Sameness and substance*. Cambridge, MA: Harvard University Press.

Wimsatt, William. 1980. Units of selection and the structure of the multi-level genome. *PSA 1980 Volume Two: Symposia and Invited Papers* 47(2): 122–183.

{8}

Individuation, Individuality, and Experimental Practice in Developmental Biology

Alan C. Love

8.1 Individuality: (Evolutionary) Theory and Biological Practice

Natural philosophers have long been fascinated by individuality and the relationship between parts and wholes. Empirical examples from animals and plants where intuitive notions of individuality seem to break down have been especially provocative (e.g., siphonophores or quaking aspen). Increased attention to complex life cycles of terrestrial and marine invertebrates led to an explosion of competing perspectives in the nineteenth century on how to conceptualize individuality, a debate that carried forward into the twentieth century (Huxley 1926; Elwick 2007; Godfrey-Smith 2016; Lidgard and Nyhart 2017). The development of immunology in the twentieth century brought an additional set of considerations to debates about biological individuality (Loeb 1937; Pradeu 2012; Anderson and Mackay 2014). More recently, individuality has been of interest to philosophers because of its crucial role in different aspects of evolutionary biology. These debates include whether species are individuals or classes, what counts as a unit of selection, and how transitions in individuality occur evolutionarily. These discussions often rely on prior accounts of individuality to determine *whether* the concept is applicable, as in the case of species, and *when* the concept is applicable, as in cases where there is a question about whether a group can exhibit properties that are not simply an aggregate sum of its individual member organisms. Recent empirical discoveries that some organisms are genetic or immunological chimeras (Ross, French, and Orti 2007; De Tomaso 2014) and that most are composed of substantial genetic and cellular material from their microbiota (Gilbert, Sapp, and Tauber 2012; Booth 2014) have continued to spur conceptual reflection on the nature of biological individuality (Guay and Pradeu 2015*a*; Love and Brigandt 2017).

For many philosophers, individuality is a pressing issue for biology. Ellen Clarke offers a perspicuous and forceful articulation: "there is a real problem of biological individuality, and an urgent need to arbitrate among the current plethora of solutions to it" (Clarke 2010: 312). Biological individuals are ubiquitous in life science inquiry, and conceptions of individuality undergird a variety of basic concepts.

> All these concepts—generation, trait, and phenotype—only have meaning in relation to the more primitive concept "individual." Biologists implicitly invoke a particular characterization of the individual any time they use these notions, and their choice of one concept over another will have consequences, acknowledged or otherwise, for their theories and explanations. (Clarke 2010: 313)

Whether or not the concept "trait" has meaning only in relation to a preexisting conception of "individuality" is debatable. However, most would agree that conceptualization choices have consequences. For example, comparisons in terms of homology across diverse taxa are sensitive to what units are understood as parts or wholes: "adopting one definition of organism *versus* another gives a different answer to a biological question" (Pepper and Herron 2008: 624–625). This sensitivity is especially poignant in taxa that exhibit complex life cycles in which individuals at one developmental stage may comprise the parts of another individual at a later stage, such as in pelagic tunicates (Bone 1998). In addition to comparison, other individuation practices include the activities of detecting, counting, and tracking.

The problem of individuality is real for Clarke because the different characterizations of individuality pick out non-overlapping groups of items as individuals—"there is a choice to be made about which definition, or how many definitions to accept" (2010: 314)—and it is not immediately obvious which one (or set) should be adopted. Clarke identifies thirteen candidates: (1) displaying the capacity of reproduction, (2) exhibiting a life cycle, (3) having a unique genotype, (4) engaging in sex, (5) having a single-cell bottleneck during the life cycle, (6) exhibiting a separation of germ and soma, (7) displaying mechanisms that police for cheaters, (8) having identifiable spatial boundaries and contiguity or physiological integrity, (9) exhibiting histocompatibility or an immune self identity, (10) being under selection as a unit of fitness maximization, (11) having cooperation among but not conflict between parts, (12) exhibiting a common evolutionary fate or co-dispersal, and (13) being an entity that bears adaptations. These different characterizations yield nonequivalent partitions of biological systems into individuals (Table 8.1). The situation calls for adjudicative action: "there is an urgent need for the concept to be cleaned up" (p. 323).

Clarke's solution to the problem of biological individuality relies on the concept of multiple realization (Clarke 2013). The individuating mechanisms that underlie

TABLE 8.1 Different candidate accounts of biological individuality yield conflicting answers about whether particular biological entities are individuals

Candidate	Lobster Claw	Aspen Grove	Puppy	Bee Colony	Man o' War	Bacterium
Common sense	N	N	Y	N	Y	Y
Sex	N	?	Y	N	?	N
Bottleneck	N	?	Y	Y	Y	Y
Germ–soma	N	N	Y	Y	Y	N
Spatial bounds	Y	?	Y	N	Y	Y
Histocompatibility	N	?	Y	?	Y	Y
Adaptations	Y	?	Y	Y	Y	Y

Y, individual; N, not an individual; ?, undetermined. Redrawn from Clarke 2010.

individuality can be instantiated in many ways; the particular way in which a mechanism is instantiated is not as crucial (but see Sterner 2015). The two core types of mechanisms on this account—policing and demarcation—are distilled from isolating the shared outcomes of concrete causal processes that operate in contexts pertinent to the thirteen candidates (Clarke 2013: 421, 424).

- A policing mechanism is any mechanism that inhibits the capacity of an object to undergo within-object selection
- A demarcation mechanism is any mechanism that increases or maintains the capacity of an object to undergo between-object selection

There are many ways to police or demarcate, but both types of mechanisms must be present. Notably, both mechanism types are defined in terms of the operation of natural selection. Thus, physical boundaries contribute functionally—not structurally—to demarcation in virtue of their effect on fitness: "Spatial boundaries or barriers around a collection of objects *can facilitate higher-level selection* by preventing mixing or migration between collections" (p. 425; emphasis added). A biological individual is any object that exhibits both kinds of mechanisms simultaneously.

A cognate approach to individuality is found in the work of Peter Godfrey-Smith. Godfrey-Smith takes reproduction and individuality to go hand in hand because of how Darwinian evolution operates:

> Reproduction is at the center of Darwinism. . . . All talk about heritability and fitness requires that we know whether one organism is the parent of another. . . . In a case of reproduction we have: (i) the production of a new individual, (ii) primarily via the causal role of some specific pre-existing individual(s), and where (iii) the "parent" individuals are of the same kind (broadly understood) as the new individual. (Godfrey-Smith 2009: 69)

Akin to Clarke, Godfrey-Smith is sensitive to the multiple realization of reproductive entities. He takes this into account via dimensions of individuality for collective reproducers, which are defined in terms of having parts that reproduce independently of the whole. Collective reproducers can be characterized in terms of three parameters: bottleneck, germline, and functional integration. These correspond to several features from the thirteen accounts set forth by Clarke and can be classified as contributing to either policing or demarcation mechanisms, though the capacity for reproduction is central. These three parameters are differentially fulfilled in different taxa and thereby yield *dimensions of individuality*. When considered in the context of levels of organization, the three parameters contribute to the "de-Darwinization" of lower levels such that higher levels exhibit greater individuality (i.e., the ability to undergo a Darwinian process of variation and selection). However, since individuality can vary along the three parameters, marginal cases or degrees of individuality in different dimensions are recognizable, such as sponges that lack a bottleneck process and germ–soma separation but that have a degree of functional integrity.

The primary considerations in both Clarke's account of what a biological individual is and Godfrey-Smith's dimensions of biological individuality derive from evolutionary theory.

> It is in an evolutionary context that the notion of the individual really does a lot of work. . . . The biological individual is the entity that population biologists count. . . . The notion of the biological individual is inextricably bound up with the notion of fitness. . . . Measuring fitness requires that biologists count biological individuals. (Clarke 2010: 313)

Clarke's use of the definite article ("the" biological individual) and her claim that biological individuals are "inextricably bound up" with concerns about survival and reproduction implies that the perspective of evolutionary theory is somehow fundamental for all of biology when evaluating the nature of individuality. Godfrey-Smith is explicit in this regard when criticizing Griesemer's account of reproduction (Griesemer 2000), where he indicates the inextricable linkage between reproduction and individuality:

> A concept of reproduction that requires material overlap is too narrow to work well in a foundational description of Darwinian processes. . . . it understands the concept of reproduction too narrowly for the foundational role that is relevant here. . . . The link between "individuality" and reproduction is in some ways inevitable. Reproduction involves the creation of a new entity, and this will be a countable individual. (Godfrey-Smith 2009: 83, 84, 86)

Godfrey-Smith's notion of a Darwinian individual has a criterion of adequacy that the concept of reproduction is inextricably linked to the role it plays in foundational theory. The "theory" in view is evolutionary theory, inclusive of population and quantitative genetics (or "evolutionary genetics"), which is utilized in abstract theoretical representations found within the context of population biology.[1] This assumption is sometimes made explicit:

> The problem of biological individuality is a central conceptual issue in evolutionary biology, concerning our ability to delineate the biological units to which fitness can be properly attributed and which participate in the evolutionary process. . . . There is a genuine multiplicity of concepts of the organism and this multiplicity is potentially damaging to the discussion and application of evolutionary theory. (Clarke 2013: 413, 417)

Although evolutionary theory is in focus, the perspective offered is meant to undergird all notions of individuality across theory and practice in the life sciences; it is "a monistic account of organismality" (p. 429; cf. Love and Brigandt 2017).[2]

Clarke is careful to circumscribe the nature of her conceptual question: "The problem I treat here is one of finding necessary and sufficient conditions for membership of the class, rather than of finding identity conditions for particular organisms" (Clarke 2013: 414–415). The latter issue is akin to the traditional problem of personal identity (Olson 2010) and involves reidentifying the same individual over time. When juxtaposed with her own account of what counts as an individual and Godfrey-Smith's dimensions of individuality, Clarke's clarification helps to make salient a critical methodological assumption that lies behind their approaches to the problem of individuality. The necessary and sufficient conditions for individuality or the mapping of an exemplar onto the three-dimensional space of individuality are exercises in fundamental theory.[3] Many philosophers of science assume that our best fundamental theories inform us about the basic furniture of the world. This fundamental theory then governs scientific practice; that is, it tells scientists what to count or how tightly to associate their preferred model system with different dimensions of individuality (Love and Brigandt 2017). The theoretical account of what an individual is determines how biologists go about

---

[1] Although Griesemer holds that reproduction and individuality are linked, it is not because evolutionary theory is fundamental or foundational. Instead, according to Griesemer, the notion of reproduction is conceptually prior to any role it plays in evolutionary theory (Griesemer 2000). As a consequence, what counts as a biological individual when the link to reproduction is in view extends beyond the domain of evolution. However, there are implications for evolutionary theorizing, such as how bottlenecks are identified (see Griesemer 2016).

[2] "I advocate taking this as the basis for our understanding of organismality, . . . something is an organism only if it has mechanisms that give it those properties necessary for participating in a selection process. . . . Individual organisms should be defined by the possession of . . . mechanisms [that] provide the causal basis of an object's capacity to undergo selection at its own hierarchical level, rather than at any other level" (Clarke 2013: 430, 435). Godfrey-Smith (2013) recognizes a difference between evolutionary and physiological approaches by distinguishing between "Darwinian individuals" and "organisms," but only the former plays a foundational role.

[3] "Fundamental theory" does not imply reduction to the molecular or microphysical level. The relevant sense of fundamental concerns foundational *with respect to other sciences*, which may or may not involve an allied conception of reduction to a lower level of organization, and, in this particular case, these two items are decoupled.

the practice of individuation. "Biological theory demands a clear organism concept, but at present biologists cannot agree on one" (Clarke 2013: 413). Given the existence of multiple, conflicting accounts of individuality, adjudication must occur at the fundamental theoretical level or the ensuing advice on how to proceed will be contradictory and the resulting empirical inquiry pertaining to biological individuals will be muddled.

Should fundamental theory always govern scientific practice? Is it necessary to answer the question "what is an individual?" before researchers can pursue inquiry that involves individuation practices of different kinds to study biological individuals? Karen Kovaka answers these questions negatively (Kovaka 2015). She argues that practices of individuation are sensitive to but not dependent on a consensus theoretical answer to the question of what a biological individual is. As a consequence, no resolution of the problem of individuality (*sensu* Clarke) is required to carry out empirical research. While biologists' conceptions of individuals are relevant to and affect how they understand different processes in biological systems, the empirical findings gathered from investigations of these processes generate input that feeds into how biologists conceptualize individuals. Thus, instead of an asymmetrical relationship of dependence, where practices of individuation depend on a worked out theory of biological individuality, the symmetrical relationship of sensitivity means that input goes both ways and leads to modified conceptions of individuality and individuation practices. Although Kovaka's distinction is helpful and persuasive, it leaves open the question of how, in detail, practices proceed without depending on a theory, and it doesn't identify a rationale for advocating dependence on fundamental theory in the first place (i.e., why it would be prioritized over practices).

What if we reversed the methodological assumption that theory guides practice and specifically explored the sensitivity of theory to practice? Instead of "what individuality is" determining how individuation is practiced, we might explore how individuation is practiced in order to understand what accounts of individuals are supported in particular contexts of inquiry, thereby fleshing out a symmetrical relationship of sensitivity. Why might exploring this reversal be warranted? One reason is that different investigative contexts have been provocateurs of different questions about biological individuality through history. Understanding these contexts has the potential to inform why these questions were provoked apart from any consensus or commitment to a fundamental theory of individuality. A second reason is that not all of biological investigation is theory-guided (Waters 2007). In particular, molecular, cell, and developmental biology do not rely on a fundamental theory, evolutionary or otherwise, to govern their highly successful practices (Love 2014).[4] Instead of appealing to a foundational theory, we might

[4] The claim that developmental biology does not rely on a fundamental theory is not the same as a claim that there are no theories in developmental biology (see the diverse chapters in Minelli and Pradeu 2014). Theory can still *inform* developmental biology even if it does not *guide* it (Waters 2007).

scrutinize alternatives. The fact that structured scientific problems play a central role in the individuation practices of biologists suggests one possibility. This orientation animates the analysis in the remainder of this chapter. To see what this means and how it operates, we need to say more about scientific problems in the experimental contexts of developmental biology and what is required in practice to accomplish individuation.

8.2 Individuation, Tracking, and Scientific Problems in Developmental Biology

Developmental biology is the science of characterizing and explaining how a variety of interacting processes generate the heterogeneous shapes, size, and structural features of an organism as it develops from embryo to adult, or more generally throughout its life cycle (Love 2008, 2015). Although it is commonplace in philosophy to associate sciences with theories, such that the individuation of a science is often dependent on a constitutive theory or group of models, it is uncommon to find presentations of developmental biology making reference to a theory or theories of development. I have argued elsewhere that the absence of any reference to a theory of development or some set of core explanatory models should be taken at face value (Love 2014). Instead of trying to reconstruct theories out of the relevant discourse (e.g., multiple allied molecular models or differential gene expression) or holding that it is an "immature" science, developmental biology can be understood as organized and guided by structured scientific problems that provide it with ample and accessible epistemological scaffolding.[5] An erotetic structure of stable and broad domains of problems (*problem agendas*), such as differentiation or morphogenesis, direct inquiry and organize knowledge in studies of ontogeny. This guidance extends to the coordination of individuation practices relevant to experimental inquiry.

In standard textbook presentations of developmental biology (e.g., Slack 2013), clusters of questions are organized around problem agendas that reflect generally delineated phenomena: differentiation, specification or pattern formation, morphogenesis, and growth. These phenomena can be subdivided into different types, partitioned in terms of whether they occur earlier or later in development or in specific tissue layers, and combined to investigate a variety of phenomena. We can characterize these subdivisions within problem agendas in terms of at least five variables: abstraction, variety, connectivity, temporality, and spatial composition (Love 2014). The values given to these variables structure the constellation of research questions within a problem agenda. For example, research questions oriented around events in zebrafish gastrulation are structured in a way that differs from research questions oriented around vertebrate neural crest cell migration

---

[5] Cf. footnote 4.

because they involve different values for the five variables: degree of abstraction (zebrafish vs. vertebrates), temporality (earlier vs. later), spatial composition (tissue layer interactions vs. a distinctive population of cells), variety (epiboly vs. epithelium to mesenchyme transition), and connectivity (gut formation and endoderm vs. organogenesis and ectoderm/mesoderm). These configurations can be adjusted readily in response to shifts in the values for different variables. Some of these variables help to structure problems in other sciences (e.g., degree of abstraction), but this set of variables—especially the juxtaposition of temporality and spatial composition—is central to structuring developmental biological problems.[6]

Returning to the question of individuation practices, these five variables are germane to what makes something a biological individual in experimental inquiry. For example, in particular episodes of individuation, the degree of abstraction governs relationships of similarity between parts and wholes. Gastrulation can be compared across different taxa when treated abstractly despite occurring through a variety of mechanisms. Since temporality and spatial composition are distinctively important to the structuring of problems in developmental biology, it is perhaps not surprising that they play a central role in shaping individuation practices. First, individuals and their components must not only be identifiable at a time ("detected") but tracked *through time*. This tracking of individuals need not involve homology comparisons across taxa or the ability to discriminate between different individuals (e.g., for counting). The physical segregation of developing embryos from one another is adequate for detection, but it is not adequate to temporally track the properties of individuals (wholes) or their components (parts). These properties are what determine sensitivity conditions for classifying a particular entity as a member of the class of *developing* embryos in experimental contexts.

Second, spatial composition is crucial for individuation practices in developmental biology. This is especially relevant to the identification of parts of individuals, but also to the spatial boundaries that circumscribe the developing embryo as such. However, these spatial boundaries are not defined in terms of natural selection; demarcation is structural and accomplished without a concern for whether it "increases or maintains the capacity of an object to undergo between-object selection." Although embryos are sometimes tracked as living individuals, in most cases they will not survive to the next generation. The manipulative techniques and protocols applied to yield data during ontogeny often result in death (e.g., chemical fixation or sacrifice and dissection) or reproductive impairment (e.g., through mutagen exposure). This automatically suggests that functional accounts of individuality that invoke

[6] Much more could be said about each of these variables. For example, a focus on vertebrate neural crest cell migration is more abstract than zebrafish neural crest cell migration because the former requires subtracting or ignoring many details of this causal process that differ across vertebrate taxa. Note that an ontology of species or taxa as individuals, which is relevant for problem agendas within systematics, is not in play here. Problem agenda anatomy can be delineated further with three dimensions: history, heterogeneity, and hierarchy (Brigandt and Love 2012; Love 2014). I bracket these here for expositional clarity with respect to questions surrounding individuality.

intergenerational processes (e.g., bottleneck, common evolutionary fate) and behavior related to sexual maturity (e.g., reproduction, histocompatibility) are irrelevant (i.e., practices of individuation in developmental biology are insensitive to theories of individuality based on these considerations). At the same time, our focus is shifted to a characterization that is sensitive to these experimental practices: having identifiable spatial boundaries and contiguity or physiological integrity. The former is necessary for tracking embryos: "These candidates expect biological individuals to be visible and readily apprehensible phenomena, with edges and continuous existence in between" (Clarke 2010: 318). Identifiable spatial boundaries are often sufficient for tracking individuals, but if embryos must live through the experimental conditions then physiological integrity is required as well.

Tracking spatial composition is frequently done with distinctive morphological indicators, such as cell shape or relative position of a tissue, and molecular markers, such as gene expression for cell types or regions of tissue. For example, the whole embryo can be mapped onto a set of normal developmental stages based on morphological signatures. The developmental trajectory from fertilized zygote to fully formed adult is broken down into distinct temporal periods by reference to the occurrence of major events, such as fertilization, gastrulation, or metamorphosis (Minelli 2003: chapt. 4). New researchers (e.g., graduate students) learn these through observational training repeated on many embryos. These individuation practices are ubiquitous in developmental biology and have emerged from a long history of working with embryos in different investigative contexts. They enable researchers in different laboratory contexts to have standardized comparisons of experimental results on developing individuals and their components (Hopwood 2005, 2007). Molecular markers are particularly helpful for identifying distinct parts through time. Ectoderm tissue expresses specific genes that mark its identity (e.g., *BMP-4, Otx2, Chordin*, or *Nestin*). These individuation practices are so widespread in developmental biology that they have been commercialized, and new techniques arise on a relatively regular basis.

Thus far, we have seen that the structure of scientific problems helps to coordinate individuation practices in developmental biology by focusing attention on what it means to track spatially bounded individuals and their components through ontogeny. Before turning to an experimental case study to see these features in operation, it is necessary to further characterize the difference between counting and tracking. If counting individuals is the practice that scientists are engaged in, then there might be some justification for starting with fundamental theory (but see Kovaka 2015). Although there is no a priori reason to expect that one fundamental theory governs all of biology, it is possible that a theory for a particular domain is critical for telling one what to count in that domain.[7] The actual

---

[7] Parallel arguments might apply to other practices, such as detecting or comparing (cf. Pepper and Herron 2008).

counting itself may or may not be challenging, but it is crucial to know what is being counted. Scientists, however, are not only interested in counting individuals; they also want to follow them empirically. This "following" is what we have been referring to as *tracking* (Griesemer 2007).[8]

One might think that tracking is sufficiently similar to counting and therefore warrants similar treatment: develop a theory about what is being tracked before executing the tracking. The difficulty with this view is that tracking is usually a prerequisite for deriving sufficient empirical information on which to base any theory. If scientists always required a fundamental theory of what they were tracking in order to do the tracking, then a lot of science would be stopped in its tracks. There is an asymmetry between empirical practice and fundamental theory in which the former is prioritized. This is not to say that the eventual formulation of some theory of the items being tracked will be uninformative; it could refine how the tracking is done to yield better predictions or explanations. The historical development of tracking practices demonstrates that this refinement occurs, though their general stability over time belies any connection to a progressively developed fundamental theory; changes in practice and changes in theory are not tightly correlated. Regardless, here we see a critical rationale for why individuation practices will routinely be the starting point rather than theories of individuality: you simply would not be able to get started on the methodological assumption that theory precedes practice.

What is involved in tracking? This is not the place to offer a complete account of tracking in biology, and therefore we can rely on earlier discussions for the main details (Griesemer 2007). Tracking is a means to an end, with implicit and explicit commitments. The end is often to understand the causal relationships in a process of interest and characterize how aspects of this process occur under assumptions about what is a relevant process (or in what entity it is manifested). More generally, the goal is frequently to attain a causal explanation of the phenomenon of interest. The means to the end are representations that have embedded commitments about what features are relevant to track and how the tracking displays those features, such as by visualization. These assumptions determine what is tracked and how and constrain how other items are followed or represented. Different disciplines have standardized varieties of tracking practices. In developmental biology, in situ hybridization is one prominent example that tracks gene expression through the visualization of mRNA transcript presence in regions of the developing embryo (Figure 8.1). The representation contains "marks" of the feature being tracked—in

[8] Importantly, not every entity that can be counted or tracked is automatically considered an individual. However, developmental biological practice uses spatial boundaries to detect, count, track, and compare organisms and their component parts through time. It is not clear why one would want to deny these the status of individuals apart from an independent commitment to the governing role of some fundamental theory from another science, such as evolutionary biology (see discussion in Section 8.4).

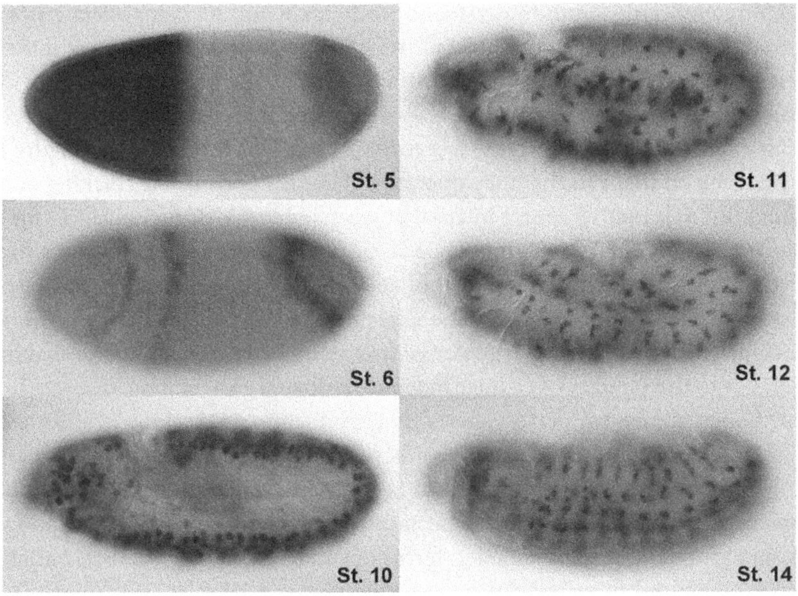

FIGURE 8.1 *Tracking embryos and their parts in a Drosophila embryo with molecular markers. Dark patches represent expression of the Hunchback gene, which plays a causal role in establishing the segmentation network cascade early in development (St. 5 and St. 6). Staging corresponds to "normal stages" marked by morphological indicators and shared across different research communities investigating ontogeny.*
https://en.wikipedia.org/wiki/File:Hunchback_in_situ.jpg

this case, RNA transcripts from a particular gene—and this permits dissecting causal relationships in the system, such as the role of a gene in establishing the segmentation network cascade early in ontogeny. Sometimes the mark is permanent and other times the mark is removable, such as by quenching a chemical reaction or because it fades over time (e.g., due to diffusion). Sometimes the mark is only mentally recorded, as in the recognition of a particular stage of embryogenesis through a morphological indicator, though this is often represented elsewhere in some capacity (e.g., normal stages).

We can formulate an abstract template for tracking that can guide a search for how individuation practices are manifested in experimental research.[9] First, a process of interest is identified (e.g., gastrulation) with a particular end state or consequence to be explained (e.g., the existence of three germ layers). Second, the process is identified with, transformative of, or located within a bounded system or individual (e.g., gastrulation involves the movement of cells and tissues in

[9] This template is adapted and modified from Griesemer's discussion of tracking (Griesemer 2007: 413).

the entire embryo). Third, the system or individual and its component parts are marked in order to track the process through time (temporality) and at different levels of organization (spatial composition), such as the gene expression of cells composing a sheet that moves during gastrulation. This marking can be more or less manipulative in that it can preserve existing causal relationships (Figure 8.1) or intervene on them so that they operate differently. Fourth, the tracking occurs through the relevant temporal interval at germane levels of organization for the specified process. Finally, an inference is made about causal relationships in the process based on the overall pattern of stasis and change of marks with respect to an end state or consequence to be explained in the bounded system (i.e., the individual). For example, a particular gene is assigned responsibility for making a difference to the rapidity of cellular movement during gastrulation.

Although somewhat perfunctory, this introduction to tracking is adequate to explicitly characterize individuation practices. They are coordinated by particular research questions within problem agendas. In developmental biology, those problem agendas—differentiation, specification or pattern formation, morphogenesis, and growth—are subdivided in terms of values taken on by several variables, of which temporality and spatial composition play special roles. Tracking via temporality and spatial composition is accomplished with molecular and morphological techniques that selectively pick out embryos as whole individuals containing various component parts undergoing developmental processes. This can be elaborated with an abstract template to help isolate the specific practices and characterize their associated assumptions. The relativity of individuation practices to particular problems is a function of tracking a preferred set of features as causes and effects in biological individuals while simultaneously ignoring other sets of features. Other features could become germane in the context of other problems, thereby modulating individuation practices and leading to different kinds of individuals. There is no need to adjudicate between different accounts of biological individuality; displaying mechanisms that police for cheaters or being under selection are simply irrelevant to the tracking procedures engaged in by most developmental biologists. These tracking procedures are oriented to spatiotemporal features of parts and wholes in development as causes and effects because of the research questions being asked. Policing and demarcation mechanisms (*sensu* Clarke 2013) are not pertinent to the problem agendas of developmental biology and therefore immaterial to its individuation practices in experimental inquiry.

8.3 The Problem of Growth and Individuation Practices

One of the central problem agendas in developmental biology is *growth*. Several central research questions revolve around the regulation of cell division: "How do our cells know when to stop dividing? How is cell division so tightly regulated?" (Gilbert 2010: 2). Another question at the cellular level is the regulation of cell size. Cell division and cell size in combination contribute to the size of anatomical elements, such as a wing or a limb, and the overall body of an individual organism. At the level of individual cells, the mechanisms of cell regulation are relatively well understood. At higher levels of organization, "the overall control of growth is still somewhat mysterious" (Slack 2013: 362). At these higher levels of spatial composition, the problem agenda of growth is subdivided into two distinct questions: (1) final organism size and (2) the coordination of relative sizes between different parts and the whole individual. This erotetic structure governs inquiry: "there must exist mechanisms that can sense the overall size of the organism and regulate the growth of individual parts accordingly" (p. 362). Further problem structure can be isolated by looking at the connections between these two questions, such as how growth maintains symmetries (e.g., forelimb length). This relates to both overall size and the coordination of growth among parts and the whole (Wolpert 2010).

Two of the primary biochemical pathways participating in embryonic growth are the insulin-TOR system and the Hippo pathway. The former is involved in gene expression related to positively stimulating cell division and metabolizing nutrients, whereas the latter negatively regulates cell division. Defects in these pathways yield distinct phenotypes: mutations that affect insulin-TOR–related gene expression often result in undergrowth; mutations that affect the Hippo pathway often result in overgrowth. These two biochemical pathways are modulated more systemically in an embryo via growth hormones, such as juvenile hormone or ecdysone. In *Drosophila*, the prothoracic gland that secretes ecdysone is critical for sensing overall size because ecdysone represses juvenile (growth) hormone. Manipulation of the size of the prothoracic gland leads to robust predictions regarding overall size: glandular size increases, which increase the amount of ecdysone, lead to smaller adults; glandular size decreases, which decrease the amount of ecdysone, lead to larger adults since juvenile hormone is not suppressed.

Concentrating on a particular example of a study devoted to one of these research questions helps to illuminate individuation practices in action. Oliveira and colleagues set out to investigate the coordination of overall body size with wing disc size under internal (physiological) and external (environmental) perturbations during *Drosophila* development (Oliveira et al. 2014). Although the developmental coordination of growth is robust to perturbations—full-grown organisms exhibit proper relationships between overall size and proportionality for component parts under a variety of physiological and environmental conditions (Kuntz and Eisen 2014)—how this is accomplished remains an open question. Its poignancy can be recognized by juxtaposing two facts: (1) this coordination involves the integration of gene expression patterns from the biochemical pathways active in component cells and tissues with global levels of the hormone ecdysone that bring about developmental transitions across the entire body, and (2) these gene expression patterns and systemic hormone levels are modulated by internal and external conditions that change through time. "The timing of these developmental events changes with environmental and physiological conditions

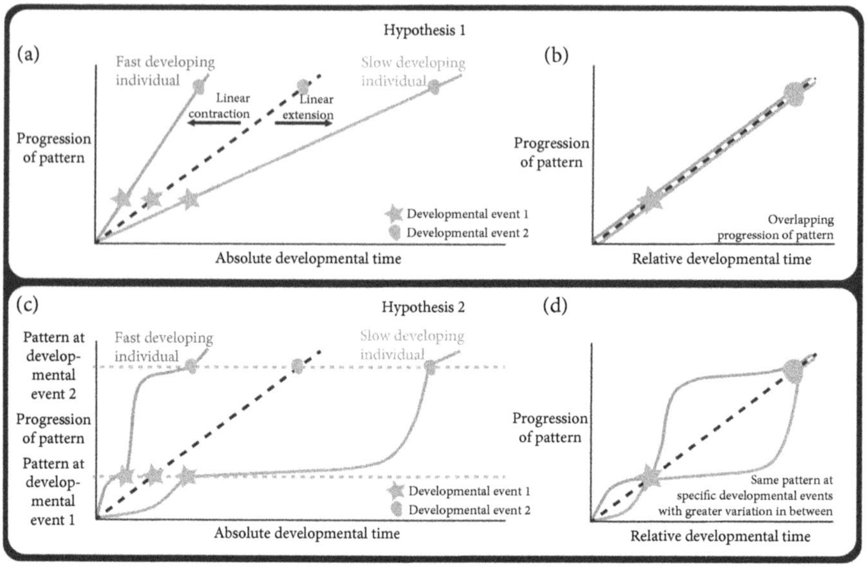

FIGURE 8.2 *Two hypotheses for the relationship between the growth of a whole organism and its component parts during development (see text for discussion).*

From Oliveira et al. 2014. Creative Commons Attribution License 4.0 (http://dx.doi.org/10.1371/journal.pgen.1004408.g001).

but how this affects tissue development is not fully understood" (Oliveira et al. 2014: 1).

The aim of the study was to distinguish between two different hypotheses about how this coordination could occur (Figure 8.2). The first possibility is that whole-body development (WBD) and tissue-specific patterning (TSP) progress at the same rate. Thus, whether an individual develops slowly or quickly, depending on variables such as temperature, different hypotheses can be distinguished by examining patterns in the progress of WBD and TSP over time. If WBD and TSP progress at the same rate, then relationships of progression would be linear when plotted against absolute developmental time (Figure 8.2A) and remain the same if developmental time is relativized to particular events during ontogeny (Figure 8.2B). The second possibility is that WBD and TSP do not progress at the same rate but are coordinated at key developmental events ("milestones"), such as molting. On this hypothesis, one would expect a large degree of variability in trajectories between milestones under altered developmental time courses (Figure 8.2C) and not expect the overall pattern of progression to match except with respect to the developmental milestones (Figure 8.2D).

To test these hypotheses, the researchers needed to track the process of growth for WBD and TSP with respect to the progression of particular parts. The former was tracked with elements from the standard periodization of *Drosophila* development (Figure 8.3).

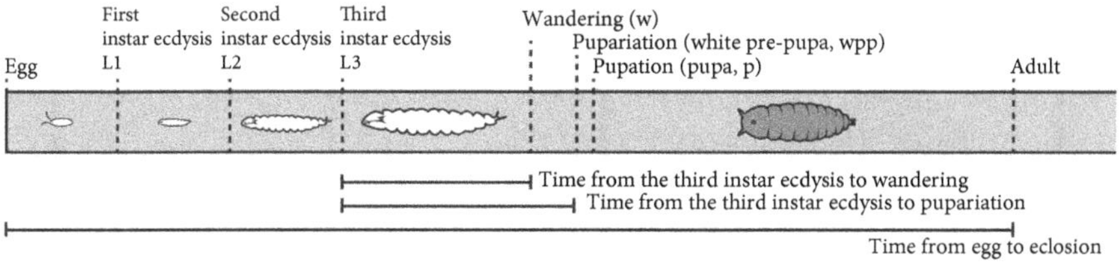

FIGURE 8.3 *A representation of the standardized periodization of Drosophila development (see text for discussion).*

From Oliveira et al. 2014. Creative Commons Attribution License 4.0 (http://dx.doi.org/10.1371/journal.pgen.1004408.g005).

> The juvenile period comprises three larval moults. This is followed by a wandering stage where larvae leave the food and search for a pupariation site. Larval development ends with pupariation, whereupon the fly metamorphoses into its adult form. These events provide useful markers of whole-body development. (Oliveira et al. 2014: 2)

Discernable pulses of ecdysone regulate these developmental events or milestones (e.g., third instar ecdysis). In the latter case (i.e., the progression of parts), the researchers tracked the imaginal discs from which the wings develop. These were good candidates because ecdysone pulses have been shown to regulate gene expression in the wing disc (Mirth, Truman, and Riddiford 2009) and therefore might be involved in coordinating TSP with WBD. Here, the researchers could not draw on existing tracking practices but needed to generate a new scheme. This was necessary because they needed to compare WBD via developmental event with some other set of marks to see if TSP progression in the wing disc exhibited variability between milestones (in accord with the second hypothesis) or was proportionally similar throughout (in accord with the first hypothesis).

A wing-disc staging scheme was created using immunocytochemistry to mark the expression of eleven genes across eleven time points, inclusive of three developmental milestones. In doing this, the researchers attended to temporality through the selection of the eleven time points and to spatial composition by defining the locations and patterns of gene expression within the wing (Figure 8.4). The expression patterns for six of the genes are displayed in the columns of Figure 8.5 and the staging scheme derived from this expression is shown in Figure 8.6. Not all of the genes tested were helpful in constructing the staging scheme. Thus, Figure 8.5 shows only the six useful markers from which the scheme was built and leaves out five other genes whose expression also was tested. In fact, the expression of two of the genes (*Achaete* and *Senseless*) could assign wing discs to nine of the eleven stages in the scheme. This meant there was differential usefulness among the markers. Once completed, the staging scheme was calibrated by categorizing simulated disc patterns to determine if it classified the same patterns from the sampled discs similarly.

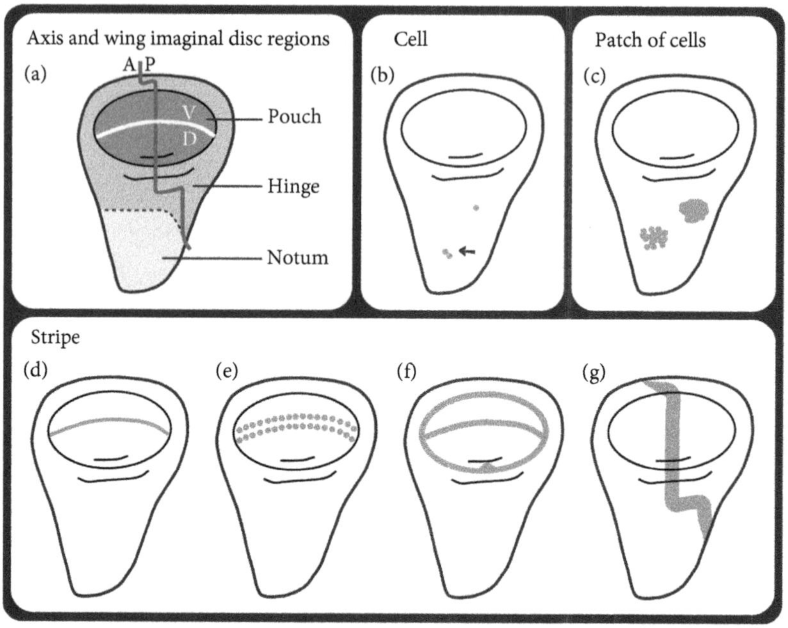

FIGURE 8.4 *Diagrammatic representation of the Drosophila wing disc and how different gene expression patterns are scored.* **A**: *Anterior (A) and posterior (P) axis; dorsal (D) and ventral (V) domains.* **B**: *Cell = one or two round dots of expression.* **C**: *Patch of cells = delimited or diffuse clustered pattern.* **D–G**: *Four different manifestations of a stripe pattern. Combinations of these elements describe all of the observed patterns for each gene tracked.*

From Oliveira et al. 2014. Creative Commons Attribution License 4.0 (http://dx.doi.org/10.1371/journal.pgen.1004408.s001).

With this tracking scheme in hand, Oliveira and colleagues then experimentally altered the developmental rate internally by modifying ecdysone production and externally by changing the temperature. Ecdysone production was increased or decreased, with the former shortening the duration of the third larval instar (from Stage A to K; Figure 8.6) and the latter lengthening its duration. Temperature was adjusted upward (from 25°C to 29°C), which shortens the time to adult eclosion and therefore reduces overall body size, as well as downward (from 25°C to 18°C), which lengthens the time to adult eclosion and consequently increases the overall body size (cf. Figure 8.3). When the patterns of WBD to TSP in wing development were compared, the second hypothesis was supported because the progression of patterning was variable between developmental events but synced with two particular milestones. Future research is needed to test whether this is true for other aspects of TSP in addition to the wing and whether different tissues coordinate with the same subset of developmental milestones.

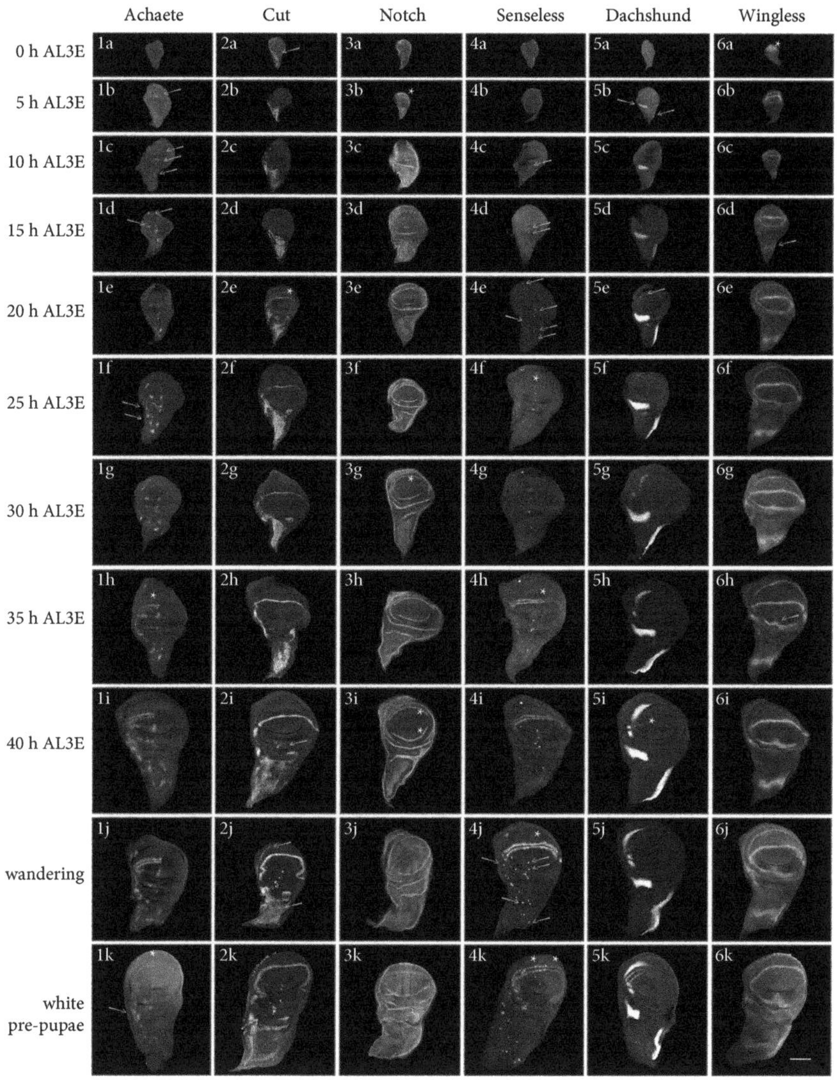

FIGURE 8.5 *Tracking of the expression for six genes (Achaete, Cut, Notch, Senseless, Dachsund, and Wingless) in wing disc development across eleven temporal stages (AL3E = after third star ecdysis).*

From Oliveira et al. 2014. Creative Commons Attribution License 4.0 (http://dx.doi.org/10.1371/journal.pgen.1004408.g002).

If we return to the abstract template formulated in Section 8.2, we can distill the main features of the individuation practices involved in tracking growth-related events for the developmental experiments that facilitated discriminating between these two hypotheses.

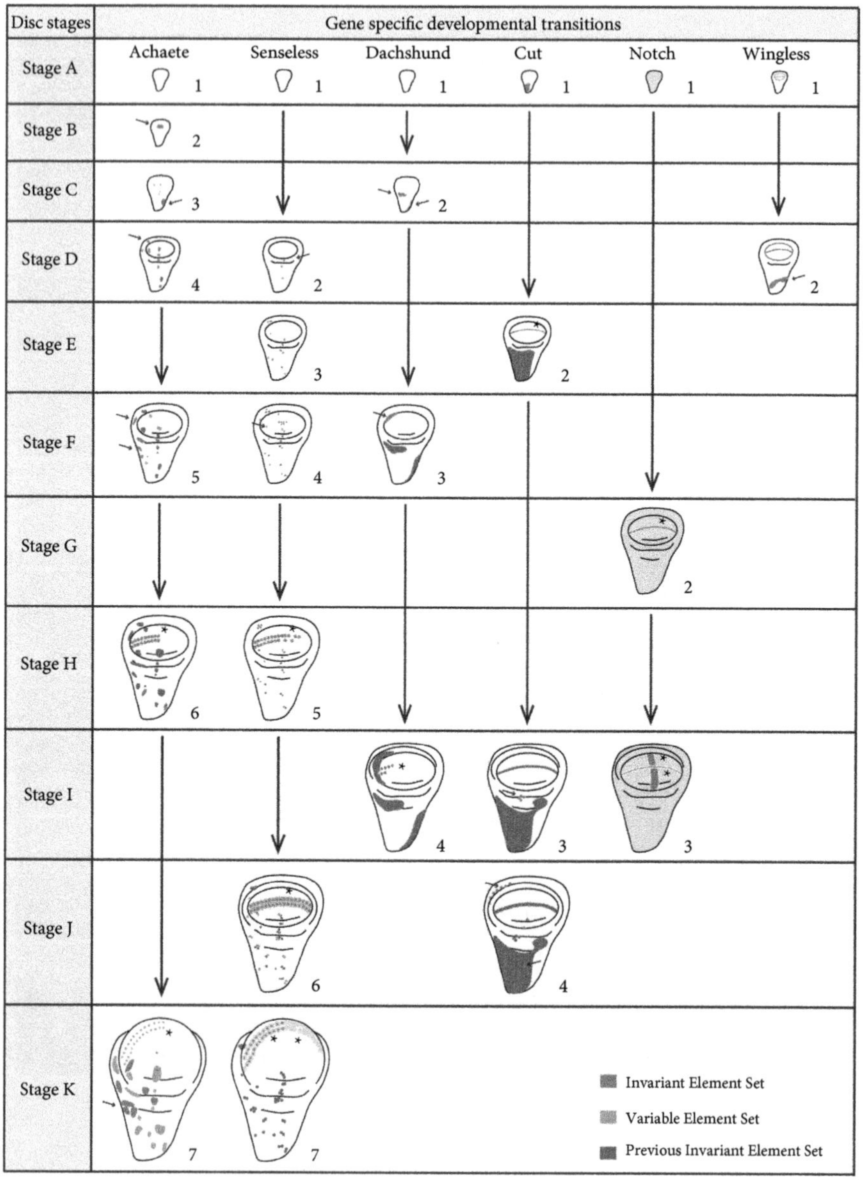

FIGURE 8.6 *Standardized staging scheme for the wing disc based on gene expression depicted in Figure 8.5.*

1. A process of interest is identified—growth—with a particular end state or consequence to be explained—the coordination of growth between the whole-body and the wing disc.
2. The process is identified with, transformative of, or located within a bounded system or individual—the growth of the whole embryo and the wing contained within it.
3. The system or individual and its component parts are marked in order to track the process through time (temporality) and at different levels of organization (spatial composition). The individual embryo is tracked through developmental milestones (WBD), and the wing disc is tracked through a staging scheme related to gene expression patterns (TSP).
4. The tracking occurs through relevant temporal intervals—third instar ecdysis to wandering, third instar ecdysis to pupariation, and egg to eclosion (Figure 8.3)—at germane levels of organization for the specified process—whole-body and the wing disc.
5. An inference is made about causal relationships in the process based on the overall pattern of stasis and change of marks with respect to an end state or consequence to be explained in the individual: "patterning aligns with whole-body development at specific developmental milestones . . . and shows greater variation between these milestones" (Oliveira et al. 2014: 11).

What we have achieved in this section is a concrete explication of how individuation practices are coordinated by particular research questions within problem agendas. For the problem agenda of growth and the research question of how relative sizes between different parts and the whole individual are coordinated, particular tracking methods that are keyed to temporality and spatial composition via molecular and morphological techniques selectively pick out embryos as whole individuals along with their component parts undergoing growth processes. The individuation practices are driven by particular questions that foreground a set of features as causes and effects in developing organisms. The resulting spatially bounded individuals and their component parts are tied distinctively to the practices involved in tracking them. At no point was it necessary to decide between competing theoretical accounts of what an individual is in order to accomplish successful individuation in this experimental context (cf. Kovaka 2015).[10]

[10] This does not mean there are no interestingly relevant theoretical conceptions of individuality to probe in this vicinity. One of these is *genidentity*: "the identity through time of an entity *X* is given by the continuous connection of states through which *X* goes. . . . the individual *X* is never presupposed or given initially, because the starting point is the decision to follow a specific and appropriate process *P*, and the individual *X* supervenes on this process. . . . what we single out as an 'individual' is always the by-product of the activity that is being followed, not its prior foundation" (Guay and Pradeu 2015*b*: 317–318). A full comparison and analysis is beyond the scope of this chapter (see also Murray and Crowther 2016).

8.4 Reconfiguring the "Problem of Individuality"

Where does this leave the "problem of individuality?" Did we simply solve it differently than Clarke or Godfrey-Smith, such that our accounts are competing and must be adjudicated? Clearly, the answer is negative. Sections 8.2 and 8.3 illustrate that the methodological assumption privileging fundamental theory is *unnecessary* (Kovaka 2015); it is not a prerequisite to executing scientific practice. Individuation practices in developmental biology proceed without a fundamental theory of the individuals tracked in its experimental procedures. There was no need to tell scientists what to count or how tightly to associate their preferred model organism with different dimensions of individuality. The priority of fundamental theory is also *unwarranted* because this requirement would prevent the successful testing of hypotheses, such as whether WBD is coordinated with TSP in particular ways. Thus, a theoretical account of what an individual is does not determine how developmental biologists go about their individuation practices. This is dramatically exemplified by the staging schemes involved. Neither the WBD scheme drawn from the normal stages of *Drosophila* development nor the TSP scheme for the wing disc is presumed to mark out WBD for all organisms or TSP for any and all component parts. There is no prior commitment by experimental developmental biologists as to what individuals are apart from the local dictates of their individuation practices. The so-called problem of individuality is reconfigured from arbitration among alternative accounts via fundamental theory to the characterization of pertinent problem structure that provides criteria of adequacy for individuation in practice.

The ability to conceptualize individuals differently in distinct contexts of inquiry, as defined by different problems, makes it possible to answer research questions that would otherwise remain unaddressed. Criteria of adequacy vary across these contexts and account for variation in individuation practices.[11] In many areas of evolutionary inquiry, measurements of fitness are important for explaining changes in populations. With respect to this aim, an account of what an individual is has the reliable measurement of fitness via counting as a criterion of adequacy.[12] Similarly, some considerations for decomposing individuals into parts are motivated and constrained by the need to account for evolutionary transitions, which require the suppression of selective dynamics among constituent parts. In developmental inquiry, in contrast, spatial boundaries of parts are important for understanding changes during ontogeny. Therefore, a way of tracking individuals or decomposing individuals into parts has spatial separation through time as a criterion of adequacy. In physiological inquiry, functional

---

[11] This goes beyond previous acknowledgments that different criteria of individuality may be needed for different taxa (Clarke 2012, 2013; Wilson and Barker 2013).

[12] It would be a mistake to infer that there is only one concept of individuality that serves this aim (see Ereshefsky and Pedroso 2015; Sterner 2015).

interconnections are central to tracking activity in different systems (e.g., the immune system). In such an epistemic context, counting individuals may require taking into account mutualistic relationships and decomposing individuals into parts that transgress specific spatial boundaries and diverge from what is relevant to selective dynamics. Tracking individuals in this context occurs via methodologies similar to and different from those observed in developmental biology. Parallel considerations apply to practices of comparing individuals phylogenetically (Pepper and Herron 2008).

Before drawing a more general conclusion from this reconfiguration of the problem of individuality based on the fact that fundamental theory is neither necessary nor warranted for the individuation practices of experimental biologists, let me briefly consider two objections. The first is that both Clarke and Godfrey-Smith can recover the same individuals and component parts that are tracked experimentally in the individuation practices of developmental biologists. What differs is the basis for why something is taken to be an individual. For their accounts, it is functional properties related to being a unit that can undergo selection; for developmental biologists, it is structural properties (both molecular and morphological) that have crystallized out of long-established historical practices of individuation in experimental contexts. However, this difference is crucial because developmental biologists do not rely on a theory of biological individuality based on natural selection to track developing embryos and their component parts. A fundamental theory does not actually govern (and never has governed) scientific practice in these investigative contexts. Instead, structured problems play this role. If we want to describe correctly what biologists do, comprehend why they do it, and make explicit what standards apply to their practices, then the appeal to fundamental theory from evolutionary biology gets it wrong even if there is no substantial difference in what ends up being understood as a biological individual in these experimental contexts.[13]

A second objection that could be raised is that Clarke's account (and, by implication, Godfrey-Smith's account) is, in fact, problem-based. As Clarke notes,

> The problem . . . which motivates the different accounts I have described, is the problem of identifying those properties a living unit must have in order that measurements of its fitness will support valid evolutionary inferences. In other words, so that counting the units enables us to predict and explain how the traits of such units are changing over time, under the action of natural selection. (Clarke 2013: 429)

[13] However, there are novel types of individuals and component parts in developmental biological practice that would not come under the aegis of a fundamental theory based on evolutionary considerations, such as explants, organoids, and interspecies chimeras (Eiraku et al. 2011; Fatehullah, Tan, and Barker 2016; Hong et al. 2012; Sive, Grainger, and Harland 2007).

One indicator that suggests Clarke is not oriented toward individuation practices guided by problem contexts is that she eschews the task of finding identity conditions for particular individuals (Clarke 2013: 414–415). However, in order to avoid double counting, one must either mark individuals (a form of tracking) or have some way to reidentify the same individual at a later time. Even if we assume (for a moment) that Clarke's account is only intended to apply in the specialized problem context of some aspects of evolutionary theory and not govern individuation practices elsewhere in biology, we should not interpret Clarke's account as being responsive to a research question. The difficulty is signaled by Clarke's use of the modal verb: "the problem of identifying those properties a living unit *must* have in order that measurements of its fitness will support valid evolutionary inferences" (p. 429; emphasis added). What is the basis for this claim? It derives from a philosophical commitment about the nature of the world and how our knowledge should correspond to it. The commitment is a form of monism and is discernable in a subsequent passage that clarifies the "problem" that motivates Clarke's account: "This description of the problem implies that the solution must involve finding *the* hierarchical scale at which selective change is going to take place" (p. 429; emphasis added). The monism is signaled by the definite article; "valid evolutionary inferences" must comply with these strictures.

We can clarify this commitment by elaborating terminology from an earlier discussion of monism and pluralism in the context of debates about units of selection (Waters 2005: 313–314).

- *Genic theory*: The theory that models evolutionary situations in terms of selection for individual alleles
- *Genic monism*: Genic theory alone provides the correct account of all situations involving natural selection
- *Multilevel theory*: The theory that models evolutionary situations in terms of selection at one or more levels
- *Multilevel monism (strong)*: For a given situation, the multilevel theory will provide a uniquely correct model of the selection situation that identifies the true level of selection in that situation
- *Multilevel monism (weak)*: In some situations, the multilevel theory will provide uniquely correct models of the selection situation that identify the true levels of selection in those situations
- *Multilevel pluralism (weak)*: In some situations, an evolutionary process can be modeled correctly as selection for units at one level (or multiple levels) and alternatively modeled correctly as selection for units at another level (or multiple levels)
- *Multilevel pluralism (strong)*: For a given situation, an evolutionary process can be modeled correctly as selection for units at one level (or multiple levels) and alternatively modeled correctly as selection for units at another level (or multiple levels)

Clarke is a multilevel monist, most likely of the strong variety: "individuating mechanisms make evolutionary change more likely to take place at one level rather than another, and so determine at which level/s we should choose to sample change in order to gain an informative and undistorted perspective on the overall pattern of change" (Clarke 2013: 429). A multilevel pluralist rejects the assumption that there is a uniquely undistorted perspective on the overall pattern of change: "the causes in evolutionary situations are sufficiently complicated that they can be described truthfully in multiple ways" (Waters 2005: 313). However, the weak variety of multilevel pluralism is consistent with there being a fact of the matter about selection operating simultaneously on multiple levels in some situations. (In other words, weak multilevel monism and weak multilevel pluralism are not mutually exclusive.) Recalling the role of research questions and problem agendas, it is not always the goal of evolutionary research to provide a uniquely correct model of any given selection situation—multilevel selection can be correctly modeled in a plurality of ways for some explanatory purposes. Yet from the perspective of other research aims, there can be genuine debate over whether unique models of multilevel selection processes are correct. And, empirically, we are not always in a position to discriminate whether selection is operating on multiple levels because some studies are not powerful enough to detect selection at different levels.

Identifying Clarke as a multilevel monist is relevant in this context because her account of individuality is defined in terms of selection processes. If there is an undistorted fact of the matter about the level at which selection acts, then we can unambiguously say that whatever entity is being selected on at that level is a genuine individual. This animates the need to "clean up the concept" of biological individuality; multilevel monism demands it. We can extend the preceding taxonomy to include individuality and encapsulate the results of our earlier analysis.

- *Individuality monism (strong)*: For a given situation, there is a uniquely correct account that identifies the genuine individual in that situation
- *Individuality monism (weak)*: In some situations, there is a uniquely correct account that identifies the genuine individual in that situation
- *Individuality pluralism (weak)*: In some situations, individuality can be modeled correctly in more than one way
- *Individuality pluralism (strong)*: For a given situation, individuality can be modeled correctly in more than one way

On the strong variety of individuality monism, a fundamental theory is necessary because this is what guides the scientist to detect or count genuine individuals for any given situation in practice—a uniquely correct account is required. For the weaker variety, any theory that is necessary would not be fundamental because it would only govern "in some situations." On the weak variety of individuality pluralism, fundamental theorizing is not required and individuation practices—detecting, comparing, counting, or tracking—are guided by distinct research questions with more or less input from particular theoretical commitments. These yield different criteria of adequacy for individuals in different contexts of investigation; in some situations, the same biological entity or system may require different construals of individuality. Biological individuals are "sufficiently complicated that they can be described truthfully in multiple ways," and empirical findings may alter conceptualizations of individuality (Kovaka 2015). However, this is not as encompassing as the strong variety of individuality pluralism, which demands that individuality can be modeled correctly in more than one way for any given situation. This may be true but it is far from established empirically.

Clarke has given us good reason to think she holds something like the stronger variety of individuality monism ("a monistic account of organismality"; Clarke 2013: 429). I think Godfrey-Smith adheres to the weaker variety, as indicated in his recognition of a difference between evolutionary or Darwinian individuals and physiological individuals or organisms (Godfrey-Smith 2013). Probing the application and further implications of this interpretive taxonomy is necessary but must remain a task for another day. Importantly, the adoption of either form of individuality monism would be an obstacle to successful practices in the experimental inquiry we find in developmental biology (and in many other places). The requirement of having a theory of individuality in place prior to tracking individuals accurately in scientific practice is descriptively false. Individuation practices that are not shackled by this requirement make it possible to test hypotheses about causal mechanisms and reevaluate our understanding of how various processes operate in biological individuals.

The "problem of individuality" articulated by Clarke (and others) is an artifact of monistic philosophical assumptions about the nature of scientific knowledge and the world we attempt to describe and explain. Her functional definition may outstrip others in generality—an important criterion in fundamental theorizing—but this is purchased with monistic assumptions that are a barrier to its applicability in particular scientific practices. The problem-relative nature of biological individuation reconfigures this so-called problem and accounts for many of the standard, ubiquitous, and successful investigative strategies in experimental inquiry that involve heterogeneous individuation practices. These practices are part of tracking procedures that mark out particular kinds of individuals and their component parts. In developmental biology, this tracking is reliant on specific variables from the erotetic structure of its problems agendas, such as temporality and spatial composition, and can be seen in the testing of hypotheses related to how developing organisms coordinate WBD and TSP. A pluralist stance (Kellert, Longino, and Waters 2006) on individuality is more apposite in capturing how individuation practices contribute to the diverse goals of manipulation, prediction, and explanation for diverse kinds of biological individuals. Philosophers should resist the "urgent need to arbitrate" and instead focus their attention on comprehending the concrete contexts of inquiry where biologists individuate.

Acknowledgments

Earlier versions of this chapter were presented to the 2014 Taiwan Conference on Individuation (National Chung Cheng University), the HPS department at the University of Pittsburgh (2015), and the 2015 Society for Philosophy of Science in Practice Conference (Aarhus). I am grateful to these audiences for constructive and critical feedback that improved the chapter substantially. Three anonymous referees and the editors provided a variety of helpful comments that also aided me in revising the chapter. I am indebted to participants of the 2016 Winter School on "Philosophical and Historical Dimensions of Biological Individuality" at the University of Sydney, where a variety of issues surrounding individuality were fruitfully debated and discussed. The research and writing of this chapter was supported in part by grants from the John Templeton Foundation: "Integrating Generic and Genetic Approaches to Biological Phenomena" (46919) and "From Biological Practice to Scientific Metaphysics" (50191).

References

Anderson, Warwick, and Ian R. Mackay. 2014. Fashioning the immunological self: The biological individuality of F. Macfarlane Burnet. *Journal of the History of Biology* 47: 147–175.

Bone, Quentin (ed.). 1998. *The biology of pelagic tunicates*. New York: Oxford University Press.

Booth, Austin. 2014. Symbiosis, selection, and individuality. *Biology & Philosophy* 29: 657–673.

Brigandt, Ingo, and Alan C. Love. 2012. Conceptualizing evolutionary novelty: Moving beyond definitional debates. *Journal of Experimental Zoology (Mol Dev Evol)* 318B: 417–427.

Clarke, Ellen. 2010. The problem of biological individuality. *Biological Theory* 5: 312–325.

Clarke, Ellen. 2012. Plant individuality: A solution to the demographer's dilemma. *Biology & Philosophy* 27: 321–361.

Clarke, Ellen. 2013. The multiple realizability of biological individuals. *Journal of Philosophy* CX: 413–435.

De Tomaso, Anthony W. 2014. Allorecognition, germline chimerism, and stem cell parasitism in the colonial ascidian, *Botryllus schlosseri*. *Biological Theory* 9: 423–430.

Eiraku, Mototsugu, Nozomu Takata, Hiroki Ishibashi, Masako Kawada, Eriko Sakakura, Satoru Okuda, et al. 2011. Self-organizing optic-cup morphogenesis in three-dimensional culture. *Nature* 472: 51–56.

Elwick, James. 2007. *Styles of reasoning in the British life sciences: Shared assumptions, 1820–1858*. London: Pickering & Chatto.

Ereshefsky, Marc, and Makmiller Pedroso. 2015. Rethinking evolutionary individuality. *Proceedings of the National Academy of Sciences USA* 112: 10126–10132.

Fatehullah, Aliya, Si Hui Tan, and Nick Barker. 2016. Organoids as an *in vitro* model of human development and disease. *Nature Cell Biology* 18: 246–254.

Gilbert, Scott F. 2010. *Developmental biology*, 9th ed. Sunderland: Sinauer.

Gilbert, Scott F., Jan Sapp, and Alfred I. Tauber. 2012. A symbiotic view of life: We have never been individuals. *Quarterly Review of Biology* 87: 325–341.

Godfrey-Smith, Peter. 2009. *Darwinian populations and natural selection*. New York: Oxford University Press.

Godfrey-Smith, Peter. 2013. Darwinian individuals. In *From groups to individuals: Evolution and emerging individuality*, eds. Frédéric Bouchard and Philippe Huneman, 17–36. Cambridge, MA: MIT Press.

Godfrey-Smith, Peter. 2016. Complex life cycles and the evolutionary process. *Philosophy of Science* 83: 816–827.

Griesemer, James. 2000. Development, culture, and the units of inheritance. *Philosophy of Science* 67: S348–S368.

Griesemer, James. 2007. Tracking organic processes: Representations and research styles in classical embryology and genetics. In *From embryology to Evo-Devo*, eds. Jane Maienschein and Manfred Laubichler, 375–433. Cambridge, MA: MIT Press.

Griesemer, James. 2016. Reproduction in complex life cycles: Toward a developmental reaction norms perspective. *Philosophy of Science* 83: 803–815.

Guay, Alexandre, and Thomas Pradeu (eds.). 2015*a*. *Individuals across the sciences*. Oxford: Oxford University Press.

Guay, Alexandre, and Thomas Pradeu. 2015*b*. To be continued: The genidentity of physical and biological processes. In *Individuals across the sciences*, eds. Alexandre Guay and Thomas Pradeu, 317–347. Oxford: Oxford University Press.

Hong, Ni, Songlin Chen, Ruowen Ge, Jianxing Song, Meisheng Yi, and Yunhan Hong. 2012. Interordinal chimera formation between medaka and zebrafish for analyzing stem cell differentiation. *Stem Cells and Development* 21: 333–2341.

Hopwood, Nick. 2005. Visual standards and disciplinary change: Normal plates, tables and stages in embryology. *History of Science* 43: 239–303.

Hopwood, Nick. 2007. A history of normal plates, tables and stages in vertebrate embryology. *International Journal of Developmental Biology* 51: 1–26.

Huxley, Julian. 1926. The biological basis of individuality. *Philosophy* 1: 305–319.

Kellert, Stephen H., Helen E. Longino, and C. Kenneth Waters. 2006. Introduction: The pluralist stance. In *Scientific pluralism*, eds. Stephen H. Kellert, Helen E. Longino, and C. KennethWaters, vii–xxix. Minnesota Studies in the Philosophy of Science (Volume 19). Minneapolis: University of Minnesota Press.

Kovaka, Karen. 2015. Biological individuality and scientific practice. *Philosophy of Science* 82: 1092–1103.

Kuntz, Steven G., and Michael B. Eisen. 2014. *Drosophila* embryogenesis scales uniformly across temperature in developmentally diverse species. *PLoS Genetics* 10: e1004293.

Lidgard, Scott, and Lynn K. Nyhart. 2017. The work of biological individuality—concepts and contexts. In *Biological individuality: Integrating scientific, philosophical, and historical perspectives*, eds. Scott Lidgard and Lynn K. Nyhart. Chicago: University of Chicago Press.

Loeb, Leo. 1937. The biological basis of individuality. *Science* 86: 1–5.

Love, Alan C. 2008. Explaining the ontogeny of form: Philosophical issues. In *The Blackwell companion to philosophy of biology*, eds. Anya Plutynski and Sahotra Sarkar, 223–247. Malden: Blackwell Publishers.

Love, Alan C. 2014. The erotetic organization of developmental biology. In *Towards a theory of development*, eds. Alessandro Minelli and Thomas Pradeu, 33–55. Oxford: Oxford University Press.

Love, Alan C. 2015. Developmental biology. In *The Stanford encyclopedia of philosophy*, ed. Edward N. Zalta. http://plato.stanford.edu/entries/biology-developmental/.

Love, Alan C., and Ingo Brigandt. 2017. Philosophical dimensions of individuality. In *Biological individuality: Integrating scientific, philosophical, and historical perspectives*, eds. Scott Lidgard and Lynn K. Nyhart, 318–348. Chicago: University of Chicago Press.

Minelli, Alessandro. 2003. *The development of animal form: Ontogeny, morphology, and evolution*. Cambridge: Cambridge University Press.

Minelli, Alessandro, and Thomas Pradeu (eds.). 2014. *Towards a theory of development*. Oxford: Oxford University Press.

Mirth, Christen K., James W. Truman, and Lynn M. Riddiford. 2009. The Ecdysone receptor controls the post-critical weight switch to nutrition-independent differentiation in *Drosophila* wing imaginal discs. *Development* 136: 2345–2353.

Murray, Christopher M., and Brian I. Crother. 2016. Entities on a temporal scale. *Acta Biotheoretica* 64: 1–10.

Oliveira, Marisa M., Alexander W. Shingleton, and Christen K. Mirth. 2014. Coordination of wing and whole-body development at developmental milestones ensures robustness against environmental and physiological perturbations. *PLoS Genetics* 10: e1004408.

Olson, Eric T. 2010. Personal identity. In *The Stanford encyclopedia of philosophy*, ed. Edward N. Zalta. http://plato.stanford.edu/archives/win2010/entries/identity-personal/.

Pradeu, Thomas. 2012. *The limits of the self: Immunology and biological identity*. Oxford: Oxford University Press.

Pepper, John W., and Matthew D. Herron. 2008. Does biology need an organism concept? *Biological Reviews* 83: 621–627.

Ross, Corinna, Jeffrey A. French, and Guillermo Ortí. 2007. Germ-line chimerism and paternal care in marmosets (*Callithrix kuhlii*). *Proceedings of the National Academy of Sciences USA* 104: 6278–6282.

Sive, Hazel L., Robert M. Grainger, and Richard M. Harland. 2007. *Xenopus laevis* Keller explants. *Cold Spring Harbor Protocols* 2007(6): pdb.prot4749.

Slack, Jonathan M. W. 2013. *Essential developmental biology*, 3rd ed. Chichester: Wiley-Blackwell.

Sterner, Beckett. 2015. Pathways to pluralism about biological individuality. *Biology & Philosophy* 30: 609–628.

Waters, C. Kenneth. 2005. Why genic and multilevel selection theories are here to stay. *Philosophy of Science* 72: 311–333.

Waters, C. Kenneth. 2007. The nature and context of exploratory experimentation. *History and Philosophy of the Life Sciences* 29: 275–284.

Wilson, Robert A., and Matthew Barker. 2013. The biological notion of individual. In *The Stanford encyclopedia of philosophy*, ed. Edward N. Zalta. http://plato.stanford.edu/entries/biology-individual.

Wolpert, Lewis. 2010. Arms and the man: The problem of symmetric growth. *PLoS Biology* 8: e1000477.

{9}

Experimental Individuation

CREATION AND PRESENTATION

Ruey-Lin Chen

9.1 Introduction

This chapter examines the problem of individuation from the perspective of experimental practices. Experimentation is a particular research activity using apparatus or mechanisms to manipulate and intervene in objects of inquiry. This feature makes experimental individuation distinct from individuation in the contexts of traditional philosophical and scientific theories.

From the perspective of traditional philosophy, individuation is understood both as a way that makes something an individual (metaphysical individuation) and as a way a knower discerns an individual (epistemological individuation). As metaphysician Jorge Gracia says:

> The double understanding of individuation raises two distinct problems of individuation: identifying the causes of metaphysical individuation, and of epistemological individuation. In both cases the causes are referred to as the *principle of individuation*. Attempts to settle the metaphysical and epistemological problems of individuation presuppose an understanding of the nature of individuality. (Gracia 1999: 424)

This traditional interpretation of individuation assumes the following: (1) individuation presupposes individuality, and (2) the concept of individuality with its attendant criteria is universal, being valid for all individuals in our universe. However, the two assumptions are not *a priori* true. Some philosophers of science, based on new theories, findings, and practices in physics, biology, and other sciences, have challenged the belief that all metaphysical or epistemological criteria of individuality are universal (see a number of articles in Guay and Pradeu 2016). Exploring individuation in scientific contexts implies that the understanding of individuality may be grounded on individuation. The two points bring forth a new approach: investigating individuation in scientific contexts and establishing new criteria of individuality based on those scientific practices that are a kind of individuation. This chapter aims to justify this new approach, focusing on individuation in experimental contexts. It does this by extending the arguments from my experimental account of individuality to experimental individuation. In the previous work (Chen 2016), I suggest a conception of experimental individuality (defined by separability, manipulability, and maintainability of structural unity) extracted from experimental individuation whose process and conditions in turn are the topic of this chapter. The fact that I extract a conception of individuality from experimental practices suggests the priority of individuation to individuality in the experimental contexts.

This chapter is thus developed along the following line. Section 9.2 argues that recent philosophy of science challenges traditional criteria of individuality, motivating a turn to consider individuation. Individuation in scientific contexts raises the problem of *scientific individuation*, which includes *theoretical* and *experimental* approaches. Section 9.3 considers different meanings of "individuation" used by philosophers, uncovering the practical meaning of "individuation" in which experimental individuation is to be understood. Following the semantic analysis, I explore experimental individuation and distinguish its two modes: the *experimental creation* and the *experimental presentation* of individuals. I call the former the *ontological* mode and the latter the *epistemological* mode of experimental individuation and, based on the two modes, formulate a group of particular questions (namely, how to create, maintain, compose, and present the individuality of experimented entities). In the Sections 9.4 and 9.5, I use three experimental cases (the creation of Bose-Einstein condensates (BECs) and fermionic condensates and the presentation of individual genes) to develop an answer. I argue that effective experimental mechanisms and operative conditions for maintaining the structural unity of experimented entities when those entities are separated and manipulated offer a synthetic answer. Section 9.6 offers the conclusion.

9.2 From Individuality to Individuation

Ever since the twenty-first century arrived, interest in the topic of biological individual and individuality has grown explosively (Clarke and Okasha 2013). Journals' articles (Griesemer 2000; Dupré and O'Malley 2007, 2009; Dupré 2010, Mossio and Moreno 2010; Pradeu 2010; Clarke 2010, 2013), monographs, and edited collections explore different dimensions of the topic (Michod 1999; Wilson 1999; Godfrey-Smith 2009; Calcott and Sterelny 2011; Pradeu 2012, 2013; Bouchard and Huneman 2013; Guay and Pradeu 2016*a*). Interest is not confined to philosophy of biology as many philosophers of physics hold a strong and durable interest in the topic of physical identity and individuality (French and Readhead 1988; French 1989; French and Krause 2006; Saunders 2006; Ladyman and Don Ross

2007; Ladyman and Bigaj 2010; Morganti 2009, 2013; French 2014). However, the style of exploring the issue of individuals in the philosophy of biology is quite different from that in the philosophy of physics. Philosophers of biology are typically concerned with diverse living entities occupying various "levels" in the hierarchy of the living world, while most philosophers of physics focus on fundamentally physical objects such as quantum particles.

For the issue of individuality in the philosophy of biology, "organism" and "biological individual" are two key concepts. Organisms are usually taken as the paradigmatic case, but they are not the only kind of biological individuals. Genes, gene networks, genomes, cells, organisms, superorganisms, lineages, groups, populations, and even species also count as biological individuals from the perspective of evolutionary theory. This is because they all are units of evolutionary hierarchy on which natural selection acts, with successive levels arising through major evolutionary transitions (Maynard-Smith and Szathmáry 1985). With the rise of the idea about evolutionary individuality, the relationship between the category of organism and that of biological individual has now been reconsidered.

Peter Godfrey-Smith (2009, 2013) argues that some organisms are not Darwinian (or evolutionary) individuals because of their inability to reproduce while some Darwinian individuals are not organisms because of the absence of metabolic functions. Thomas Pradeu (2010, 2012, 2013) argues that immunity is not only an adequate criterion for identifying an organism and delineating its boundary but also can be extended to other biological individuals, even in an evolutionary perspective. Ellen Clarke (2013) proposes functional interpretations of policing and demarcation mechanisms that are used to define organisms and suggests a new definition of biological individuality based on her redefinition of *individuating* mechanism. A few articles in *From Groups to Individuals*, edited by Bouchard and Huneman (2013), explore how entities such as biological groups, communities, superorganisms, and the like count as biological individuals.

By and large, most philosophers of biology address the problem of defining organisms or offer criteria for biological individuals, but, as I will argue in Section 9.3, their positive accounts suggest that some conceptions of *individuation* or *individuating* are presupposed. Moreover, the exploration of individuation is inevitably involved in two interrelated questions: First, *how is an individual organized from or composed of its parts or components?*[1] Second, *how does biological individuality maintain itself?* The two questions arise because biological individuals are always composed of parts and always gain their individuality when they are born at some temporal point and lose it when passing away at a later temporal point.

---

[1] Dupré and O'Malley formulate this question addressed to organisms in terms of the following statement: "Answering the question, 'What is an organism?', requires seeing that there is a great variety of ways in which cells—sometime genomically homogeneous, sometimes not—combine to form integrated biological wholes" (2007: 842).

Philosophers of physics, by contrast, focus on the issue of the identity and individuality of quantum objects. Ever since several philosophers have demonstrated that Leibniz's Principle of Identity of Indiscernibles is violated by some quantum objects (French and Readhead 1988; French 1989), the problem of whether quantum objects such as bosons, fermions, elementary particles, and quasi-particles are individuals is a key debate topic. Some philosophers argue that the concepts of objects and individuals in the subatomic scale can be eliminated. They endorse structural realism, arguing that only ontic structures are real (French and Krause 2006; Ladyman and Don Ross 2007; Ladyman and Bigaj 2010; French 2014), with French extending eliminativism about individuals to biological objects (French 2014). To save the intuition of quantum particles as individuals, some philosophers argue that weak discernibility can serve for this purpose (Saunders 2006).[2] Other philosophers invoke the conception of "primitivity" or "primitive thisness" to save the concepts of individual and object in the quantum domain (Morganti 2009, 2013). Those debates concerning quantum objects are potentially directed toward the problem of whether there are *indivisible* (undividable) or *ultimate* individuals,[3] which is not a topic of concern in this chapter.

I instead wonder how scientific objects composed of quantum objects become and maintain themselves as individuals, given that the individuality of quantum objects is at issue. Beyond the domain of quantum physics, a few philosophers of physics (Ladyman and Ross 2007; Guay and Pradeu 2016*b*; Saunders 2016) have considered the question of whether other objects posited in physical theories are individuals. According to what criterion, however, can we identify macroscopic objects such as waves, rays, chemical molecules, crystals, minerals, clouds, air masses, rivers, tributaries, mountains, mountain ranges, and a great variety of artifacts as individuals? This is a genuine and serious problem.

A difficulty emerges when one wishes to delineate a clear boundary between two adjacent rays, clouds, air masses, minerals, mountains, mountain ranges, and the like. Metaphysical criteria of individuality, such as discernibility and delineability, are not available to those objects. The difficulty in identifying some kinds of physical macroscopic objects by observation is similar to that in identifying some kinds of organisms. Unfortunately, we do not have a theory of natural selection that can help define the individuality of physical objects. We thus face the criterion problem of physical macroscopic objects.

---

[2] Saunders (2006) distinguishes the principle of indiscernibility into three versions with different degrees of strength: absolute discernibility, relative discernibility, and weak discernibility. Which version should individuality be based on? There are many debates over this; see Ladyman (2016).

[3] The Latin origin of "individuals" means "in-dividable (undividable) entities." If "individual" refers to the kind of undividable entities, then, by intuition, they seem to be invariant, unchangeable, indecomposable, and immortal. "Atoms" in the ancient Greek sense refer to this very kind of undividable individuals. This meaning seemingly remains in the thoughts of some metaphysicians and philosophers of physics because they attend only to some entities that have these features, such as elementary particles and quantum objects in contemporary physics.

The following instance presents the criterion problem more clearly and deeply. Consider a pile of mechanical parts, a workable machine, and a statue made of mechanical parts. Under some intuition, a pile of mechanical parts does not constitute an individual (although this intuition is controversial), while both a machine assembled out of parts in a specific way and a mosaic statue reassembled out of mechanical parts in a different way are individuals. What distinguishes the former from the latter two? Both the machine and the statue are composed of mechanical parts, while there is no composition in a pile of mechanical parts. One may wonder: *By what way in general is an individual composed of its parts?* This question presents the composition problem of physical individuals, paralleling to the organization problem of biological individuals.

Like biological individuals, almost all physical macroscopic things come into being and cease to be at different temporal points. A wave rises and soon passes away. A river occurs when water flows through a strip of land and vanishes when the water dries up. A water molecule forms under some conditions and dissociates in others. An artifact is present when it is made and ruined when broken down or taken apart. Whenever a physical thing forms and vanishes, it always undergoes a period of time through which it maintains itself as an individual. *What maintains the individuality of physical things during this period?* This question is exactly similar to the maintenance problem of biological individuality. Considering all biological and physical things together, I call the questions related to the composition of individuals and the maintenance of individuality *the core problems of individuation in general.* Here, the term "individuation" means the process by which objects form into and maintain themselves as individuals or through which objects get their individuality.

The purpose of this chapter is not to provide a solution to the core problems of individuation in general, which can be fully accounted only if we have examined the problems in the particular (i.e., scientific theoretical and experimental) contexts of individuation. Given that individuation in experimental contexts has received less attention, this chapter aims to explore *experimental individuation* whose problems can be formulated by referring to individuation in general. I hope that such an exploration will help develop a full account of individuation in general.

9.3 From Individuation to Experimental Individuation

Metaphysicians and philosophers of physics usually use "individuation" to mean invariant principles referring to natures, essences, or intrinsic properties. Those principles provide certain logical or semantic criteria for identifying an entity as an individual and thus offer answers to the question of what makes individuals *qua* individuals. Philosophers of biology typically use this term "individuation" differently. As we discussed in the previous section, "individuation" may well mean the *process* in which the biological individuality *forms* and *continues.* When philosophers of biology propose various "principles" of individuation, such as action of natural selection, reproductive mechanisms, policing mechanisms, metabolic functions, immunity or immune mechanisms, and the like, as criteria for identifying biological individuals, they really mean *changing processes* rather than invariant principles. This is not only because the composition and maintenance of individuals make up a temporal process, but also because those actions, functions, or mechanisms that cause composition and maintenance are *in process*.

Another meaning of the term is used by philosophers who think that individuation is a cognitive operation of mind by which we sort an object out, or identify, or treat some entity as an individual (Chauvier 2016). Such an operation depends on some conception or cognitive framework that may be derived from metaphysical or scientific (physical, chemical, biological, etc.) theories that they as such are cognitive products. To sum up the discussion so far, there are at least three meanings (or extensions) of individuation.

1. "Individuation" refers to the (natural) principle of individuation, denoting synchronic essences, intrinsic properties, or logical and semantic features that provide criteria for identifying or treating entities as individuals.
2. "Individuation" refers to the (natural) process of individuating, denoting the diachronic process of the composition and maintenance of individuals through which entities become and continue to be individuals.
3. "Individuation" refers to the cognitive operation, based on a specific conception or cognitive framework by which one sorts an object out or identifies or treats entities as individuals. The cognitive operation of individuation depends on some conceptions of individuation. There can be a variety of conceptions of individuation.

The three meanings are related to one another. The specific process of individuation in an abstractive form can always be taken as a principle of individuation. Moreover, the principle or the process can be represented by a specific conception or cognitive framework of individuation. In traditional terms, epistemological individuation is closely related to metaphysical individuation.

Traditional metaphysicians uncover the principle or formulate the conception of individuation by constructing metaphysical theories based on a number of paradigmatic cases (observed or intuited) or by making a logical and semantic analysis of the terms "individuals," "individuation," and even "nonindividuals" (Lowe 2016). Philosophers of physics uncover the principle or formulate the conception by examining criteria from metaphysical theories against the best physical theories—say, the theory of relativity or quantum mechanics. Philosophers of biology tend to uncover the natural process of individuation because they investigate the diversity of the living world. They find the process of individuation or establish criteria of individuality from the best biological theories; namely, the best physiological theory, the best immunological theory, or the theory of evolution by natural selection (Godfrey-Smith 2009; Pradeu 2012; Clarke 2013).

In formulating the conception of individuation or uncovering the principle or process of individuation, most philosophers attend to one or more of a variety of theories—metaphysical, physical, or biological. However, a few philosophers have argued that experiments can produce or realize novel things such as bosonic condensates or cultured pluripotent stem cells that arguably satisfy criteria of individuality (Kovaka 2015; Chen 2016; Fagan 2016; Chapter 6, this volume).[4] Some also argue that the conception of individuality implicated in experimental practices is different from and/or independent of theories (Chen 2016; Love, Chapter 8, this volume). Their cases and arguments support two significant points. First, scientific experiments can individuate things via producing or realizing things that embody plural criteria of individuality. Second, experimental practices can offer a specific conception or cognitive framework of individuation that is different from and/or independent of the theoretical ones. The first point indicates that "individuation" may be used in situations in which experimenters or technicians realize (create or present) individuals via their experimental or technical practices. For instances, a technician creates a robot by combining a variety of mechanical parts or an experimenter realizes a fermionic condensate by arranging an effective mechanism. Thus, "individuation" has a fourth meaning:

4. "Individuation" refers to the practice of individuating, denoting the (artificial) practice performed in realizing (creating or presenting) an individual by scientists, experimenters, or technicians.

Although the practice through which one individuating an entity always undergoes a process in which the entity becomes and continues to be an individual, the process is not spontaneous in nature. It is a process mixed with artificial operations and interventions. Furthermore, as the second point indicates, such kind of individuation can offer a specific conception of individuation or a particular conception of individuality that is extracted from those manipulations, interventions, production, or creations (Chen 2016). As a consequence, individuation in experimental contexts produces an ontological output and an epistemological output.

First, creation has an ontological implication. Experimentally creating an individual is experimentally individuating it. Thus, one may wonder (Q1) *how experiments create an individual.* Following the first question, we would like to

[4] Fagan (2016; Chapter 6, this volume) proposes an interesting but seemingly paradoxical argument: stem cells are not biological individuals in the same way as cells of multicellular organisms, but cultured pluripotent stem cells are biological individuals in the way of multicellular organisms. Fagan convincingly argues that stem cells cannot be experimentally confirmed to satisfy the major criteria of biological individuality that cells of multicellular organisms do, such as clear physical boundaries, physiological integration, and functional autonomy, but, she argues, pluripotent stem cells in artificial cultures do satisfy these criteria in the way multicellular organisms do. In this case, Fagan may uncover another mode of the experimental individuation: transforming a type of individual into another type.

know (Q2) *how experimenters maintain the individuality of created entities through the experimenting process.* If the created individual is composed of components, then we ask a further question (Q3): *how experimenters combine components into a unitary individual.* The three questions are asked in the contexts of experimental creation, which is an ontological mode of experimental individuation.

Not all experimented individuals are created in experiments. If an entity that is not made from experimenting keeps its individuality throughout the process of experimenting, then we would also like to ask how experimenters accomplish this. We would ask this because mistaken operations or inadequate interventions may impair the experimented entity and even break down its individuality. In fact, at the start, we do not *know* whether the entity is an individual or not unless scientists do verify (present) that it is. In this sense, the process of *experimental presentation* is an epistemological mode of experimental individuation.

Experimentation is traditionally regarded as a useful instrument or method for confirming or disconfirming theoretical hypotheses. A few hypotheses posit or suppose theoretical entities that cannot be directly observed by unaided eyes (e.g., phlogiston, oxygen, ether, electromagnetic fields, atoms, electrons, quarks, and so forth). Those posited entities may be real or fictional. In order to confirm a theoretical hypothesis that posits a key entity, scientists can conduct an experiment to verify the existence of the entity. If an entity can be verified to exist by an experiment, then it is usually presented as an individual sample or an individual in the particular experiment. If an entity can be presented as an individual in a particular experiment, then it must be able to satisfy the criteria of experimental individuality. As a consequence, experimentation offers a means of verifying the individuality of entities as long as the criteria of experimental individuality are satisfied. Stated otherwise, we can use experiments to realize the individuality of experimented entities in the sense of presentation or verification. As I argue later, for example, transgenic technology can present that a number of genes are individuated entities. A related question naturally arises: (Q4) *how do experimenters present that an experimented entity is an individual?* In the next sections, I will answer questions Q1–Q4.

Before moving into the next section, let me answer two possible questions. One may wonder whether the statements in the foregoing paragraph presuppose that a conception of experimental individuality is prior to the practice of individuation. The answer is negative because the idea here is that the conception of experimental individuality and its attendant conditions are extracted from the process of experimental practices in which scientists identify objects as individuals (see Chen 2016). Thus, this conception is finally grounded on the practice of individuation. What I am doing in this chapter is to explicitly make the ordering between individuality and individuation and to generalize the conception of experimental individuality and its attendant conditions to similar cases, but this does not mean that the conception of individuality is necessarily prior to any practice of individuation. In addition, one may think that this chapter is addressing the special composition problem rather than the problem of individuation. It is true that the former is involved in this chapter, but, from the perspective of individuation as a process, the special composition problem is part of the individuation problem because individuation includes the process of the composition and maintenance of individuals.

9.4 Two Modes of Experimental Individuation: Creation and Presentation

In my previous work (Chen 2016), I proposed a conception of experimental individuality that is at work in physical and biological experiments. This conception is defined by three properties: manipulability, separability, and maintainability of structural unity. These properties can be realized, respectively, by three experimental operations: separation, manipulation, and maintenance of structural unity. I call the production of individuals by means of these operations and in accordance with their corresponding properties the experimental realization of individuality. The successful performance of the three operations constitutes the realization conditions of individuality. At the point of producing an individual, the process in the experimental realization of individuality can be called *experimental individuation* as well.

As I have argued, a scientist is said to experimentally produce or realize an individual if and only if she could maintain its structural unity when manipulating and separating it (Chen 2016). The maintenance of structural unity is a requirement for the successful manipulation and separation of experimental individuals and vice versa. In order to understand experimental individuation based on the three realization conditions, two questions have to be answered: (1) *How do scientists maintain an experimental entity's structural unity when manipulating and separating it?* (2) *Why is the structural unity of an experimental entity capable of being maintained when being separated and manipulated?*

A preliminary answer consists of two points. First, scientists have to make an *effective* mechanism to maintain the entity's individuality when manipulating and separating it because an effective experiment always contains an effective mechanism. Second, the mechanism is so designed and constructed as not to impair the structural unity of the entity through the overall process of experimenting. However, the answer would be void if we do not know what the mechanisms are, how they are effective, and how they function to maintain individuality. Thus, we have to find what conditions characterize such an effective mechanism.

Many definitions of mechanisms have been proposed, and debates over them continue (Glennan 1996, 2002; Machamer, Darden, and Craver 2000; Woodward 2002; Darden 2002; Fagan 2012; Levy 2013). I will not enter into those debates. I simply adopt Machamer, Darden, and Craver's definition because it is very suitable for the purpose of this chapter. The definition states: "Mechanisms are entities and activities organized such that they are productive of regular changes from start or set-up to finish or termination conditions" (Machamer, Darden, and Craver 2000: 3). However, most mechanisms on which philosophers have so far tended to focus are natural.

Experimental mechanisms are artificial. They are designed and constructed by scientists. In constructing an experimental mechanism and realizing it, scientists arrange and organize experimented objects and apparatuses in order to produce activities and interactions—among those objects and apparatuses. Every experiment has a starting point and a termination point. When an experiment is performed, its starting circumstances are the set-up conditions of the experimental mechanism and its finishing circumstances are the termination conditions. Thus, one may say that an experimental mechanism is the very experimenting process as well. To match with the artificiality of experimentation, I define "experimental mechanisms" as "experimenters organize experimental entities (including experimented objects and experimental apparatuses) and activities (including experimenters' operations and interventions) such that those entities and activities are productive of changes and interactions from start or set-up to finish or termination conditions."

Applying this definition to the experimental production of individuality, mechanisms created by scientists enable experimented entities that satisfy the criterion of individuality to be realized via producing some specific conditions. If the conditions enable the realized individuals to be manipulated and separated so that their inner structures are maintained in a changeless state, then such a mechanism can effectively operate to maintain the experimental individuality.

Provided the individuality of an entity may form under some conditions and vanish under others, I call the former the *forming condition* and the latter the *vanishing condition*. Between the forming and the vanishing of the entity's individuality, an experimental mechanism has to operate under a further set of constraints such that the individual is maintained. The specific *operating conditions* define the *effectiveness* of the experimental mechanism and characterize *experimental individuation*. In some cases, the operating conditions may include the forming and vanishing conditions. If we know those conditions, then we can answer the question of why an entity's individuality can be produced or created in experimenting. If we know how scientists produce those conditions in their experiments, then we can answer the question of how to create and maintain the entity's individuality.

As the individuality of an entity is maintained by some experimental mechanism, it may or may not appear at any temporal point in the entire process from the start to the termination. In some cases, the forming condition of individuality appears at a temporal point *after* the start and *before* the termination. Experimental mechanisms operate and maintain the individuality of entities from their forming to their vanishing. These requirements define the *experimental creation* of individuality, which has been elegantly embodied in the realization of BECs. Although entities such as BECs vanish at the termination of the experiment, not all experimented entities necessarily do so. They may continue to exist and maintain their individuality after the completion of experiments.

In some other cases, experimented entities are not produced by the experiments that use them as target objects, and therefore neither the forming condition nor the vanishing condition of the entities' individuality are produced in the course of the experiments. In such a situation, experimenters do not test whether the entities are individuals before performing the operation. However, the individuality of the entities can be verified when the experiments are successfully finished because the entities remain the same before, during, and after the experiments. This indicates that individuality is maintained through the overall process of experimenting. Transgenic technology exemplifies the alternative, epistemological mode of experimental individuation.

In what follows, I discuss three cases that use different experimental mechanisms with different forming, vanishing, and operating conditions. The realization of bosonic atomic, molecular, and fermionic condensates exemplifies the experimental creation of individuality under distinct sets of operative conditions, while the realization of transgenes exemplifies the experimental presentation of individuality.

9.4.1 THE CREATION OF BOSE-EINSTEIN CONDENSATES

The first case is the creation of BECs that are described in detail by Chen (2016). Here, I offer a simplified version, specifying the forming, vanishing, and operating conditions of BECs. The first BEC in the laboratory was composed of bosons. Bosons—atoms with integer spins—can condense into a collective in which all bosonic atoms act uniformly as an individual (namely, a BEC) because they all occupy the lowest energy level and collapse into the degenerate state when the environmental temperature reduces to a ultracold state.

In 1995, American scientists Eric Cornell and Carl Wieman, in the Joint Institute of Laboratory Astrophysics (JILA) at the University of Colorado and the National Institute of Standards and Technology, used sophisticated new techniques to create BECs in which there were about 2,000 indistinguishable rubidium atoms that condensed into a single individual and existed for approximately fifteen seconds (Anderson et al. 1995). Cornell and Wieman used three key techniques—laser cooling, optical and magnetic trapping, and evaporative cooling—to make an experimental mechanism. The laser cooling technique makes a setting in which the atoms absorb photons out of laser light that is cast in the direction opposite to the atoms' motion, thereby slowing down the atoms and cooling them. Laser light is used not only to cool the atoms, but also to trap them, keeping them away from the room-temperature walls of the glass box that contains a tiny amount of the experimental gas. Six vertically intersected laser beams are arranged to keep the slowing atoms in a restricted space, as if they have been fixed. These trapped atoms are at an extraordinarily low temperature, but not low enough to enter into Bose-Einstein condensation.

Cornell and Wieman continued to cool those atoms by using magnetic trapping and evaporative cooling. At this stage, they hold the atoms in a magnetic field by carefully controlling the shape and strength of the field in which the atoms move around and collide with one another. The most energetic atoms escape from the magnetic bowl and carry away more than their share of the energy, leaving the remaining atoms colder. When the energy is reduced to a level below the critical temperature, a BEC comes into being (Cornell and Wieman 1998: 40–43). When the temperature around the condensate goes up to the critical point, the degenerate state that constitutes the internal unity of parts (atoms) within the individual is broken down, and the structured entity disappears. According to my argument, a BEC is an experimental individual because it can be separated and manipulated without losing its structural unity (Chen 2016). As a result, the experimental mechanism not only creates but also maintains the individuality of BECs.

In the mechanism for producing and maintaining BECs, rubidium atoms, laser beams, photons, and a magnetic field whose shape and strength can be controlled are working *entities*. Cooling, photon-absorbing, speed-slowing, trapping, separating air from the glass wall, keeping the slowing atoms in a restricted space, evaporating, and (atoms) escaping are *activities* in which working entities are participating. The individuality of BECs forms and vanishes during experimenting. The forming condition is that the atoms' temperature decreases to a level below the critical temperature, and the vanishing condition is that the temperature is just at the level above the critical point. If scientists wish to maintain the individuality of BECs, then they have to keep the temperature at always below the critical point—this is the operating condition for maintaining the individuality of BECs.

9.4.2 THE CREATION OF FERMIONIC CONDENSATES

After BECs were created from Bose gases in 1995, physicists turned to pursue the next goal: the creation of condensates out of Fermi gases. The production of condensates from Fermi gases is more difficult because, according to Pauli's principle of exclusion, fermions—atoms with half-integer spins (1/2, 3/2, 5/2, etc.)—are excluded from occupying the same energy levels. In order to overcome this difficulty, physicists use paired fermionic atoms to conform to the condition of integer spins. However, different bounds and interactions between fermionic atoms produce different kinds of fermionic pairs: from bosonic molecules of tightly bound fermionic atoms and closely interactive fermionic atom pairs to loose-knit Cooper pairs (see Cho [2003] and Collins [2003] for a brief introduction).[5]

[5] How does one distinguish bosonic molecules of fermionic atoms from fermionic atom pairs and from Cooper pairs? A figurative description is helpful. Both bosonic molecules of fermionic atoms

Physicists usually view this series of ultracold states as a continuum from BECs to Bardeen-Cooper-Schrieffer (BCS) superfluids (the BCS is an extreme type, with atomic BECs at the other extreme). Some physicists are quite interested in exploring the BEC–BCS crossover by means of experimental intervention, but there are no clear boundaries throughout the BEC–BCS crossover. Accordingly, "[i]n discussing condensation of a Fermi gas throughout the BCS–BEC crossover, terms such as Cooper pairs, molecules, BEC, and fermionic condensates often have ambiguous meanings" (Regal, Greiner, and Jin 2004: 040403-1). Deborah Jin's team defines "condensation of fermionic atom pairs," or equivalently, "fermionic condensates" as "condensation (i.e., the macroscopic occupation of a single quantum state) in which the underlying Fermi statistics of the paired particles play an essential role" (Ragal, Greiner, and Jin 2004: 040403-1). Some other scientists use terms such as "superfluidity in resonantly interacting Fermi gases" in place of "fermionic condensation" (Kinast et al. 2004).

BECs are nonetheless obviously different from fermionic condensates in both theoretical and experimental aspects. In the theoretical aspect, members in a BEC occupying the same quantum state are described by Bose statistics, while members in a fermionic condensate are described by Fermi statistics. In experimental observations, a BEC appears as a round point, whereas a fermionic condensate exhibits a cigar shape (O'Hara et al. 2002: 2180; Pitaevskii and Stringari 2002: 2145; Kinast et al. 2004: 150402-1). Naturally, molecular BECs out of Fermi gases are still BECs rather than fermionic condensates. One wonders whether or not BECs out of bosonic atoms are a different type from molecular BECs out of paired fermionic atoms. I will return to this question later.

In 2003, three teams (D. Jin's team at JILA and W. Ketterle's team at MIT, both in the United States, and R. Grimm's team at Universität Innsbruck in Austria) almost simultaneously reported that they had created molecular BECs out of Fermi gases (Greiner, Regal, and Jin 2003; Zwierlein et al. 2003; Jochim et al. 2003). Soon after, in early 2004, Jin's team declared that they had created fermionic condensates (Regal, Greiner, and Jin 2004). Three months later, J. E. Thomas's team at Duke University in the United States reported the same achievement (Kinast et al. 2004).[6]

In creating BECs from bosonic molecules and fermionic condensates from fermionic atom pairs, scientists use techniques and instruments similar to those

---

and fermionic atom pairs dance the tango. In bosonic molecules, every fermionic atom keeps the same partner, whereas in fermionic atom pairs the atoms continuously exchange partners. Atoms in Cooper pairs also dance, but not in the close style of a tango. Atoms in Cooper pairs interact with each other over a longer distance and produce a loose connection, whereas atoms in fermionic pairs interact with each other over a shorter distance and produce a stronger bond.

[6] In 2002, J. E. Thomas's team had produced a highly degenerate, strongly interacting Fermi gas of atoms that does not reach condensation, but the boundary line between the superfluid of strongly interacting Fermi gas of atoms and the superfluid of resonantly interacting Fermi gas is not very clear. In general, physicists are mainly concerned about the superfluidity of various kinds of condensed matter rather than their individuality.

used in creating BECs from bosonic atoms. The process of producing those condensates is also similar to that of producing BECs from Bose gases. In order to create molecular BECs, scientists use the specific technique of magnetic field Feshbach resonances in addition to standard techniques.[7] After a Fermi gas (e.g., an isotope of Potassium ^{40}K) has been cooled to a high degree of quantum degeneracy by laser and evaporative cooling, scientists tune a magnetic field sweeping across a Feshbach resonance to cause a two-body effect. As a result, most of the fermionic atoms are converted into bosonic molecules adiabatically, and few atoms remain free. If the temperature of the initial atom gas is sufficiently low (in Jin team's experiment, T/TF = 0.17, where TF is Fermi temperature), then molecular BECs will be realized.[8] In order to create fermionic condensates in the BEC–BCS crossover, scientists carefully tuned the magnetic field and the resonance to produce cooperative many-body effects that can give rise to the condensation of fermion pairs (Regal, Greiner, and Jin 2004).

The experimental mechanisms for producing and maintaining molecular BECs and fermionic condensates are similar to those for atomic BECs. The forming and vanishing conditions of both condensates are a critical temperature plus a specific magnetic field and a resonance effect—for molecular BECs, the resonance effect is a two-body type, whereas for fermionic condensates, the resonance effect is a many-body type. The operating conditions of both mechanisms are to keep the temperature below the critical point and to keep the magnetic field and the resonance in working order.

9.4.3 THE PRESENTATION OF PARTICULAR GENES AS INDIVIDUALS

The third case is one of presenting the individuality of antibiotic resistance genes via experiments involving recombinant DNA and cell transformation. It is an epistemological mode of experimental individuation different from the creation of bosonic and fermionic condensates. The individuality of genes is not produced in experimenting, but it can be presented by successfully transplanting a gene—a kind of experiment that produces transgenic organisms. This case is also described in detail in my previous work (Chen 2016), but there I do not characterize it as a mode of experimental individuation alternative to experimental creation.

Stanley Cohen and Herbert Boyer conducted the first complete experiment of recombinant DNA in 1973 and 1974, in which they transferred two different

---

[7] A Feshbach resonance is a feature of many-body systems in which a bound state is achieved if the systems are between at least one internal degree of freedom and the reaction coordinates. Herman Feshbach was an American physicist who first gave a precise analysis of this type of resonance.

[8] This process is more complicated, as D. Jin's team states: "The molecular condensate is not formed by any active cooling of the molecules, but rather merely by traversing the predicted BCS–BEC crossover regime" (Greiner et al. 2003: 537).

DNA segments encoding proteins for ampicillin and tetracycline resistance, respectively, into *Escherichia coli*, thereby realizing the transformation of this bacterium. Both DNA segments are called "antibiotic resistance gene." In performing the experiment, Cohen and Boyer used small circular plasmids (extrachromosomal pieces of DNA) as vectors to transfer a foreign DNA segment into a bacterial cell. The plasmids were made by cutting out a (supposed) antibiotic resistance gene from other bacteria with the restriction enzyme *Eco*RI, linking the gene into a plasmid by using another enzyme, DNA ligase, and transferring the plasmid into an *E. coli* cell lacking the ability to resist antibiotics. The result, a modified *E. coli* cell, was able to resist antibiotics and contained the antibiotic resistance gene. In this experiment, the antibiotic gene was separated from its original bacteria and then was manipulated (i.e., linked and transferred). Its structural unity was not broken down, thereby allowing it to be expressed in the other kind of bacteria throughout the process. Scientists thus identify it as *a* gene—an individual biological entity.

In performing the experiment, Cohen and Boyer arranged an experimental mechanism in which *E. coli*, plasmids from other bacteria, DNA segments encoding a protein conferring antibiotic resistance, restriction enzyme *Eco*RI, and DNA ligase are working entities, and cutting DNA segments out, linking DNA segments with plasmids, transferring DNA segments into bacterial cells, and recombining DNA segments with transformed bacteria's genomes are activities. The individuality of genes is not created in the experiment, nor does it vanish when the experiment is terminated. Of course, this does not imply that individual genes have no processes from formation to disappearance. Rather, the forming and vanishing conditions do not happen in the process of cell transformation experiments.

These kinds of experiments offers the strongest evidence verifying the individuality of genes because the structural unity of genes is not lost when they are separated and manipulated; otherwise, if the structural unity of a transgene were not maintained, it would not be expressed on the new genomes. In such a situation, scientists could not warrant some DNA segment that they cut to be a gene. As a result, even if that DNA segment is a genuine gene, scientists could not verify that it is! This leads to the conclusion that the individuality of that segment as a gene would not be presented. Even the individuality of that DNA segment as such would not be verified because it might disappear in the cellular environment of some other kind of bacteria. However, at least the individuality of *some* genes has been experimentally presented if they can be expressed successfully. Moreover, they should count as individuals only if they have been experimentally presented as individuals. In this sense, the individuality of those genes epistemologically depends on the practice of experimental individuation. Moreover, we should note that the experiments of transgenes cannot present *all* located genes as individuals; what they have verified is that some particular transgenes can be warranted to count as individuals.

9.4.4 TWO MODES OF EXPERIMENTAL INDIVIDUATION

The three cases exemplify two modes of experimental individuation: one is ontological and the other epistemological. In the ontological mode, the individuality of Bose and fermionic condensates is created and maintained during a period of time within the experiments. In other words, scientists use a special experimental mechanism to combine a large amount of bosonic atoms or fermionic atoms into a Bose or a fermionic condensate and maintain its individuality in the process of experimenting. In the epistemological mode, the individuality of transgenes is presented due to the successful performance of transgenic experiments because the structural unity of those genes is maintained through the experimental process. This view implies a conjunctive claim: *only those genes that are experimentally presented or verified should count as individuals,* and, *therefore, not all posited genes can be identified as individuals.* The individuality of those genes, which had not been nor cannot be experimentally individuated, has not been warranted or cannot be known either: *no individuation of a particular gene, no knowledge of the gene as an individual.*

Each of the three cases illustrates that *effective* mechanisms can maintain the individuality of experimented entities. The "effectiveness" is specified by the operating conditions of maintaining mechanisms in experimenting. Therefore, I offer a synthetic answer to the questions of creation and presentation in terms of effective mechanisms and their operating conditions. The next section takes up the question of composition.

9.5 Structural Unity and Its Maintenance

How do experimenters combine components into a whole individual? A necessary condition is that the experimenters who create the individual have to endow a structural unity to all components. To do so, the experimenters must design a mechanism to make and maintain the structural unity of the individual. In other words, the formation and maintenance of structural unity explain the composition problem of experimented individuals. The composition and the maintenance problems are two faces of a coin in the context of experimental creation.

As we have seen, a BEC is realized when all experimented bosonic atoms occupy the same quantum state and act uniformly and collectively as an individual. The Bose condensate is composed of Bose gases that are bosonic atoms in the phase of air. The components of molecular BECs are bosonic molecules that are composed of paired fermionic atoms, and the components of fermionic condensates are fermionic atom pairs.

Are these condensates created in the laboratories different kinds from one another? Physicists have not yet provided a clear answer. It looks as if there are three or more general kinds of condensates, in a loose sense of the word "kind," because scientists use different kinds of components with different inner structures to constitute different condensates. However, these condensate *kinds* may not be distinct kinds because scientists have not yet learned what is happening within each kind of component with a specific inner structure when each condensate is realized. By definition, each condensate is the thing in which all components (bosonic atoms or bosonic molecules or fermionic atom pairs) occupy the same quantum state and become indistinguishable with one another—this overall state implies a set of specific operative conditions that is used to realize the structural unity of all components that constitute a condensate. If the set of specific operative conditions for a particular condensate cannot be maintained, then the condensate would decompose immediately.

Different sets of operative conditions are of course needed for different kinds of elements and for different components with different chemical structures. However, no component kind is relevant to the individuality of condensates because either we have no knowledge about how the components' inner structure engages with the composition of condensates, or the chemical structures of different component kinds are irrelevant to the form of composition. If one wants to verify that an experimented entity is an individual, then one needs to only present that its structural unity is maintained when being separated and manipulated. Separating and manipulating experimented entities involve intervening in their structures. Interventions risk impairing or even breaking down the structures of the things, causing them to be no longer individuals—they decompose into many material fragments that may or may not be still individuals in some way. If these risks can be avoided in experimenting, then the maintenance of the structural unity of experimented entities can be verified.

So far, I say little regarding the notion of structural unity. What is structural unity? What is a structure? Does every entity have a structure? Does an elementary particle have a structure? I leave aside the question of whether there are genuine *atomic* entities that have no inner structures. However, it is obvious that an absolute majority of things are composed of components in a great variety of ways. This accounts for the significance of the general composition problem in traditional metaphysics (van Inwagen 1995). By "structure," I refer simply to various forms of composition. A form of composition is the whole of all connections among all components. Every specific form of composition is a specific structure. This chapter is not the place to discuss the notion of structure fully,[9] but it is helpful to list some general forms of material composition.

[9] "Structure" is a term widely used in philosophy of science. "Theory structure" is used in the discussion about the structure of scientific theories. Sometimes, theoretical models are taken as representing theory structures. In the debate over the realism issue, structural realism is proposed as an alternative position to scientific realism and anti-realism. Speaking roughly, structural realists claim that the theoretical or mathematical structures of mature scientific theories reflect the real structure of the world. In this chapter, the meaning of "structure" is not involved in those contexts.

(s1) *Linear structure*: The form in which a component connects another adjacent component in a linear order. A typical instance is the sequence of nucleotide bases comprising a gene.

(s2) *Radial structure*: The form in which a component connects some other components in a multidimensional way. It may be two-dimensional or three-dimensional. Chemical structures of many molecules exemplify this form.

(s3) *Reticulate (network) structure*: The form in which every component connects some other components with one another. Every component may be a node or a connection. Connections among nodes may be two-dimensional or three-dimensional. Spider webs and the steel structures of buildings are two typical examples.

(s4) *Uniform structure*: The form in which all components are the same and behave in a uniform way. Fish groups and troops in military parades are two candidates for this form.

(s5) *Functional structure*: The form in which every component is connected with some other components through functional relationships. For example, bodies of organisms are composed of various organs with different functions.

(s6) *Composite structure*: Different forms of structures are connected into a single individual with a global structure that is composed of heterogeneous substructures. A machine may be composed of module parts. Different module parts have different forms of composition.

A structure is thus a compositional structure. "Structural unity" means the unity of compositional structures or the formal unity of composition. Impairing or breaking the structural unity of an individual is to break its form of composition. Breaking the compositional form of an individual is cutting off the connections between components, or decomposing it. Conversely, maintaining the structural unity of an individual is combining its components into a whole in a persisting state. This indicates that another distinctive feature of compositional structures is *invariant during a period of time*.[10] To understand the relation between structural unity and individuality, one may consider nonstructural or formless things. Mere aggregation or random connections of components, such as air, fluid, and mass, are nonstructural and formless. They are so because the connections among components cannot remain invariant during a period of time. They are not structurally unitary, either. People as well as metaphysicians usually ascribe these nonstructural and formless things to the category of nonindividuals (Lowe 2016).

[10] Although all compositional structures remain invariant for a period of time, they are still changeable. They change in a gradual way. For example, the structure of a multicellular organism largely remains invariant for the length of its life. However, many of the organism's cells are renewed continuously. This chapter leaves a fuller discussion of this point for another time.

This ascription indicates that the unity of compositional structure is necessary for individuality.

To sum up, the maintenance of structural unity is a necessary condition of experimental individuation. It offers a general answer to not only the composition problem of individuals, but also to the maintenance problem of individuality in the context of experimental practice.

9.6 Conclusion

I have argued herein that there are two basic modes of individuation in experimental situations. In the ontological mode, by constructing an effective experimental mechanism and finding effective operative conditions, an experimenter can create a new individual via combining its components into a unitary whole under specific conditions. In the epistemological mode, experimenters can present the individuality of experimented entities via designing an effective procedure through which the structural unity of those entities is not lost when they are separated and manipulated. In both modes of experimental individuation, experimented entities are separated and manipulated and their structural unity is maintained. However, the former produces an ontological outcome and the latter an epistemological outcome.

Acknowledgments

I deeply thank two anonymous reviewers. Their valuable reviews have helped improve the contents and arguments of this chapter. The draft was presented in the Taiwan Conference on Scientific Individuation in the Physical, Biological and Experimental Sciences at National Chung Cheng University in 2014. Many thanks go out to Yi-Wei Liu, Alexandre Guay, Otávio Bueno, Melinda Bonnie Fagan, and other members of the audience for their valuable comments. This research is funded by Taiwan's Ministry of Science and Technology (MOST 105-2410-H-194-072-MY3).

References

Anderson, M. H., J. R. Ensher, M. R. Matthews, C. E. Wieman, E. A. Cornell. 1995. Observation of Bose-Einstein condensation in a diluted atomic vapor. *Science* 269 (July): 198–201.

Bouchard, Frédéric, and Philippe Huneman. 2013. *From groups to individuals: Evolution and emerging individuality*. Cambridge, MA: MIT Press.

Calcott, Brett, and Kim Sterelny. 2011. *The major transitions in evolution, revised*. Cambridge, MA: MIT Press.
Chauvier, Stéphane. 2016. Why individuality matters. In *Individuals across the sciences*, eds. Alexandre Guay and Thomas Pradeu, 25–45. New York: Oxford University Press.
Chen, Ruey-Lin. 2016. The experimental realization of individuality. *Individuals across the sciences*, eds. Alexandre Guay and Thomas Pradeu, 348–370. New York: Oxford University Press.
Cho, Adrian. 2003. Ultracold atoms spark a hot race. *Science* 301: 750–752.
Clarke, Ellen. 2010. The problem of biological individuality. *Biological Theory* 5(4): 312–325.
Clarke, Ellen. 2013. The multiple realizability of biological individuals. *The Journal of Philosophy* 110(8): 413–435.
Clarke, Ellen, and Samir Okasha. 2013. Species and organisms: What are the problems? In *From groups to individuals: Evolution and emerging individuality*, eds. Frédéric Bouchard and Philippe Huneman, 55–75. Cambridge, MA: MIT Press.
Collins, Graham. 2003. The next big chill. *Scientific American* (December): 26.
Cornell, Eric, and Carl Wieman. 1998. The Bose-Einstein condensate. *Scientific American* (March): 40–45.
Darden, Lindley. 2002. Strategies for discovering mechanisms: Schema instantiation, modular subassembly, forward/backward chaining. *Philosophy of Science* 69 (Supplement): S354–S365.
Dupré, John. 2010. The polygenomic organism. In *Nature after the genome*, eds. S. Parry and John Dupré, 19–31. Oxford: Blackwell Press.
Dupré, John. 2012. *Processes of life: Essays in the philosophy of biology*. Oxford: Oxford University Press.
Dupré, John, and Maureen O'Malley. 2007. Mategenomics and biological ontology. *Studies in the History and Philosophy of the Biological and Biomedical Science* 38: 834–846.
Collected in John Dupré. 2012. *Processes of life: Essays in the philosophy of biology*, 188–205. Oxford: Oxford University Press.
Dupré, John, and Maureen O'Malley. 2009. Varieties of living things: Life at the intersection of lineage and metabolism. *Philosophy and Theory in Biology* 1: 1–25.
Collected in John Dupré. 2012. *Processes of life: Essays in the philosophy of biology*, 206–229. Oxford: Oxford University Press.
Fagan, Melinda B. 2012. The joint account of mechanistic explanation. *Philosophy of Science* 79(4): 448–472.
Fagan, Melinda B. 2016. Cell and body: Individuals in stem cell biology. In *Individuals across the sciences*, eds. Alexandre Guay and Thomas Pradeu, 122–143. New York: Oxford University Press.
Fagan, Melinda B. Individuality, organisms, and cell differentiation. This volume.
French, Steven. 1989. Identity and individuality in classical and quantum physics. *Australasian Journal of Philosophy* 67(4): 432–446.
French, Steven. 2014. *The structure of the world: Metaphysics and representation*. Oxford: Oxford University Press.
French, Steven, and Décio Krause. 2006. *Identity in physics: A historical, philosophical and formal analysis*. Oxford: Oxford University Press.

French, Steven, and Michael Redhead. 1988. Quantum physics and the identity of indiscernibles. *British Journal for the Philosophy of Science* 39(2): 233–246.

Glennan, Stuart. 1996. Mechanism and the nature of causation. *Erkenntnis* 44: 49–71.

Glennan, Stuart. 2002. Rethinking mechanistic explanation. *Philosophy of Science* 69(Supplement): S342–S353.

Godfrey-Smith, Peter. 2009. *Darwinian populations and natural selection*. New York: Oxford University Press.

Godfrey-Smith, Peter. 2013. Darwinian individuals. In *From groups to individuals: Evolution and emerging individuality*, eds. Frédéric Bouchard and Philippe Huneman, 17–36. Cambridge, MA: MIT Press.

Gracia, Jorge J. E. 1999. Individuation. In *Cambridge dictionary of philosophy*, ed. Robert Audi, 424–425. Cambridge, UK: Cambridge University Press.

Greiner, Markus, Cindy A. Regal, and Deborah S. Jin. 2003. Emergence of a molecular Bose–Einstein condensate from a Fermi gas. *Nature* 426(December): 537–540.

Griesemer, James. 2000. The units of evolutionary transition. *Selection* 1: 67–80.

Guay, Alexandre, and Thomas Pradeu. 2016*a*. *Individuals across the sciences*. New York: Oxford University Press.

Guay, Alexandre, and Thomas Pradeu. 2016*b*. To be continued: The genidentity of physical and biological processes. In *Individuals across the sciences*, eds. Alexandre Guay and Thomas Pradeu, 317–347. New York: Oxford University Press.

Jochim, S., M. Bartenstein, A. Altmeyer, G. Hendl, S. Riedl, C. Chin, J. Hecker Denschlag, R. Grimm*. 2003. Bose-Einstein condensation of molecules. *Science* 302: 2101–2103.

Kinast, J., S. L. Hemmer, M. E. Gehm, A. Turlapov, and J. E. Thomas. 2004. Evidence for superfluidity in a resonantly interacting Fermi gas. *Physical Review Letters* 92(15): 150402-1-4.

Kovaka, K. 2015. Biological individuality and scientific practice. *Philosophy of Science* 82:1092–1103.

Ladyman, James. 2016. Are there individuals in physics, and if so, what are they? *Individuals across the sciences*, eds. Alexandre Guay and Thomas Pradeu, 193–206. New York: Oxford University Press.

Ladyman, James, and Tomasz Bigaj. 2010. The principle of the identity of indiscernibles and quantum mechanics. *Philosophy of Science* 77: 117–136.

Ladyman, James, and Don Ross. 2007. *Every thing must go: Metaphysics naturalized*. Oxford: Oxford University Press.

Levy, Arnon. 2013. Three kinds of new mechanism. *Biology and Philosophy* 28: 99–114.

Lowe, E. Jonathan. 2016. Non-individuals. In *Individuals across the sciences*, eds. Alexandre Guay and Thomas Pradeu, 49–60. New York: Oxford University Press.

Machamer, Peter, Lindley Darden, and Carl Craver. 2000. Thinking about mechanisms. *Philosophy of Science* 67(1): 1–25.

Maynard-Smith, John, and Eörs Szathmáry. 1985. *The major transitions in evolution*. Oxford: Oxford University Press.

Michod, Richard E. 1999. *Darwinian dynamics: Evolutionary transitions in fitness and individuality*. Princeton: Princeton University Press.

Mossio, Matteo, and Alvaro Moreno. 2010. Organisational closure in biological organisms. *History and Philosophy of the Life Sciences* 32: 269–288.

Morganti, Matteo. 2009. Inherent properties and statistics with individual particles in quantum mechanics. *Studies in the History and Philosophy of Modern Physics* 40: 223–231.

Morganti, Matteo. 2013. *Combining science and metaphysics. Contemporary physics, conceptual revision and common sense*. Houndmills, Basingstoke: Palgrave Macmillan.

O'Hara, K. M., S. L. Hemmer, M. E. Gehm, S. R. Granade, J. E. Thomas*. 2002. Observation of a strongly interacting degenerate Fermi gas of atoms. *Science* 298: 2179–2182.

Pitaevskii, Lev, and Sandro Stringari. 2002. The quest for superfluidity in Fermi gases. *Science* 298: 2144–2146.

Pradeu, Thomas. 2010. What is an organism? An immunological answer. *History and Philosophy of Life Sciences* 32(2–3): 247–267.

Pradeu, Thomas. 2012. *The limits of the self: Immunology and biological identity*. Oxford: Oxford University Press.

Pradeu, Thomas. 2013. Immunity and the emergence of individuality. In *From groups to individuals: Evolution and emerging individuality*, eds. Frédéric Bouchard and Philippe Huneman, 77–96. Cambridge, MA: MIT Press.

Regal, C. A., M. Greiner, and D. S. Jin. 2004. Observation of resonance condensation of fermionic atom pairs. *Physical Review Letters* 92(4): 040403-1-4.

Saunders, Simon. 2006. Are quantum particles objects? *Analysis* 66: 52–63.

Saunders, Simon. 2016. On the emergence of individuals in physics. In *Individuals across the sciences*, eds. Alexandre Guay and Thomas Pradeu, 165–192. New York: Oxford University Press.

van Inwagen, Peter. 1995. *Material beings*. Ithaca: Cornell University Press.

Wilson, Jack. 1999. *Biological individuality: The identity and persistence of living entities*. Cambridge: Cambridge University Press.

Woodward, James. 2002. What is a mechanism? A counterfactual account. *Philosophy of Science* 69(Supplement): S366–S377.

Zwierlein, M. W., C. A. Stan, C. H. Schunck, S. M. F. Raupach, S. Gupta, Z. Hadzibabic, and W. Ketterle*. 2003. Observation of Bose-Einstein condensation of molecules. *Physical Review Letters* 91(25): 250401-1-4.

{ 10 }

Emergent Quasiparticles

OR, HOW TO GET A RICH PHYSICS FROM A SOBER METAPHYSICS

Alexandre Guay and Olivier Sartenaer

10.1 Introduction

Among the very architects of the recent reemergence of emergentism in the physical sciences, Robert B. Laughlin certainly occupies a prominent place. Through a series of works beginning as early as his Nobel lecture in 1998, a lecture given after having been awarded, together with Störmer and Tsui, the Nobel prize in physics for its contribution in the elucidation of the fractional quantum Hall effect (FQHE), Laughlin openly and relentlessly advocated a strongly anti-reductionistic view of physics—and, more particularly, of the interface between condensed matter and particles physics—which culminated in what can be considered his emergentist manifesto: *A Different Universe. Reinventing Physics from the Bottom Down* (2005). In spite of this prominent role in the vindication of emergentism, rare are the philosophers who have paid serious attention to Laughlin's insights, even among those sympathetic to the idea of emergence. The subtleties of his view—it is true, often concealed in many technicalities—have accordingly, and somewhat unfortunately, mainly passed unnoticed.[1]

The starting point of this chapter is a willingness to somehow remedy this situation by taking Laughlin's emergentism seriously. More specifically, reflecting on Laughlin's insight, according to which "one of the things an emergent phenomenon can do is create new particles" (1999: 863), we propose an exploration of the way in which emergence can shed light on the ontological status of some quantum entities—more particularly, so-called quasiparticles—that would come into being on the occasion of certain physical transformations. On the occasion of such an

---

[1] Two recent notable exceptions are Lancaster and Pexton (2015) and Lederer (2015).

exploration, we will be in a position to claim that a precise and metaphysically kosher way of capturing Laughlin's insightful emergentism exists, one that allows for vindicating the existence of a rich physical world—made of diverse, ontologically irreducible, yet connected physical domains—within the boundaries of a metaphysically sober worldview. Incidentally, we will show that a natural interpretation of quantum electrodynamics turns out to be, perhaps surprisingly, rather hospitable to (a version of) ontological emergence, in favor of the nonreductive physicalists' greed, *pace* Kim (2005: 22), for "having the cake and eating it too."

Here is our battle plan. In Section 10.2, we open the discussion by unpacking Laughlin's intuition that pertains to the emergence of quasiparticles. More particularly, after having explicated what we think is the philosophical core of Laughlin's emergentism (Subsection 10.2.1), we describe what constitutes his most convincing and well-articulated exemplification of an emergent phenomenon—the one that has actually driven him toward emergentism in the first place—namely the FQHE (Subsection 10.2.2), on the occasion of which anyons are claimed to emerge. In Section 10.3, we turn to an exposition of one of the rare, explicit philosophical reconstructions of Laughlin's emergentism available on the current philosophical market, namely Carl Gillett's "mutualism," and contend that such a reconstruction, though perhaps successful, involves an unnecessary excess of metaphysics. In support of such a claim, we show in Section 10.4 that a metaphysically cheaper alternative way of understanding emergence, embedded in Paul Humphreys's recent "transformational" framework, is available and better captures Laughlin's emergentist insights. Finally, in Section 10.5, we build on our previous reflexion in order to explicate, and subsequently evaluate the plausibility of, the claim according to which anyons are to be considered as emergent quasiparticles.

10.2 Laughlin on Emergent Quasiparticles

10.2.1 LAUGHLIN ON EMERGENCE

In a nutshell, there is emergence in Laughlin's view as soon as physical entities are organized together in a very specific "emergence-engendering" way, to the effect that the transition between these entities—the "parts"—and the specifically organized collective configuration they give rise to—the "whole"—is governed by new physical "principles of organization" (hereafter [POS]; see, e.g., Laughlin 2005: xiv). Because these [POS] are at the core of Laughlin's emergentism, let us proceed to unpack them.

First, [POS] are claimed to be "physical" in the *via negativa* sense that they are not tied to anything—forces, powers, or entities—that one would be ready to qualify as nonphysical (e.g., mental or social). In spite of the fact that, as we will explicate later on, [POS] "transcend" the laws and rules of microphysics—microphysics being understood here as the physics that pertains to the "ultimate parts," viz. the most basic constituents of natural things, whatever they are—[POS] still "grow out of the microscopic rules" (Laughlin et al. 2000: 29–32), thereby guaranteeing some minimal nomological continuity or self-sufficiency for the physical realm broadly construed. It is in this last sense that Laughlin's emergentism rejects both the view that physics simply reduces to a set of "master rules" and the opposite view, according to which it is rather an "open frontier" (Laughlin 2005: 6).

[POS] are also inherently "organizational," to the extent that they appear upon the organization of entities into systems of entities and regulate the passage from the former to the latter. Accordingly, Laughlin's emergence primarily has a collective, holistic, or hierarchical flavor in the way that it is mainly about a transition between "levels"—the (lower or microphysical) level of the parts and the (higher or macrophysical) level of the whole—in the spirit of the traditional slogans: "The whole is more than the sum of its parts" or "More is different" (Laughlin and Pines 2000: 29).

Finally, [POS] are "novel." As with any emergentist view, dealing with the crucial tenet of novelty turns out to be very delicate. In order not to let too much ambiguity bear on this tenet, it can prove helpful to first stress the possible senses in which Laughlin's [POS] are *not* (necessarily) supposed to be novel. To begin with, the type of novelty considered here is to be contrasted with what can be referred to as *degenerate* novelty, that is to say, [POS] are not merely *different* from other physical principles or rules that govern the evolution of emergent systems or their composing parts (and, ultimately, their most basic constituents). Second, and more interestingly, the novelty of [POS] does *not* also reduce to mere *epistemic* novelty (i.e., unexpectedness relative to a given state of knowledge or cognitive/technological access at a given moment in time). Third and finally—let's call this *cosmological* novelty—Laughlin's [POS] are not (necessarily) novel in the sense that their advent and action is unprecedented in the history of the whole evolving universe.

Actually, it turns out that [POS] are novel in a sense that is at the same time more radical than mere degenerate or epistemic novelty and that doesn't have the unique and contingent character of cosmological novelty. It essentially amounts to the idea—let's refer to it as *ontological* novelty—that [POS] are *sui generis* or fundamental, in the sense that they appear as brute empirical laws that aren't already in existence, even potentially or implicitly, outside the specific empirical context that gave rise to them.[2] Correlatively, their appearance and action could not have

[2] That Laughlin actually endorses this last point is, of course, a matter of exegesis. According to us, it is the only way to make sense of the radical epistemic cutoff that Laughlin's emergence inexorably entails (see later discussion). Furthermore, ontological novelty seems a reasonable interpretation of expressions like: emergent states are "*fundamentally* different" from basal states (Laughlin 1999: 864, emphasis added), emergent principles are "transcendent," and their existence shows us "that for at least some *fundamental* things in nature the Theory of Everything is irrelevant" (Laughlin and Pines 2000: 28–29, emphasis added), or "collective organizing principles . . . are *in a real sense* independent of them [the microscopic rules]" (Laughlin et al. 2000: 32, emphasis added).

been predicted theoretically before the appropriate experimental configuration was reached (spontaneously or not), nor could they have been deduced from "first principles"; namely and ultimately, the laws of microphysics (Laughlin and Pines 2000: 30).

Now, it seems to be the case that, in Laughlin's implicit ontology where laws or principles govern the behavior of entities, the emergent status of [POS] entails the emergence of entities that [POS] are about, along the exact same lines. Accordingly, upon emergence, new entities arise that are (1) physical (for they are made up of the physical entities that constitute their emergence basis), (2) high-level (for they only exist at the level of an organized collection of their basal entities), and (3) ontologically new (for they are as fundamental as their ultimate basal entities are). This being said, there is a similar sense in which, in Laughlin's view, properties, processes, behaviors, or forces can also be said to be emergent as soon as some [POS] are at play in the relevant way.

Elizabeth Barnes recently proposed a way to characterize emergence that nicely fits with Laughlin's implicit construal of the concept, and this can accordingly be used to put some philosophical flesh on Laughlin's emergentist bone. According to Barnes (2012), emergents are entities that are fundamental and yet dependent. As we will see later on, emergent entities like anyons are indeed to be conceived of, in Laughlin's view, as fundamental as electrons—both are parts of the building blocks of reality—and yet the former depend on the latter (and not the other way around, a fact that justifies the idea that anyons are "higher level" than electrons).[3] It should be clear from this that, as it is the case in Barnes's account, Laughlin's emergence is primarily an ontological matter—emergence ascriptions are to be taken realistically as saying something about the way the world is—even if, concomitantly, they are systematically accompanied by principled epistemic effects like unpredictability or nondeducibility from "first principles."

10.2.2 EMERGENCE PAR EXCELLENCE: THE FRACTIONAL QUANTUM HALL EFFECT

Among the phenomena that Laughlin is ready to call "emergent" in the sense just described, the FQHE certainly occupies a prominent place. If we take seriously Laughlin's claim that the discovery of this effect was the "opening movement of the age of emergence" (Laughlin 2005: 76), then it is rather natural to focus on this

[3] Should it turn out that electrons themselves are not among the basic building blocks of reality, this would not modify Laughlin's view on emergence in any significant way. What he would probably say in this context is that electrons are themselves to be considered as emergent on the whatever-fundamental-entity-there-is-below in the exact same sense as anyons are to be considered emergent on electrons.

peculiar quantum phenomenon in order to provide some empirical concreteness to Laughlin's emergence ascriptions.[4]

To understand this effect and, more importantly, its idiosyncrasies, it is worthwhile to briefly describe its underlying physics.[5] To begin with, the Hall effect was first discovered by Edwin H. Hall in 1879. In an experimental set-up consisting of a conductor lying in a magnetic field and in which flows an electric current orthogonal to the field, the effect consists in the advent of a Hall potential V_H and a corresponding Hall resistance R_H perpendicular to both the magnetic field and the current. Such a phenomenon is due to the accumulation of the electrons on one side of the conductor, given the fact that the electrons are deflected by the magnetic field. In this purely classical effect, the Hall resistance is proportional to the intensity of the magnetic field and inversely proportional to the electronic density (as can be seen through the straight dotted line in Figure 10.2).

In 1980, Klaus von Klitzing discovered that the Hall resistance doesn't actually vary in linear fashion with the intensity of the field, but rather exhibits plateaus as a function of the field (see Figure 10.1). Furthermore, these plateaus occur at some values of the Hall resistance that are insensitive to the nature of the material involved in the experimental set-up but are a function of some constants—namely, the electric charge e and Planck constant h—divided by an integer (called the *filling factor*; see later discussion). This phenomenon, known as the *integer quantum Hall effect* (IQHE), can be explained through the machinery of the quantization of cyclotron orbits of electrons in magnetic fields (called the *Landau quantization*). Three facts are relevant here for providing such an explanation. First, the energy levels (called the *Landau levels*) are discrete and equidistant, with a spacing proportional to the intensity of the field. Second, these levels are highly degenerate (that is, a lot of electrons can occupy each level) and such a degeneracy is also itself proportional to the intensity of the magnetic field. Third, electrons are fermions. Accordingly, following the exclusion principle, they cannot occupy the same states as other electrons. It follows from these facts that, starting from an experimental situation where a given Landau level is completely filled by electrons (let's say, with a field of approximately $5T$, when the third Landau level is fully occupied; see Figure 10.1), decreasing the intensity of the field, and consequently decreasing the amount of available states in each Landau levels, will lead electrons, of which there is a fixed density, to begin to fill the next Landau

[4] Especially in his 2005 book *A Different Universe*, Laughlin generalizes his emergentism to other domains of physics and, by extension, of science, to the effect that his emergence becomes rather mundane, if not almost universal. By sticking to the specific case of the FQHE, we don't want to commit ourselves to such generalizations. We indeed think that some idiosyncrasies of the FQHE are responsible for it being emergent in an interesting sense (see Section 10.4.2), but we don't know whether other phenomena, especially in less formalized sciences like biology or psychology, share enough of these idiosyncrasies to also be plausibly qualified as emergent.

[5] For a more complete treatment and some philosophical insights, see Lederer (2015).

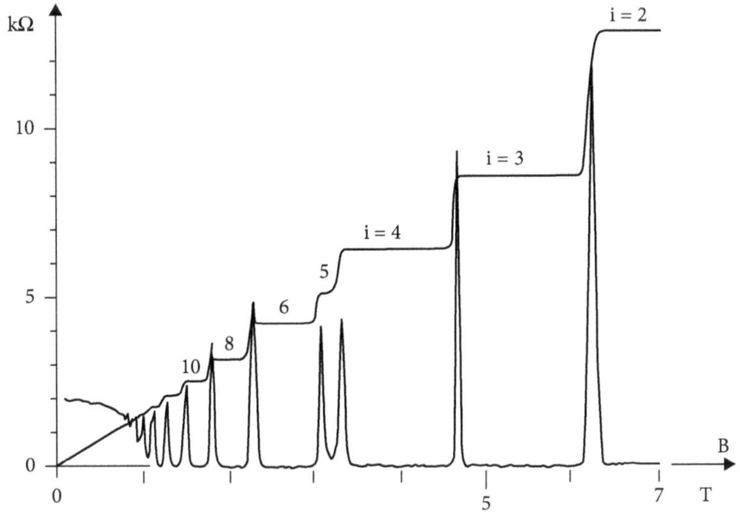

FIGURE 10.1 *The integer quantum Hall effect. The hall resistivity shows different plateaus as a function of the magnetic field. The Ohmic resistance, represented by the lower curve, drops to zero for each plateau.*

From the 1998 "Nobel Press Release" (October 22, 2015). http://www.nobelprize.org

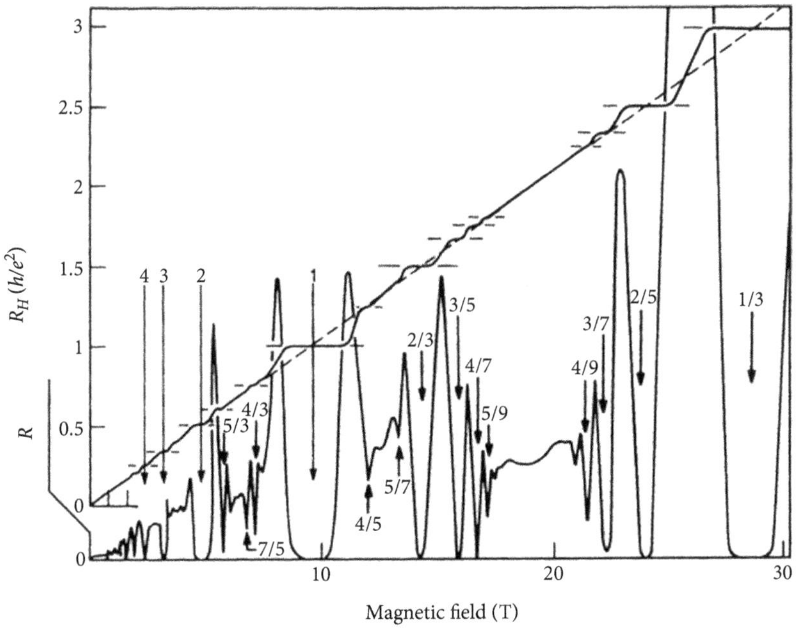

FIGURE 10.2 *The fractional quantum Hall effect. The straight dotted line captures the dependence relation between the Hall resistance and the field in the classical limit. Beside plateaus of integer filling factors (for values of the field below ±10T), plateaus with fractional filling factors can be seen.*

From the 1998 "Nobel Press Release" (October 22, 2015). http://www.nobelprize.org

level (here the fourth), leading to a transition to a next plateau of Hall resistance. Conversely, by raising the intensity of the field and, concomitantly, the number of possible occupation states in each Landau level, there will be a point (here around ±6*T*) when the lower Landau level (here the second) will have enough states available to house all the electrons, hence leading to a transition to an adjacent plateau of Hall resistance. It is of common use to ascribe each Landau level, and correspondingly each of the Hall plateaus, with a filling factor ν that can only take integer values. Given this, the IQHE is an indirect way of making manifest a purely quantum phenomenon, viz. the quantized nature of electronic orbitals in a magnetic field.

In 1982, through experiments involving lower temperatures and a stronger magnetic field, Störmer, Tsui, and their co-workers discovered other steps in the Hall resistance than those involved in the IQHE. In particular, these new steps can also be expressed through the same constants (*e* and *h*), but it turns out that—in sharp contrast with the IQHE—they occur at values of the filling factor that are fractions over an odd integer (e.g., 1/3, 2/5, 3/7, 2/3, etc.; see Figure 10.2). To account for such an effect, naturally referred to as the "fractional" quantum Hall effect, the Landau machinery proves inadequate. A new approach had to be envisioned that would take into account something that had so far been neglected; namely, the electron–electron interactions inside the experimental set-up. It was in this context that Laughlin made his insightful proposal that led him to be awarded the 1998 Nobel prize in physics.

It is not the place here to enter into the details of Laughlin's approach, which involves many technicalities that turn out to be irrelevant for our present purpose. Moreover, we will lay down some aspects of the theoretical underpinnings of Laughlin's work in a slightly different perspective (namely, a quantum field theoretical perspective rather than a nonrelativistic quantum mechanical perspective) in due time (see Section 10.4.2). For the time being, let us content ourselves with stressing Laughlin's main insight: the fractional occupation of the Landau levels corresponds to fractionally charged excitations obeying fractional statistics, which correspond, at least according to Laughlin himself, to new (types of) particles, namely *quasi*particles (famously referred to by Wilczek as "anyons"). These new particles are "the elementary excitations of a distinct state of matter" (Laughlin 1999: 863), called a *topological state of matter*, and exert a new type of force, namely "a long-range velocity-dependent force . . . a gauge force—which is unique in the physics literature in having neither a progenitor in the underlying equations of motion nor an associated continuous broken symmetry" (Laughlin 1999: 871).

It is noteworthy that what makes the FQHE a case of emergence, in contrast with the IQHE, is the fact that, in the case of the former but not (presumably) of the latter, electron–electron interactions occur, leaving room for the advent of collective effects of the electron gas, viz. anyons. What matters for us here is that anyons are to be considered emergent in Laughlin's sense (Laughlin and Pines

2000: 29) for they satisfy the three basic ingredients identified earlier: namely, dependence (anyons are sustained by underlying electrons), hierarchy (anyons are in some sense collective effects of electrons), and fundamentality or ontological novelty (anyons are fundamental particles in their own right, not simply "second-class electrons").[6] This last point is worth emphasizing, for it constitutes the core of Laughlin's intuition about emergent quasiparticles: namely, that a physical process involving emergence is able to create new particles out of "old" ones. Such an intuition is doubly vindicated in Laughlin's work. First, from an empirical point of view, anyons in the experimental set-up and electrons in the vacuum are on a par, for they behave in qualitatively similar ways: "They carry momentum, energy, spin, and charge, scatter off one another according to simple rules, obey Fermi or Bose statistics depending on their nature" (Laughlin and Pines 2000: 29). Second, they are also theoretically indistinguishable, in the sense that "there is no logical way to distinguish a real particle from an excited state of the system that behaves like one" (Laughlin 1999: 863). This being said, the only thing that could significantly differentiate anyons from electrons in terms of ontological status is that the conditions of existence of the former are more restrictive than the ones of the latter—hence the former and not the latter are called *quasi*particles.[7]

Considering anyons as emergent quasiparticles is philosophically significant on three different levels. First, from the *ontological* point of view we've just stressed, it makes the FQHE the place for the advent of new entities that are to be considered as fundamental or ultimate as the other fundamental entities they are emerging from.[8] Put differently, with the FQHE, a new physical domain is made accessible. Second, as an *epistemological* correlate of this, this new physical domain is representationally broken off the "old" one, in the sense that the epistemic resources of the latter are irrelevant or inadequate to account for the former. Accordingly, as

[6] The corresponding [PO] that is responsible for such an emergence is "Anderson's localization" (Laughlin 1999; see also Anderson 1958), which governs the fact that electrons removed from or added to a filled Landau level are localized by the impurities in the conductor and hence do not contribute to electric transport (a fact that accounts for the constant conductivity that is typical of the Hall plateaus). As an anonymous referee rightly pointed out to us, considering localization as the only relevant [PO] at stake here cannot possibly be the whole story for getting at emergent anyons, notably because the very same principle is also supposed to explain the IQHE. Complementing Laughlin's proposal in this respect is beyond the scope of this chapter, especially given the fact that the reconstruction of Laughlin's insights that we propose in Section 10.4 doesn't need the invocation of any [PO].

[7] Gelfert (2003) considers quasiparticles as unreal because, according to his intuition of "entity realism," derivative or composite things cannot be real. Of course, such an intuition merely amounts to a denial of the very possibility of ontological emergence, something that can prove to be empirically falsified, as we claim it is in Section 10.4.2. It is also somewhat ironic that the Nobel press release relative to Laughlin's shared prize refers to anyons as *not* being particles but entities that merely *result from* a "dance of electrons."

[8] It is noteworthy that Laughlin's implicit construal of fundamentality certainly doesn't amount to something like "being noncomposite." A plausible interpretation is then to consider Laughlin's fundamentality as meaning something like "being subject to fundamental laws," such laws being those deemed fundamental by practicing physicists.

a case of Laughlin's emergence, the FQHE marks a dividing line between particles physics and condensed matter physics. This obviously bears on the debates relating to the possible unity of science. In Laughlin's own words, rather than a "Theory of Everything" from which every phenomenon could be deduced—including the advent of anyons—science is made of "a hierarchy of theories of things" that are epistemologically autonomous from one another (Laughlin and Pines 2000: 30).[9] In addition to these ontological and epistemological stakes, Laughlin's emergence also suggests a *methodological* prescription for physics. Pure theoretical physics, characterized by a high weight given to deductive knowledge, should be abandoned in favor of a form of experimentalism that grants preference to inductive knowledge. According to Laughlin, "the central task of theoretical physics in our time is no longer to write down the ultimate equations but rather to catalogue and understand emergent behavior in its many guises" (Laughlin and Pines 2000: 30).

Now that what constitutes the core of Laughlin's implicit emergentism has been described and exemplified, with a special focus on the ontological status of emergent quasiparticles, let us turn to one of the rare philosophical reconstructions of Laughlin's emergence that has been attempted in the recent literature pertaining to the metaphysics of emergence.

10.3 Gillett's Philosophical Reconstruction

In 2010, Carl Gillett put forward a complete reconstruction of the metaphysical underpinnings of a view he himself dubbed "scientific emergentism," of which he took scientists like Anderson and Laughlin to be the prominent figures. According to Gillett, such an emergentism is committed to a strong version of emergence that is incompatible with "compositional reductionism," a view according to which, when it comes to part-whole configurations, high-level property instances are causally inert, all of their potency being systematically and completely preempted by their lower level, realizer property instances.

Gillett's main contention is that scientific emergentists have recently provided us with some tools to oppose such a reductionism without having to embrace a dualistic stance by pointing toward a "missed option" in the way parts and whole can come to be related in some empirical situations. Such an option, which has been overlooked by philosophers in the recent debates, can be delineated on the basis of two interconnected ingredients: namely, a new view of aggregation—the "conditioned view"—and a new determinative relationship—the so-called *machresis*.

[9] In Laughlin's layered view, this brings about autonomous theories at different "levels." In the perspective that we embrace in Section 10.4, such a picture will be reconsidered. The possible irreducibility of theories will be construed in a diachronic sense, with respect to which the theories at play are to be considered as being about domains of fundamentality that are temporally, rather than hierarchically, separated.

Unpacking these two notions requires some preliminaries, among which are (any variant of) the causal theory of properties together with a notion of natural hierarchy built in terms of levels of mechanism. On such a basis, one can capture the relationship between higher level and lower level property instances through Gillett's own notion of "dimensioned realization," on the following model (Gillett 2010: 29, with slightly modified notations): a set of property instances *B* realizes an instance of a property *X*, in an individual *s* under condition $, if and only if *s* has powers that are individuative of an instance of *X* in virtue of the powers contributed, under $, by *B* to *s* or *s*'s constituents, but not vice versa.

This being said, the conditioned view of aggregation, in contrast with the "simple view" that inexorably leads to compositional reductionism, allows that lower level realizer property instances only contribute certain powers on the condition they are realizing a certain higher level property instance. What is relevant here for emergence is that this view of aggregation, unlike the simple one, allows for higher level properties, though realized, to make a genuine difference in the world, for it allows for the existence of a peculiar species of downward determination—the so-called *machresis*—holding between the higher level realized property instances and their lower level realizer property instances. *Machresis*, which is distinct from the more usual, dualistically inclined determinative relation that is downward causation, is a relation through which a realized property instance can be efficacious by determining synchronically, noncausally and noncompositionally, the contributions of powers by property instances in its own realizers (and not by contributing powers itself).

According to Gillett, this "mutualist" determinative dynamics (see Figure 10.3) adequately captures the metaphysical picture that scientists like Laughlin have implicitly in mind when making emergence ascriptions. Should this be true, considering the emergence of anyons in the peculiar case of the FQHE would lead Laughlin to be committed to at least these three facts: (1) anyons (or their

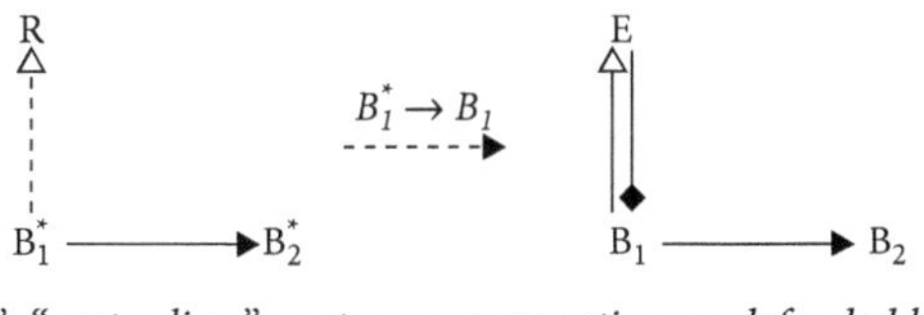

FIGURE 10.3 *Gillett's "mutualism" or strong emergentism as defended by scientific emergentists like Laughlin. Left: The dynamics of the simple view of aggregation. A realized property instance R is merely resultant from, or compositionally reducible to, its basis* B_1^*, *insofar as its defining powers are preempted by the powers contributed by* B_1^*. B_1^* *causes* B_2^* *at* t_2 *regardless of it realizing R at* t_1. *Right: The dynamics of the conditioned view, which is reached from that of the simple view as soon as the relevant basis* B_1 *obtains. In this case, the higher level realized property instance E is strongly emergent for it machretically determines the powers contributed by* B_1. *Put differently,* B_1 *would never cause* B_2 *at* t_2 *if not realizing E at* t_1.

properties) are realized by an underlying, simultaneous basis (probably made of electrons or their properties), to the effect that there is a sense in which anyons are higher level than such a basis (in a mechanistic sense of level); (2) realizing anyons (or their properties) is the appropriate conditions for the basis to contribute new powers that would never have appeared otherwise; and (3) anyons are then causally efficacious by machretically determining the powers contributed by their underlying realization basis. It is noteworthy that, together, these three facts cover the three ingredients that are defining of Laughlin's emergence as presented in Section 10.2: namely, dependence (here cashed out in terms of realization), hierarchy (here cashed out in terms of levels of mechanism), and ontological novelty (here cashed out in terms of higher level causal efficacy).

This being said, it is important to note that applying Gillett's reconstruction of Laughlin's emergentism to the case of the FQHE is to be done with some caution. Indeed, Gillett's analysis turns out to be mainly inspired from emergence ascriptions that are tied to a different [PO] than the one at play in the FQHE: namely, spontaneous symmetry breaking (instead of localization; see, e.g., Gillett 2010: 32), and nothing a priori requires that the determinative dynamics at play in these different cases should be the same. This leaves us with three possible options. First, Gillett's reconstruction does apply *mutatis mutandis* to the case of the FQHE, as it is suggested by Gillett himself (2016: 346, n14).[10] Second, Gillett's reconstruction does *not* apply to the FQHE, and therefore Gillett is mistaken in arguing that his mutualism captures the implicit thought of scientific emergentists like Laughlin (remember that, according to Laughlin, the FQHE is the paradigmatic example of an emergent phenomenon). Third, Gillett's reconstruction does *not* apply to the FQHE, but it applies adequately to other emergence ascriptions made by Laughlin (like the one relative to, say, superconductivity), to the effect that Gillett's analysis is merely incomplete. Laughlin's emergentism turns out to be richer than it first seemed, in the sense that it allows a single template to accommodate very different kinds of emergence.

Which one of these options is really the case doesn't actually matter to us, for knowing exactly what Gillett and Laughlin have in mind is irrelevant to what we want to claim in what follows: namely, that it is possible to recover Laughlin's important insights that pertain to the idea of emergent quasiparticles without the metaphysical inflation put forward by Gillett. To put it more bluntly, we contend that it is possible to get Laughlin's rich physics without Gillett's convoluted metaphysics.

[10] A possible way to accomplish this extension is to adopt a model of the FQHE where symmetry breaking plays a role. In this class of models, the FQHE is explained by the fact that a gas of composite-bosons condenses spontaneously (Rao 2001).

10.4 A Metaphysically Cheaper Alternative

In this section, we concisely describe an alternative account of emergence, called "transformational emergence," which we have developed more extensively elsewhere (Guay and Sartenaer 2016) on the basis of an idea first proposed by Paul Humphreys (2016). By exemplifying such an emergence in the case of the FQHE, we claim to recover Laughlin's strong intuition about emergent quasiparticles without having to embrace Gillett's heavy metaphysical machinery. In the next section, we draw some conclusions with respect to the idea of emergent particles.

10.4.1 ON TRANSFORMATIONAL EMERGENCE

As with any other account of emergence, transformational emergence [TE] is devised in order to make sense of two ideas that are *prima facie* in tension: namely, that a putatively emergent entity *depends on* its emergence basis and that, at the same time, the former is *novel* with regard to the latter. More precisely, considering the case of a physical system S at two successive stages of its evolution (S_1 at t_1; S_2 at t_2), one will say that S_2 transformationally emerges from S_1 when the ideas of dependence and novelty are met in the following sense:

- (DEP) S_2 is the product of a spatiotemporally continuous process going from S_1 (e.g., causal, and possibly fully deterministic). In particular, the "realm" R to which S_1 and S_2 commonly belong (e.g., the physical realm) is closed, to the effect that nothing outside of R participates in S_1 bringing about S_2. And yet:
- (NOV) S_2 exhibits new entities, properties, or powers that do not exist in S_1 and that are furthermore *forbidden* to exist in S_1 according to the laws governing S_1. Accordingly, different laws govern S_2.

It is noteworthy that the "forbidden" clause in (NOV) introduces a modality aspect to [TE], according to which, upon emergence, a physical domain that was previously barred—including entities and their properties, subject to specific laws—becomes accessible. Put differently, a [TE] ascription is essentially a modal claim. It is about impossible phenomena (according to the pre-emergence laws) that, upon emergence, become possible and even actual (due to the advent of new laws that reconfigure the space of possibilities).[11]

[TE] shares some commonalities with Gillett's emergence, among which the fact that the advent of emergent entities makes a genuine causal difference in the

[11] It should be noted that transformational emergence does not rely on a specific metaphysical conception of natural laws. However, laws have to be able to circumscribe the domain of physically possible states.

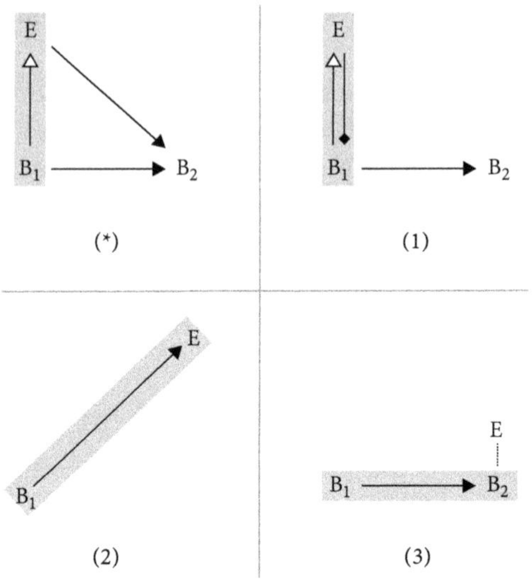

FIGURE 10.4 *Some possible ways (1–3) to define emergence without facing exclusion worries. Diagram (*) represents Kim's (1999) emergence that is subject to exclusion as it combines some synchronic upward determination with downward causation. Diagram (1) captures Gillett's (2010) strong emergence, which avoids exclusion by conceiving of downward determination as noncausal and synchronic. Diagram (2) is a representation of Humphreys's (1997) fusion emergence, which avoids exclusion by replacing synchronic upward determination by a diachronic, interlevel fusion operation. Diagram (3) captures [TE], which avoids exclusion by combining diachrony and "level egalitarianism" (i.e., E can be identical to B_2, though it is of course not necessary).*

world without breaking causal closure or entailing a rejection of physicalism.[12] What sharply distinguishes [TE] from Gillett's proposal, though, is that the former, in contrast with the latter, is a *diachronic* variety of the concept; that is, there always necessarily is a temporal delay between a putative transformational emergent and its emergence basis. Another contrast that is worth stressing is that [TE] is essentially *nonholistic* in the sense that it can perfectly tolerate a thorough egalitarianism when it comes to the idea of a discrete hierarchy of levels in nature. It is on the basis of these two distinguishing ingredients—diachrony and "level egalitarianism"—that [TE] actually turns out to be immune to exclusion-style worries (see Figure 10.4) and that the account captures, instead of the traditional

[12] For these reasons, we are inclined to consider [TE] as a form of "weakly ontological" emergence, while Gillett prefers to refer to his account as "strong emergence," "ontological emergence" being reserved for anti-physicalistic versions of the notion. This difference is only terminological.

emergentist slogans, the ideas that "*after* is different" or "the whole *is* the sum of the *transformed* parts."[13]

10.4.2 TRANSFORMATIONAL EMERGENCE AND THE FQHE

Because no one can claim to have a direct and privileged access to the ontology of natural systems, exemplifying [TE] in a concrete phenomenon, and in particular in the FQHE, requires one to identify what would be the traces that [TE] is supposed to leave in the formal constructs we use to investigate these natural systems. The two clues we are considering are the following:

- (C_1) M_1 and M_2, which both describe the same system S at two successive stages S_1 and S_2 of its evolution, are models of one and the same nontrivial theory T. And yet:
- (C_2) M_2 is not derivable from M_1 as a matter of principle, for M_2 contains features that are forbidden in M_1 according to the laws governing it. Accordingly, S_2's dynamics as described by M_2 is not continuously deformable into S_1's dynamics as described by M_1.

In a nutshell, what this operational formulation of [TE] entails is that, although transformationally emergent states are the products of their bases and can accordingly be accounted for in the light of one and the same theoretical framework, the former are epistemologically broken off the latter in the sense that there is an unbridgeable representational gap between them. Such a gap actually constitutes the best—and even also the only—evidence we could ever have to ground [TE] ascriptions.

This being said, one can make the case that there is [TE] at play in the FQHE. More particularly, we argue that the experimental production of the FQHE is an instance of a physical transformation between two states—a pre-transformation state S_1 (four-dimensional fermions) and a post-transformation state S_2 (three-dimensional particles)—to the effect that the clues (C_1) and (C_2) obtain, and, consequently, (DEP) and (NOV) hold.[14] It is not the place here to enter into the details of how to justify such a claim (for some details, see Guay and Sartenaer 2016).

[13] It is in this last respect that [TE] makes sense of an idea that Gillett's emergence also captures—though differently—namely, that "parts behave differently in wholes." For the reader who is perhaps not well-acquainted with Kim's debated "causal exclusion argument" (see, e.g., 2005: chapt. 2), "exclusion-style worries" refer to the fact that not all combinations of (DEP) and (NOV) yield acceptable forms of ontological emergence. Typically, versions of emergence that combine a construal of (DEP) in terms of supervenience with a construal of (NOV) in terms of downward causation lead to a vexing causal over-determination (when the basal level is supposed to be causally closed).

[14] Note that we do not claim that the initial state is a three-dimensional conductor in a classical regime. Contrary to the FQHE, there is no necessity to model the conductor before the effect as to be in four dimensions. In consequence, we prefer to be careful about the dimensionality of the model in the pre-transformation phase.

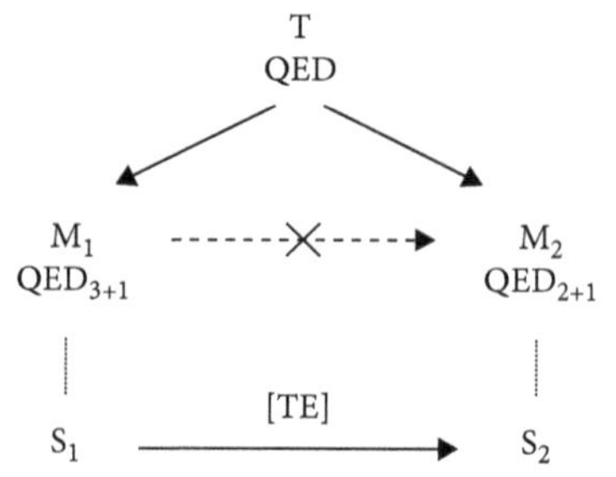

FIGURE 10.5 *[TE] in the FQHE.*

Suffice it to mention the key elements of such a justification in the following way (see also Figure 10.5):

1. It can be shown that four-dimensional quantum electrodynamics (QED_{3+1}) and three-dimensional quantum electrodynamics (QED_{2+1}) are both models of one and the same nontrivial theory, namely quantum electrodynamics (QED).
2. QED_{3+1} and QED_{2+1} are the adequate models for describing the universal properties of the dynamics of S_1 and S_2, respectively.
3. The experimental production of the FQHE can be seen as a transformation of S going from a state S_1 described by QED_{3+1} to a later state S_2 described by QED_{2+1}.[15]

As such, 1, 2, and 3 entail that (C_1) obtains to the effect that we have good reasons to believe that (DEP) holds. Already at this stage, the FQHE is very surprising for it involves the production, in our four-dimensional world, of a system that can be best described by QED_{2+1} (and, as we will see, that *cannot*, even in principle, be described by QED_{3+1}).

4. QED_{2+1} contains features that are forbidden in QED_{3+1} according to QED. In particular, the respective Lagrangian densities are analogous (*modulo* a difference in dimensionality), with the notable exception that $\mathcal{L}_{QED_{2+1}}$ contains an additional term—a so-called Chern-Simons term—that cannot exist in $\mathcal{L}_{QED_{3+1}}$. The presence of this term radically changes the accessible quantum states of matter.[16]
5. Accordingly, QED_{2+1} is nonderivable from QED_{3+1}, though both are models of one and the same general theory: namely, QED. Derivability

[15] We are not claiming anything about the actual dimensionality of the physical system. This is a claim about how to best describe the dynamics.

[16] The main new aspect is not the presence of parastatistics, since fractional statistics could in principle exist in a four-dimensional world. The real novelty is the access to new topological quantum states (Wen 2004: chapt. 8).

should be understood here as the fact that it is not possible to continuously deform one model into the other because there is no analog to the Chern-Simons term in four dimensions.

From 3 and 5, it can be seen that (C_2) obtains for the transformation involved in the FQHE.[17] Of course, what matters from a physical point of view, and what concomitantly justifies why we think thesis (NOV) also holds in this case, is the set of empirical consequences of the presence of the Chern-Simons term that captures the very distinguishing features of the dynamics of the Hall states. Some notable empirical facts, which the Chern-Simons term codes for, are that "photons" (more precisely, spin 1 excitation states of the gauge field) acquire a topological mass and that "charged particles" are endowed with magnetic fluxes and have fractional statistics. Such facts are *forbidden* to happen in a four-dimensional world described by QED_{3+1} and yet they are produced in a continuous way from a four-dimensional system adequately described by QED_{3+1}. This modal claim captures the very essence of what is *really* surprising about the FQHE: namely, to borrow the words of Laughlin himself, that the effect "was unanticipated by any theory and not analogous to anything previously known in nature. . . . [The discoverers of the effect found something] which should have been impossible" (Laughlin 2005: 77).

10.5 Transformationally Emergent Anyons?

Transformational emergence implies that the transformationally emergent state is impossible according to the very laws that govern the pre-transformation state. In a certain manner, all entities in such an emergent state are themselves emergent since they participate in an emergent state. This trivial fact is not what we are aiming at in this section. The question we will answer is, rather, do we encounter a new kind of individual in the context of the FQHE?

As early as 1984, the presence of anyons has been invoked to explain the surprising empirical behavior of the FQHE (Arovas, Schrieffer, and Wilczek 1984). Anyons are three-dimensional "particles" that exhibit very unusual statistics, interpolating between fermions and bosons (Wilczek 1982). They apparently can explain the FQHE. They are very probably no anyonic solutions in QED_{3+1} (Guay and Sartenaer 2016). In consequence, they seem to be good candidates as emergent entities.

[17] In Laughlin's own phrasing: "The fractional quantum Hall state is *not* adiabatically deformable to any noninteracting electron state" (Laughlin 1999: 869; emphasis in the original).

10.5.1 THE PROBLEM WITH QUANTUM INDIVIDUALS IN GENERAL

Quantum physics is generally considered hostile to the notion of the individual. In fact, since Schrödinger, the doxa is that if there are particles, they cannot be individuals. Since the defense of this claim, almost every position has been argued, from the fact that the metaphysics is undetermined by physics (French and Krause 2006) to the case that individuals are an emergent phenomenon in nonrelativistic quantum mechanics (Saunders 2016). In all cases, the notion of an individual used in classical physics has to be deeply revised. In this chapter, we will put aside this important debate and focus on the possibility of emergent particles (individuals or not). However, this simplification is not a panacea. The concept of "particle" is not always well-defined in quantum physics. Moreover, we have good reasons to believe that there is no such a thing as a coherent particle concept applicable to all quantum models (Ruetsche 2011). At best, only a local concept could be defined. This said, in this section, we will try to answer the following question: Are emergent particles (understood loosely as relatively stable field quanta) involved in the FQHE?

10.5.2 SOME RELEVANT THEORETICAL RESULTS

In this subsection, we will expose some theoretical results that are relevant to the discussion of the status of anyons. These results, we believe, strongly limit the number of ontological options available.

Do we have good reasons to believe that anyons are field quanta that are sufficiently stable to be understood as particles? It seems unlikely that we could obtain a simple relativistic quantum field theory where fundamental field quanta are themselves anyons (Khare 1997). However, this does not exclude that anyons are quanta of another kind. For example, Fröhlich and Marchetti (1989) proposed a model in which the FQHE could be represented by a relativistic quantum field theory, with a Chern-Simons term, where topological solitons (vortices) are quantized. According to them, such solitons (anyons) should be considered as particles since they behave as such. It seems that anyons are good candidates for being considered as emergent particles since they appear in the emergent context of the FQHE.

As already mentioned in a previous work (Guay and Sartenaer 2016), in the context of the FQHE, the dependence between electrons and anyons could go both ways. As discussed by Zee (2010: 326–327), it is possible to conceive anyons as a collective behavior of electrons. Nevertheless, it is also possible to model electrons as bond objects composed of anyons. This may be surprising but not unexpected in a quantum field theory where particles and quasiparticles are only types of field configurations. This result is a very strong challenge to the position understanding anyons as being at a higher ontological level than electrons because, in this metaphysical context, the dependence relation—e.g., supervenience or realization—does not allow for such a symmetry between putative higher entities (anyons) and lower ones (electrons). For that reason, we will claim that anyons and electrons are at the same ontological level. They still can ontologically differ in a significant way.

Mund has demonstrated (1998) that, under mild assumptions, no free model of relativistic anyons is possible. What Mund means by "free" is that the basic fields create a one-particle state out of vacuum. It does not exclude the other possible meaning of "free"; namely, that the anyons' S-matrix leads to a trivial cross-section. Apparently, according to quantum field models, electrons and anyons are different kinds of beasts. The autonomy of electrons is stronger since they can, in principle, exist alone. Anyons do not. From this we should conclude that, even if they occupy the same ontological level, anyons are not on a par with electrons. Anyons are dependent entities in the sense that, to be produced, appropriate conditions must be fulfilled. Anyons are therefore context-dependent entities, hence they are *quasi*particles. It is important to notice that this conclusion relies on the fact that free electrons and the electrons involved in the FQHE are essentially the same. This is not obvious. If the topology change is more than a mere idealization, then the electrons after the transformation are to a great extent different from the ones before. In consequence, the electrons interacting with anyons could also be context-dependent and therefore quasiparticles. Even if you do not take seriously the topological change of the dynamics, you should worry about the ontological continuity between electrons before and after the transformation. According to Haag's theorem (Earman and Fraser 2006), the interacting field theory is unitarily inequivalent to a free theory.[18] If our conception of an electron as a particle comes basically from a theory without interaction, electrons in interaction in the FQHE cannot be the same (Ruetsche 2011). Moreover, electrons in QED_{3+1} cannot be the same as free electrons. The simple possibility of interaction changes the ontology. Apparently, all particles seem, in a certain way, context-dependent.[19]

10.5.3 ANYONS EQUAL EMERGENT PARTICLES?

Let us begin this last subsection with two recapitulative remarks:

- *Are anyons collective behaviors of electrons?* This position, quite natural if one works in a nonrelativistic quantum model of the FQHE, is not convincing in quantum field theory. Even if anyons, which are solitons, are an apparently different kind of quanta than electrons, there is no reason to believe that the former are subordinated to the latter. They are

[18] To avoid this concern, we could limit ourselves to nonrelativistic models since they fall outside the theorem's domain of application. Indeed, the FQHE could be represented by an effective nonrelativistic quantum field model. Nevertheless, in this chapter, we want to aim at the general case and anticipate a broader set of objections.

[19] For a similar idea, see, for instance, Falkenburg (2015) or Wallace (2012). Wallace actually also believes that electrons, like other particles, are themselves to be considered emergent (see, in particular, Wallace 2009).

both quantas of the fundamental fields. They are both equal candidates to be particles.[20]

- *Do anyons supervene on electrons?* As mentioned earlier, Zee provided a strong reason for believing that, at least in the context of the FQHE, electrons and anyons are on the same ontological level.

These remarks exclude the possibility that anyons would synchronically emerge from an electron basis after the transformation occurred. Anyons are unmediated products of the transformation. However, this is not the end of the story. The status of anyons is strongly dependent on the interpretation of the topological change between the representations of the states before and after the transformation. The dynamics of the system are best described by a three-dimensional quantum field model, but, from this fact, should we infer that three-dimensional "particles"—namely anyons—must be involved?

The first possibility is that there is some kind of topological change involved in the FQHE. In this case, since anyons are well-defined quanta of QED_{2+1}, they are emergent quasiparticles. However, they are not the only ones. The three-dimensional analog of the electron, a fermion, is possibly an emergent quasiparticle. This conclusion could push us to entertain the hypothesis that four-dimensional electrons are also contextual particles depending on the dynamics' topology.

The second possibility is that the topological change in the representations is an idealization. In this context, because they could well be the result of an artifact, we have to be careful about any ontological claim. To guide us, we will adopt a principle attributed to John Earman and Laura Ruetsche and reformulated by Shech (2015: 1065):

> Earman–Ruetsche sound principle—If a scientific account (theory, model) uses an idealization to predict an effect which disappears when the idealization is removed then either the effect is an artifact of the idealization or else (if experiment confirms the effect) the theory is inadequate.

The existence of the FQHE is a proof that a model of QED_{2+1}, with Chern-Simons terms, can be successfully applied. But what is the status of this so-called idealization? We agree with Shech that the topological change understood as an idealization is problematic. Any mathematical series of models starting from QED_{3+1} and going toward QED_{2+1} is inhospitable to the Chern-Simons terms. In consequence, anyonic solutions seem to exist only in 2+1. Since we claimed that this kind of transformation is emergent in a modal sense, this result was expected. To recover anyons, Shech's proposition is to describe the system using a geometrical approach, in the spirit of Arovas et al. (1984), rather than a topological one. This idealization seems, at least in the context of nonrelativistic quantum mechanics, to recover anyons appropriately. We do not know how to develop

[20] More on this question is available in Manton and Sutcliffe (2004).

a similar strategy in the context of a quantum field theory, but the recovery of anyons in nonrelativistic quantum mechanics suggests that the pathological idealization can be avoided.

Do we have a reason to choose among these possibilities? Does the FQHE involve a genuine topological change of some sort? We do not have a definite answer to this question. However, emergence seems to imply that something quite radical must happen between the states before and after the transformation. A topological change of the dynamics would be such a change.

10.6 Conclusion

We are now in a position to capture Laughlin's intuition about emergent quasiparticles in a metaphysically more sober way than what has been suggested by Gillett. By making a case that anyons can be considered transformationally emergent on four-dimensional electrons (represented by QED_{3+1})—and not, as in Gillett's case, strongly emergent on an underlying, simultaneous basis of three-dimensional "electrons"—we recover Laughlin's insights as they have been exposed in Section 10.2.

Within the transformational framework, anyons are indeed fundamental entities of a new physical domain, governed by new laws that do not exist—and that cannot exist—outside of this domain. Hence, anyons are fundamental particles in their own right, provided that there is a sense in which electrons are themselves particles in their own rights. Put differently, anyons and electrons are ontologically on a par. Emergence is seen here as some kind of mechanism that can produce (ontologically) new particles out of "old" ones.

Of course, through [TE], Laughlin's idea that what emerges has to be "higher level" or "composite" is lost, though it can be kept as a parasitic feature (in the sense that it can be entertained but has no role to play in emergence ascriptions). As a result, thinking about (ontologically) emergent anyons doesn't necessarily require one to be committed to the existence of (at least) two facts that are problematic when it comes to (micro)physics: namely, "levels of mechanism" and downward determination (under the form of *machresis* or anything else).

Acknowledgments

We would like to thank Paul Humphreys, Philippe Huneman, Joshua Johnson, Peter Verdée, and two anonymous referees for helpful comments on earlier drafts of this chapter, as well as the audiences of the "Taiwan Conference on Scientific Individuation" and the Louvain-la-Neuve conference "Emergence in the Sciences of Matter." Olivier Sartenaer also gratefully acknowledges financial support from the Belgian National Fund for Scientific Research (F. R. S.-FNRS).

References

Anderson, P. W. 1958. Random-phase approximation in the theory of superconductivity. *Physical Review* 112(6): 1900–1916.

Arovas, Daniel, John R. Schrieffer, and Frank Wilczek. 1984. Fractional statistics and the quantum Hall effect. *Physical Review Letters* 53(7): 722–723.

Barnes, E. 2012. Emergence and fundamentality. *Mind* 121(484): 873–901.

Earman, John, and Doreen Fraser. 2006. Haag's theorem and its implications for the foundations of quantum field theory. *Erkenntnis* 64(3): 305–344.

Falkenburg, B. 2015. How do quasi-particles exist? In *Why more is different: Philosophical issues in condensed matter physics and complex systems*, eds. B. Falkenburg and Margaret Morrison, 227–250. Berlin: Springer-Verlag.

French, Steven, and Décio Krause. 2006. *Identity in physics: An historical, philosophical, and formal analysis*. Oxford: Oxford University Press.

Fröhlich, J., and P.-A. Marchetti. 1989. Quantum field theories of vortices and anyons. *Communications in Mathematical Physics* 121:177–223.

Gelfert, A. 2003. Manipulative success and the unreal. *International Studies in the Philosophy of Science* 17(3): 245–263.

Gillett, C. 2010. On the implications of scientific composition and completeness. Or the troubles, and troubles, of non-reductive physicalism. In *Emergence in science and philosophy*, eds. Antonella Corradini and Timothy O'Connor, 25–45. New York: Routledge.

Gillett, C. 2016. *Reduction and emergence in science and philosophy*. Cambridge: Cambridge University Press.

Guay, A., and O. Sartenaer. 2016. A new look at emergence, Or when after is different. *European Journal for Philosophy of Science* 6(2): 297–322.

Humphreys, P. W. 1997. How properties emerge. *Philosophy of Science* 64(1): 1–17.

Humphreys, P. W. 2016. *Emergence. A philosophical account*. New York: Oxford University Press.

Khare, Avinash. 1997. *Fractional statistics and quantum theory*. Singapore: World Scientific.

Kim, J. 1999. Making sense of emergence. *Philosophical Studies* 95(1–2): 3–36.

Kim, J. 2005. *Physicalism, or something near enough*. Princeton: Princeton University Press.

Lancaster, T., and M. Pexton. 2015. Reduction and emergence in the fractional quantum hall state. *Studies in History and Philosophy of Modern Physics* 52: 343–357.

Laughlin, R. B. 1999. Nobel lecture: Fractional quantization. *Reviews of Modern Physics* 71(4): 863–874.

Laughlin, R. B. 2005. *A different universe: Reinventing physics from the bottom down*. New York: Basic Books.

Laughlin, R. B., and D. Pines. 2000. The theory of everything. *Proceedings of the National Academy of Sciences* 97(1): 28–31.

Laughlin, R. B., D. Pines, J. Schmalian, B. P. Stojkovic, and P. Wolynes. 2000. The middle way. *Proceedings of the National Academy of Sciences of the United States of America* 97(1): 32–37.

Lederer, P. 2015. The quantum hall effects: Philosophical approach. *Studies in History and Philosophy of Modern Physics* 50: 25–42.

Manton, Nicholas, and Paul Sutcliffe. 2004. *Topological solitons*. Cambridge: Cambridge University Press.

Mund, Jens. 1998. No-go theorem for 'free' relativistic anyons in d = 2 + 1. *Letters in Mathematical Physics* 43: 319–328.

Rao, Sumathi. 2001. An anyon primer. *arXiv Preprint Hep-th/9209066v3*. http://arxiv.org/abs/hep-th/9209066.

Ruetsche, Laura. 2011. *Interpreting quantum theories: The art of the possible*. New York: Oxford University Press.

Saunders, Simon. 2016. On the emergence of individuals in physics. In *Individuals across the sciences*, eds. Alexandre Guay and Thomas Pradeu, 163–192. New York: Oxford University Press.

Shech, Elay. 2015. Two approaches to fractional statistics in the quantum hall effect: Idealizations and the curious case of the anyon. *Foundations of Physics* 45(9): 1063–1100 (September).

Wallace, D. 2009. QFT, antimatter, and symmetry. *Studies in History and Philosophy of Modern Physics* 40: 209–222.

Wallace, D. 2012. *The emergent multiverse: Quantum theory according to the Everett interpretation*. Oxford: Oxford University Press.

Wen, X.-G. 2004. *Quantum Field Theory of Many-body Systems: From the Origin of Sound to an Origin of Light and Fermions*. New York: Oxford University Press.

Wilczek, Frank. 1982. Quantum mechanics of fractional-spin particles. *Physical Review Letters* 49(14): 957–959 (October).

Zee, A. 2010. *Quantum field theory in a nutshell*, 2nd ed. Princeton: Princeton University Press.

{ PART III }

Individuation in Philosophical Approaches to Science

REALISM, ANTI-REALISM, ENVIRONMENTALISM

{ 11 }

Can Quantum Objects Be Tracked?

Otávio Bueno

11.1 Introduction: Experiments with Atoms and Single Quantum Particles

From a theoretical point of view, it may seem that we would never be in a position to uniquely identify and track a quantum object. After all, on a particular interpretation of quantum mechanics, such objects are spread out in space, displaying a wave-like behavior and lacking well-defined identity conditions (French and Krause 2006). Without such identity conditions, a minimum requirement for individuality is not satisfied. After all, as examined later, identity is arguably required for individuality (and by *identity* I have in mind a deflationary conception: an equivalence relation for which substitutivity holds). Moreover, it seems that only individuals can be tracked because individuality appears to be needed in order to determine whether the *same* object is being trailed in different moments of time (i.e., the same individual) rather than some other object, very similar to (or perhaps even indistinguishable from) the target. If quantum objects are non-individuals, how is it possible, even in principle, to identify and track them? (It seems that the answer is clearly negative.)

One response to what cannot be done in theory is to actually do it in practice and then figure out what exactly has been achieved. Experimentalists have often accomplished in the context of carefully designed experiments what theoreticians alleged to be impossible. In some cases, the results did not actually establish what was claimed that they achieved. Think of cases such as cold fusion.

But, in other cases, in light of relevant experimental work, the alleged theoretical impossibility had to be abandoned and the theory correspondingly rewritten. Consider the case of the experiments with splitting atoms in the late 1920s and early 1930s. Physicists spent years puzzling over the interpretation of experimental results that suggested that atoms had been split. After systematically rejecting that any such splitting could have been accomplished, in light of the theoretical assumption that ruled out its possibility, physicists eventually gave up the theoretical constraint and embraced as actual what had previously been considered impossible.

Given these considerations, the issue arises as to which attitude should be taken toward three experiments that similarly seem to challenge deeply entrenched theoretical assumptions, in this case about the individuality of quantum objects. The first consists in trapping a positron and studying its behavior for an extended period of time; in fact, for as long as three months (Dehmelt 1990). The second experiment involves trapping an ion (in an electron beam ion trap) and performing similar studies on it (Schenkel et al. 1999). The third sort of experiment is of a different kind altogether. It involves manipulating and positioning single atoms with a scanning tunneling microscope (STM) (Eigler and Schweizer 1990).

The first experiment (trapping a positron) is the most dramatic since it offers unprecedented control over a single subatomic particle. The experiment involves trapping a particle in an ultra–high-vacuum Penning trap (Dehmelt 1990: 539–540). Two negatively charged caps are set up so that they create a magnetic field. Inside the caps, a particle (an electron) is trapped, and its simplest motion is along its symmetry axis along the line of a magnetic field. Each time the particle approaches one of the negatively charged caps, it is repelled. The experiment was then adapted so that a positron was trapped, and the trapped positron, which has remained trapped for at least three months, was given a name, Priscilla. As Dehmelt notes:

> There can be little doubt about the identity of Priscilla during this period, since in ultrahigh vacuum she never had the chance to trade places with a passing antimatter twin. The well-defined identity of this elementary particle is something fundamentally new, which deserves to be recognized by being given a name, just as pets are given names of persons. (Dehmelt 1990: 539)

The other two experiments deal with objects at the nanoscale (10^{-7} to 10^{-9} meters). Although quite impressive, they involve much "larger," more stable objects, beyond the quantum scale: atoms. The second kind of experiment involves highly charged ions (Shenkel et al. 1999). At certain conditions involving extremely high temperatures, atoms end up losing several electrons. As a result, they become positively charged. The experiment involves an electron bean ion trap, which restrains ions due to a combination of electric and magnetic fields and keep them in a vacuum. The ions are then bombarded by a carefully focused and powerful electron bean, which both ionizes the atoms further and traps them: since the electron bean is negatively charged, it attracts the ions, but positively charged electrodes placed on both sides of the trap repel them. As a result, the ions are thereby trapped (for a survey of several techniques and approaches, see Schenkel et al. 1999). At the limit, a single ion could be trapped.

In the third experiment, which involves manipulating and positioning a single Xe atom, Eigler and Schweizer (1990) used a low-temperature UHT STM (Jia, Yang, and Xue 2005: 92–94). The tip of the microscope was placed on top of the selected Xe atom, and, by increasing the tunneling current between the STM's tip and the atom, the energy barrier was overcome and the atom could be slid to a nearby region on the substrate. (Some care was needed at this step for if the increase in the tunneling current was too high, the atom would be transferred to the tip.) By moving the tip, the atom was then dragged to a selected location on the substrate and released by reducing the tunneling current. This process was repeated several times so that various atomic configurations were created, from the spelling of "IBM" using Xe atoms to the formation of a "quantum corral" with Fe atoms (see Eigler and Schweizer 1990; and for a survey and additional discussion, see Jia, Yang, and Xue 2005: 92–94).

It may be argued that these experiments put some pressure on current forms of empiricism, particularly constructive empiricism (van Fraassen 1980, 1991). As will become clear, three challenges, in particular, can be raised: (1) The experiments can be used to criticize an important form of underdetermination employed by empiricists to resist commitment to the existence of the relevant unobservable objects. (2) In the context of these experiments, the nature of quantum particles, in particular their identity and individuality, seems unproblematic. Thus, no unexpected ontological commitment to metaphysically troublesome objects emerges. In this way, a significant source of empiricist concern against the commitment to the relevant quantum objects is disarmed. (3) Finally, the experiments offer an unprecedented form of access to quantum particles, one that questions the antirealist stance toward these particles that empiricists favor.

After discussing these three arguments in Section 11.2, I offer an empiricist response to each of them. As will become clear, I do not think that the considerations raised by these arguments ultimately go through. I will examine why this is the case by considering, in Section 11.3, the failure of each in turn. Throughout the discussion, questions related to the identity, individuality, and individuation of quantum particles in experimental situations remain at center stage.

11.2 Problems for Empiricism

Experiments, such as those just mentioned, show an unparalleled level of control over atomic and subatomic objects. *Prima facie*, one could argue, it is difficult to see how familiar arguments used by empiricists to undermine realist commitments could be invoked in this experimental context. Three arguments are especially significant.

11.2.1 THE INAPPLICABILITY OF UNDERDETERMINATION ARGUMENTS

It is a fact of quantum mechanics (in particular, of its nonrelativist formulation) that there are a variety of nonequivalent interpretations of the theory, including the Copenhagen interpretations (Cushing 1994), the modal interpretations (van Fraassen 1991), and the many-worlds interpretations (Wallace 2012). These interpretations provide answers to questions regarding what goes on beyond the appearances and that are not settled by the theory alone, such as whether the world is deterministic or not, whether quantum particles have well-defined identity conditions, or what exactly happens to a quantum system between measurements.

A familiar argument, used by empiricists against realist accounts of (nonrelativist) quantum mechanics, relies on such underdetermination of interpretations of quantum theory by the theory itself. The central point of this argument is to explore the fact that there are significantly different interpretations of the same quantum-mechanical, empirical results that posit substantially diverse ontological pictures beyond the appearances, but which are all empirically adequate. (Otherwise, rather than providing distinct interpretations of quantum mechanics, the proposals would amount to rival theories.) Since there are no empirical reasons to choose one interpretation over the others, and given that a number of issues left open by the theory are settled very differently—albeit tentatively—by each interpretation, it is unclear how to decide what is really going on beyond the appearances, beyond the empirical data accounted for by quantum theory. In other words, the fact that a single interpretation of quantum mechanics cannot be uniquely chosen highlights the difficulty in ultimately settling what is the true (or even the approximately true) description of the quantum domain (if any). These considerations put some pressure on those forms of realism about quantum mechanics that are tied to the identification of the ultimate constituents of reality since these constituents are largely underdetermined by the interpretations of the theory (for additional discussion, see van Fraassen 1991; French 2014).

It is not clear, however, that this underdetermination argument goes through in the context of the two trapping experiments, mentioned earlier, involving a positron or an ion. What is striking about these experiments is that, with a single positron trapped inside the apparatus, or a single ion, one could argue that it is implausible to run the standard underdetermination argument to undermine the realist's understanding of the results. Given that a single positron (or ion) is involved, it is very difficult to see how a multiplicity of theoretical alternatives could be entertained to accommodate the phenomena in question. That there is *something* inside the trap is not in question, at least for the sake of argument. After the cooling process, one can study the various properties of the object, and the results that are obtained, given the measured values of mass and charge, are consistent with the theoretical supposition that it is a positron, and only one, trapped inside the magnetic field (or, alternatively, a single ion trapped in it). It can be argued that the best explanation of the results of the experiments is given by the fact that a single positron (or a single ion) is indeed trapped inside the apparatus. It is very unlikely that anything else could account for the experimental results, particularly given that a single object is involved.

Similarly, in the case of the manipulation of atoms at the nanoscale, it can be argued that it is difficult to see, at least in principle, which sort of underdetermination could be raised to challenge a realist understanding of the experiment. What else could be going on in this context but what is provided by the description of the experimental results: namely, the manipulation of atoms? Since atoms are also being positioned, one by one, precisely at selected locations, it is unclear what kind of plurality of interpretations regarding what is going on could be offered. Once again, it seems that the explicit account of the experiments undermines any attempt to generate a variety of interpretations of the results in question.

11.2.2 THE UNPROBLEMATIC NATURE OF ATOMIC AND SUBATOMIC OBJECTS IN THE CONTEXT OF THE EXPERIMENTS

It is important to distinguish the case of the single trapped positron from that of a trapped ion. An ion is an atom (or an aggregate of atoms) that is normally neutral but which lost or gained electrons and thus is currently either positively or negatively charged. Now, atoms are much larger and more robust objects than electrons, positrons, and other subatomic objects. In the case of subatomic particles, as I will discuss later, concerns have been raised regarding their apparent lack of identity that *prima facie* seem to jeopardize the standard scientific realist understanding of them (French and Krause 2006).

With regard to atoms (whether ionized or not), since they are more stable than subatomic objects, the issue regarding their lack of identity does not seem to arise in the same way. However, concerns about their identity still emerge. Consider, in a water molecule, the fact that it does not matter which two hydrogen atoms are part of it: as long as there are two of them for one oxygen atom, a typical water molecule is formed. The particular identity and individuality of the hydrogen atoms involved does not matter at all. In fact, it is unclear how one could individuate a single hydrogen atom, recognizing it as being the same in different moments of time. Moreover, the same aggregate of atoms can form different chemical substances depending on the arrangement of the atoms involved. This is the case of isomers. For example, the atoms of carbon, hydrogen, and oxygen that form C_2H_6O can stand for either CH_3-CH_2-OH (ethylic alcohol) or H_3C-O-CH_3 (methylic ether) depending on the different configurations among the atoms (see Krause 2005). Once again, the particular identity and individuality of the atoms in question are not relevant.

One of the main arguments against the well-defined identity conditions for subatomic particles emerges from the fact that the permutation of these particles does not change the state of the quantum system they are in (French and Krause 2006). Electrons can be permutated without ever changing the quantum state of the relevant system. In other words, from the perspective of (nonrelativist) quantum mechanics, the permutation of such particles is not observable. Since in the context of the theory there are no other distinguishing features for these particles, they are completely indistinguishable from one another. As a result, they cannot be individuated since nothing about the particles, in light of their quantum-mechanical description, allows them to be singled out.

The argument is then extended further. Since subatomic particles share the same properties (such as their mass and charge), given Leibniz's principle of identity of indiscernibles and assuming, for the sake of argument, that the principle could be applied to these particles, they would have to be the same. After all, according to this principle: $\forall P \forall x \forall y \, ((Px \leftrightarrow Py) \rightarrow x = y)$. But these particles, although indistinguishable, are numerically distinct. If it were supposed that among the properties that such particles have are spatiotemporal properties, such as their location in space and time, one could argue that it would be possible to distinguish the particles in terms of their location. However, even in orthodox nonrelativist quantum mechanics, space separation is not enough to distinguish particles since there is no perfect separation: interference cannot be fully eliminated. Moreover, it is not exactly clear that such particles have a definite location either (particles do not have positions, but position-states only). They satisfy certain probability conditions to the effect that they may be located in certain regions of space and time, but this is different from actually being located in those regions. In the end, difficulties emerge to make sense of (nonrelativist) quantum mechanics under the assumption that subatomic particles have well-defined identity conditions (for additional discussion, see French and Krause 2006).

A long time ago, Hermann Weyl and Erwin Schrödinger put forward considerations along these lines. As Weyl points out:

> The possibility that one of the identical twins Mike and Ike is in the quantum state E_1 and the other in the quantum state E_2 does not include two differentiable cases which are permuted on permuting Mike and Ike; it is impossible for either of these individuals to retain his identity so that one of them will always be able to say "I'm Mike" and the other "I'm Ike". Even in principle one cannot demand an alibi of an electron! (Weyl 1931)

Schrödinger also emphasizes the point:

> We have . . . been compelled to dismiss the idea that . . . a particle is an individual entity which retains its "sameness" forever. Quite the contrary, we are now obliged to assert that the ultimate constituents of matter have no "sameness" at all. (Schrödinger 1952: 17)

And he insists:

> It is not a question of our being able to ascertain the identity in some instances and not being able to do so in others. It is beyond doubt that the question of "sameness", of identity, really and truly has no meaning. (Schrödinger 1952: 17–18)

Given that subatomic particles, on this interpretation of quantum mechanics, lack well-defined identity conditions (for a critique, see Muller and Saunders 2008; for responses, see Krause 2010; Arenhart 2017*b*), it becomes particularly puzzling to determine their nature. Their apparent lack of well-defined identity conditions rules out their individuality. In fact, as both Weyl and Schrödinger make clear, they do not take such particles to be individuals at all. But if they are not individuals, what kind of being are they? Nothing in the quantum mechanical description of the world seems to settle uniquely this issue. One could note that subatomic particles are non-individuals, but a quantum mechanical description also does not characterize what they are. At best, one could argue that non-individuals are indistinguishable from one another; namely, those that are of the same *kind* (whether they are electrons, or protons, or photons, etc.). Note, however, the use of identity in specifying *sameness* of kind: even though non-individuals are supposed not to have well-specified identity conditions, they are assumed to be of the *same* kind as others indistinguishable non-individuals of that kind. Despite this feature, this is not enough to characterize the nature of non-individuals, since they can also be formulated in terms of alternative metaphysical settings. For instance, electrons can be specified as things that have certain *properties*, such as having a particular mass and charge and being indistinguishable from other electrons. But which of these formulations, if any, characterizes their nature: kinds, properties, or some other metaphysical setting? Without a proper account of what non-individuals are, the concern is raised regarding exactly what standard scientific realists are ontologically committed to when they are ontologically committed to quantum particles (see also Ladyman 1998; French and Krause 2006; Arenhart 2017*a*).

Of course, Weyl and Schrödinger were theorizing about subatomic particles well before any of the trapping and positioning experiments mentioned earlier had been implemented. The problem here is that, in the case of the experiments with a single quantum particle, the familiar worries about the putative lack of identity of the objects involved do not seem to arise. After all, when a single positron is trapped, it seems to behave as though it had perfectly well-defined identity conditions in the sense that, relative to the macroscopic setting in which the positron is found, it remains the same object for the duration of the experiment. (Here it is assumed that the experiment is not reset; otherwise, as will become clear, there would be no guarantee that the *same* positron had been trapped again.) That is, the experiment allows researchers to study the behavior of the positron in isolation from many sources of interference, and it seems that, throughout the experiment, there is no difficulty in identifying the positron and assigning identity conditions to it, at least relative to the macroscopic environment in which it is embedded.

The same considerations apply to the experiments with a single ion (which involves an electron beam ion trap) and with the manipulation of atoms (which uses an STM). In the case of the single ion, its properties can be studied carefully in the isolated system of the ion trap. And since there is only one ion trapped, the issue of the ion's apparent lack of identity conditions does not seem to emerge.

With just one ion involved, no concern regarding the permutation of that ion apparently arises. Similarly, in the case of the atom manipulation experiment, given that a single atom is precisely moved and positioned at a time, it is also difficult to see how the issue of its lack of identity conditions could in fact arise. With the STM, atoms are moved, one by one, carefully and systematically. Underdetermination considerations do not seem to apply. Given the precise manipulation and control over the relevant objects (positrons, ions, and various atoms), an anti-realist reading of the experiments seems highly implausible.

Furthermore, one of Ian Hacking's influential arguments in support of entity realism seems to find a powerful echo in these experiments (see the final chapter of Hacking 1983). On his view, the best argument for realism about certain entities is to use them to build an instrument that produces reliable results about certain phenomena. These are the entities one should ultimately be realist about. Positron traps, electron bean ion traps, and STMs are all machines that successfully use a variety of entities to yield consistently reliable results, as illustrated by the outcomes of the three experiments under consideration. In light of Hacking's argument, the experiments seem to support realism about the entities involved.

11.2.3 TRACKING QUANTUM OBJECTS

Despite the significant differences between the various instruments that are used to obtain information about subatomic and atomic events (such as electron beam ion traps and STMs), they all seem to share something in common. Using these instruments, researchers can *track* the objects under study. And the ability to track such objects provides additional evidence for the existence of the corresponding objects. In this way, tracking supports realism about quantum and microscopic objects. Whatever these objects turn out to be, it seems clear that we are tracking them. So, there is something real being detected by the instrument.

But what requirements are involved in *tracking*? We can think of it as a form of epistemic access satisfying two conditions, which in many contexts can only be met with the use of suitable instruments (see Bueno 2011): (a) Had the sample been different, then the output from the relevant instruments would have been correspondingly different (within the sensitivity range of the instrument). (b) Had the sample been the same, then the output from the relevant instrument would have been correspondingly the same (again, within the sensitivity range of the instrument). I will call these conditions *counterfactual tracking*. (Perhaps one could weaken condition (b) and only require the condition: had the sample been indistinguishable, then the instrument's output would have been correspondingly indistinguishable. But the problem with this suggestion is that the indistinguishability would allow for the conflation of potentially different, but indistinguishable, phenomena in the light of the instrumental apparatus. Proper evidence should rule out such cases.)

Note that the counterfactuals here should not be understood in terms of possible worlds: the actual world is complicated enough to make sense of, and the relevant changes should be thought of as changes in that world rather than in some other possible, causally isolated one (see Bueno and Shalkowski 2015). Moreover, for an instrument to be properly functioning and reliable, not only does it need to satisfy the two counterfactual conditions, but also one should know (or at least have some reason to believe) that these conditions are satisfied. Otherwise, it is unclear that one would be in a position to trust fully the outcomes of the instruments, given that the possibility of artifacts and malfunctions could not be ruled out. In other words, some form of internalism should be in place to determine the reliability of the instruments in question (see Bueno 2016). (For discussion of tracking in the context of epistemology, see Nozick 1981; for an examination of counterfactuals without possible worlds, see Woodward 2003; for further details of this approach to instruments, see Bueno 2011, 2014*a*.)

Alternatively, we can think of tracking as a form of monitoring (Azzouni 2004: 129), where (a′) the object in question is monitored in time (we can detect what the object does over time), or (b′) the object is monitored in space (we can study different aspects of the object and its properties). (I will call these conditions *spatiotemporal tracking*. As will become clear, these conditions are too strong when applied to the quantum case.)

What is the relation between counterfactual and spatiotemporal tracking? It seems that if counterfactual tracking obtains, so does spatiotemporal tracking, but not vice versa. Suppose that, using a suitable instrument, we are able to track counterfactually a certain object. This is done, in part, by examining how certain changes in the sample under study yield corresponding changes in the images generated by the instrument. But these changes are obtained by monitoring the sample over time and in space. Thus, counterfactual tracking leads to spatiotemporal tracking.

Consider now the reverse: using a certain instrument, we are monitoring a certain object over time and in space. Is condition (a) of counterfactual tracking then met? That is, can we say that, if the object were not there, we would not believe that it was? Not necessarily, since we may think we are monitoring an object when we are in fact monitoring an entirely different object—or no object at all, if we are dealing with an artifact of the instrument. With regard to condition (b), we have an intriguing situation. Suppose the object we are monitoring is indeed in the sample. We would then believe the object is in the sample since we are monitoring it. So, it seems that condition (b) would be met: if the object were there, we would believe it was. However, this is the case even though we might be monitoring a completely different object than the one we think we are monitoring. For example, suppose you want to study the geometrical configuration of the surface of DNA using an STM. You apply some DNA to the surface of some conducting material, such as carbon, and scan the sample. You might think you are then monitoring the DNA but, as it turns out, you are in fact just studying the surface of carbon. Since

DNA is not a conducting material, there is no tunneling current between the tip of the STM and the DNA itself. So, even though condition (b) is met, it might be satisfied for the wrong reasons! As a result, counterfactual tracking is more stringent than spatiotemporal tracking.

This provides an additional reason for the inclusion of the internalist requirement mentioned earlier. With this requirement in place, it is not enough that the two conditions of counterfactual tracking be satisfied. One would also need to know, or have good reasons to believe, that the conditions are indeed satisfied. This requires that one understand the particular details of the interaction of the instrument and the sample under study and figure out the proper relations between them to ensure that artifacts and other potential misleading bits of information are successfully ruled out.

Given the issue of the lack of identity conditions for quantum objects, at least on some interpretations of (nonrelativist) quantum mechanics, the issue arises as to whether we can ever be in a position to track an object that lacks such conditions. In principle, it may seem that neither counterfactual nor spatiotemporal tracking requires that an object, in order to be tracked, must display well-defined conditions of identity. After all, nothing in the formulation of these notions of tracking seems to impose any requirement regarding the identity of the objects that could be tracked. Hence, the issue regarding the identity conditions for such objects is left open.

But how could that be? If an object lacks well-defined identity conditions, how can it be tracked? How could it be determined whether *the same* object is being tracked if it cannot be settled whether it is identical (or not) with other things in its surroundings? To track an object, some contrast is required between the object to be tracked and other objects so that the former can be distinguished from the latter. Moreover, in order to determine whether the *same* object is being tracked (rather than, say, a sequence of different but indistinguishable objects), the object needs to be continuously identified over time. Or, at least, one should be able to reidentify that object at any point that is needed and have good reason to believe that it is still the same object. But it is unclear how this could be achieved in the case of objects that lack well-defined identity conditions. Without identity, there is no fact of the matter as to whether the object in question is the same as or different from any other object. As a result, such an object cannot be reidentified (determined to be the *same* over time).

For these reasons, it is unclear that microscopic objects that lacked well-defined identity conditions (in light of the interpretation of quantum mechanics that takes them to be so) could really be tracked. The point becomes particularly pressing for these objects because the only form of access to them is via suitable instruments (such as STMs and electron beam ion traps). Since there is no direct, instrument-independent form of access to microscopic objects, any lack of identity conditions for these objects jeopardizes one's ability to track them unless the instrumental set-up itself somehow helps one to overcome that difficulty (a possibility I will return to later).

It may be argued that tracking does not require that the *same* object be detected in different moments of time: it is enough that objects of the same *kind* be so detected. But this is not right. The detective could not be said to be tracking the drug dealer if at different moments in time he detects some man (who may or may not be the drug dealer in question): it is crucial that the *same* man—namely, the drug dealer—be detected. Similarly, detecting an object indistinguishable (and potentially different) from the one that is being tracked, even if the indistinguishable object is of the same kind as the target one, is not enough for one to conclude that one is tracking the object under consideration.

What happens in the case of macroscopic objects that lack well-defined identity conditions? (I take it that even some macroscopic objects, such as waves and clouds, may lack well-defined identity conditions. I remain agnostic about the nature of this lack of identity conditions; that is, whether it is an epistemic or an ontological matter.) Could we track such objects? Given the arguments just presented, the answer is negative since the points just made apply equally well to macroscopic objects for which identity is not properly defined. After all, it will be unclear whether it is the *same* object that is being tracked or some other object that is, thus, *distinct* from the one that is being tracked: any such determination ultimately requires identity. However, most macroscopic objects typically have well-defined identity conditions, and it is not surprising that they can typically be tracked so well. But some macroscopic objects, such as waves, despite having for the most part relatively clear boundaries do not always have sharp ones. Can such hazy objects be tracked? To a certain extent, they can: most of their areas are delineated clearly enough. In this case, the waves do have clearly determined identity conditions. However, when that is no longer the case (i.e., when there is too much overlap between apparently different waves), difficulties emerge, and one is no longer able to track them individually. After all, in this case, there is no fact of the matter as to whether a given wave is the same as the one that has been under examination up to a certain moment or whether it is, in fact, a distinct wave altogether that overlapped with the previous one. (Of course, tracking ensembles of waves does not pose this difficulty since the hazy boundaries of their parts do not affect the identification of the group as a whole.)

For these reasons, even though counterfactual and spatiotemporal tracking do not explicitly presuppose objects that have well-defined identity conditions, it is in practice difficult to determine how such objects could, even in principle, be tracked. So, if objects have, in fact, been tracked, it seems that they do have properly characterized conditions of identity, at least while they are being tracked. (Of course, one can still run Humean doubts about whether it is still the *same* object that one is tracking if one lost contact with it for some time; see Hume 1738/2000: I, iii, ii.)

What happens in the context of the trapping and positioning experiments then? Have the relevant positrons, ions, and atoms been tracked? In this case, the argument goes, the objects, properly constrained under the particular experimental setups, seem to display well-defined identity conditions. Moreover, the fact that they have been tracked seems to provide an indication that the objects in question have the required conditions of identity. Interestingly, in each case, a *single* item is involved: a single positron trapped, a single ion encaged, a single atom positioned. In each instance, the objects seem to display the behavior of individuals; that is, they are, minimally, entities that (1) have well-defined identity conditions and (2) whose persistence conditions are robust and stable enough over time so that the objects in question can, at least in principle, be reidentified (see Bueno 2014*b*). (Reidentification, in this context, should be interpreted as a primarily ontic trait, as the outcome of the stability and robustness of the objects' persistence conditions.)

Since a single object is involved, *prima facie*, it seems that both conditions are immediately satisfied. Given that the positron and ion that are trapped and the atom that is individually positioned is a single one in each case, the concerns about lack of identity conditions do not seem to arise. Furthermore, since there is a single object in each instance, its persistence conditions, at least in principle, should not raise difficulties: the objects should, *prima facie*, all be reidentified in each case. No *other* positron or ion is trapped, and just one atom is positioned at a time. Hence, it is not surprising that it is possible to track these objects, one by one. As a result, once again, a realist commitment to these objects becomes much more plausible.

11.3 An Empiricist Response

The arguments just considered seem to provide strong considerations for the defense of a realist view of the outcomes of the trapping and positioning experiments. The arguments can be used to mount a challenge to an anti-realist, particularly empiricist, view of these results. Is it possible to respond to the challenge within empiricism? In what follows, I argue that it is.

11.3.1 THE UNDERDETERMINATION ARGUMENT REVAMPED

Is there underdetermination in the cases of the experiments involving trapping positrons and ions and positioning atoms? In all of these cases, underdetermination is indeed involved, although it is of a slightly different sort than the usual one.

First, it should be noted that the realist still invokes an inference to the best explanation to obtain the conclusion that it is a positron (or an ion), rather than some other object, that is trapped inside the magnetic field. After all, to make sense of the result of the experiment, and to assert that it is indeed a positron (or an ion) that is at stake, the experimental outcome needs to be interpreted carefully. It is inferred that, given the information provided by the trapping instrument, a positron (or an ion, depending on the experiment) provides the best explanation of the observable outcome. Determination and measurement of masses and charges need to be carried out, and, based on them, it is inferred that we are dealing with a positron (or an ion). But to identify the object in question as a positron (or an ion) rather than some other object, is to be guided by quantum mechanics. The content of the experiments, that is, the salient information provided by the relevant instrumental apparatus alone does not determine the interpretation of the results. The informational content from the experiments gives researchers a particular empirical configuration and corresponding images. To conclude that a *positron* (or an *ion*) is trapped from this information, we need to interpret the results in light of quantum mechanics. And it is at this point that underdetermination emerges once again.

After all, although quantum mechanics identifies objects with certain charge and mass as positrons (or ions), it does not settle the issue regarding their nature. Are these positrons (or ions) individuals? Or are they non-individuals (that is, objects that, minimally, fail to have well-defined identity conditions)? Can we precisely and simultaneously determine the position and the momentum of these objects? Or is simultaneous measurement of these magnitudes intrinsically unsharp in any empirical set-up? Questions such as these are left entirely open by the experiments. After all, depending on the interpretation of quantum mechanics one adopts, dramatically different answers to them are obtained.

The interpretation favored by Weyl and Schrödinger, for instance, would insist that, in the case of the positron (or the ion), we are dealing with a non-individual. A positron (or an ion) has been trapped in the experimental apparatus, but it is just an element of an equivalence class of indistinguishable objects: any other positron (or any other ion) would yield the same empirical results. (Let us denote this equivalence class by "≡".) The fact that the positron (or the ion) seems to have well-defined identity conditions in the context of the experiments only emerges as the result of the particular configuration of the experiments. In that context, it makes perfect sense to claim that any two indistinguishable positrons (or ions) would satisfy the same properties. That is:

$$(*)\forall X\, \forall x\, \forall y\, (x \equiv y \rightarrow (Xx \leftrightarrow Xy).$$

If we now assume that a given positron (or a given ion), a, is indistinguishable from itself (that is, $a \equiv a$), which is perfectly reasonable to suppose, we then obtain the notion of identity. (It is not difficult to show that (*) plus the reflexivity of the relation ≡ entail all the properties of classical identity.) Thus, in the context of the experiments, it is perfectly plausible to explain why the positron (or the ion) seems to have well-defined identity conditions. This is the case even though the identity in general of the positron (or the ion) is left open; that is, the particles' identity outside experimental contexts is not presupposed. (For additional discussion, see Krause 2011; see also French and Krause 2006. In these works, it is argued that

(*) above does not hold for all properties; as a result, the relation of indiscernibility does not collapse into identity.)

Alternatively, consider now those who favor the individuality approach, according to which, despite appearances to the contrary, quantum particles do have well-defined identity conditions (for a discussion, see van Fraassen 1991). Most interpretations of quantum mechanics typically fall under this category. These interpretations would insist that the positron (or the ion) is indeed an individual, and, as such, it immediately satisfies the usual identity conditions. No additional explanation is needed for the positron's (or the ion's) behavior in the context of the experiments. What has to be explained is something else: How is it that quantum particles, despite having well-defined identity conditions, are still indistinguishable, in the sense that their permutation is observationally invariant? Needless to say, different interpretations of quantum mechanics answer this question very differently.

The crucial point here is that the trapping experiments do not settle which (if any) of these rival interpretations is true. The results of the experiment are compatible with all of them. However, without answering the questions raised earlier regarding the nature of a positron (or an ion), it is not clear exactly how we should interpret the results of the experiments. In other words, it is not transparent what the experiments actually tell us about what is going on beyond their immediate experimental content. In order to make any claims beyond that content, one needs to invoke an interpretation of quantum mechanics, and, at that stage, underdetermination considerations emerge again. (For additional discussion, see Krause 2011.)

The same point applies to the positioning experiment: ultimately it depends on inference to the best explanation and on particular interpretations of the empirical results. In order to position atoms with an STM, a tunneling current is established between the sample and the tip of the microscope. Something is then dragged on that surface. But *what* exactly is being dragged? The STM output will not provide an answer to *that* question. One can *interpret* the result, assuming quantum mechanics, as an atom being dragged through the sample's surface. This provides the best account of the experimental result. But, clearly, this is an inference to the best explanation. As is well known, this is a very contentious form of inference in the context of an argument for a realist interpretation of the experiment in light of the difficulties raised against this inference by anti-realists (see, in particular, van Fraassen 1989).

Moreover, depending on the interpretation of quantum mechanics that is adopted, different answers regarding the question about the nature of such atoms emerge. Do such atoms have well-defined identity conditions? Are they individuals or non-individuals? Once again, different interpretations of quantum theory generate incompatible responses to these questions, and underdetermination with regard to the relevant empirical results emerges. In light of such underdetermination, the realist commitment to the objects at stake remains problematic. After all, the nature of the objects is substantially unsettled. And without a definite answer on this front, it is simply unclear what the realist is ultimately committed to; that is, what the content of realism about these experiments amounts to. (For additional discussion of what is involved in specifying the content of realism, see Chakravartty 2007.)

11.3.2 THE PROBLEMATIC NATURE OF ATOMIC AND SUBATOMIC OBJECTS EVEN IN THE CONTEXT OF THE EXPERIMENTS

It is important to separate the notions of individuality and distinguishability. Individuality can be thought of as a metaphysical notion, whereas distinguishability can be considered an epistemological one (the distinguishability of certain objects can be taken as a necessary condition for one to determine their individuality). Moreover, individuality and distinguishability are logically independent notions. Individuality can obtain without distinguishability being the case: a universe with only one object illustrates this point (assuming that such a universe, which would not have a spatiotemporal structure to prevent distinguishability, is possible). Alternatively, a universe with exactly two objects (say, two perfectly indistinguishable spheres) would also do the trick (assuming, again, the possibility of a universe without spatiotemporal structure). Each sphere is an individual, but one cannot be distinguished from the other (they have exactly the same mathematical properties after all, and there is no underlying spatiotemporal structure). However, there are still two of them.

It is harder, however, to establish that we can have distinguishability without individuality. Consider, for instance, the case of two drops of water. One could claim that the two drops could be distinguished from each other but not as an individual. We often use distinguishability as a necessary condition to identify individuals. It is because I am able to distinguish a tiger from an elephant (or, for that matter, from another tiger) that I am able, in part, to consider them as individuals.

Consider an *entity* to be a Quinean object: an object that has well-defined identity conditions. And let us accept, for the sake of the argument, that an *individual* is an entity that can be uniquely singled out; it can be uniquely determined which object it is. The understanding of quantum mechanics that emerges from Schrödinger and Weyl is one in which quantum objects (such as electrons and positrons) are not entities, since they lack well-defined identity conditions, and hence they are not individuals either, since they cannot be uniquely singled out. However, quantum objects can be thought of as members of an equivalent class of indistinguishable objects. In each case, say of electron or positron, the conditions that define the equivalence class are well-defined: these are the conditions provided by quantum mechanics. The trouble, however, is that these conditions are not sufficient to individuate (single out) each object in the class since (on the interpretation advanced by Schrödinger and Weyl) nothing allows us to distinguish the corresponding objects (for a critique, see Muller and Saunders 2008; responses have been provided by Krause 2010; Arenhart 2017*b*). As a result, we do not have individuals (or entities) since we do not have objects with well-defined identity conditions. (The facts that there are mathematical objects that can always be distinguished and that perhaps the quantum objects in question could be distinguished in a different structure are beside the point. After all, arguably the objects under consideration are not mathematical, and they are being considered in a given structure, in light of a particular interpretation; in a different interpretation, the situation would, as one would expect, be described differently.) But we still have objects, in the sense that there are things that satisfy the conditions of the equivalence class in question (although these things are not distinguishable).

So far, I am separating entities (things that have well-defined identity conditions) and objects (things that are being talked about whether or not they have such conditions). A more radical proposal would give up completely the reference to objects. On this view, we would only have certain fields and energy, and, when we detect a positron or an electron, we only detect (roughly speaking) energy at a certain region of space. We do not detect an object. Positrons or electrons would only be the result of certain interactions with the field. In other words, the detection does not go through and becomes substantially harder in the context of a quantum field theory.

This raises then the following dilemma for the experiments of the trapped ion or trapped positron. Let us focus on the case of the positron since it is the most relevant one. Could we measure continuously the trajectory of the positron as it moves around inside the apparatus? If we could not, how could we know that there is only one positron trapped inside the instrument? If it is said that the cardinal, in this case, is 1, then we are simply asserting the very point in question, rather than providing a reason for it. (One would need to rule out the possibility that one is measuring, in different instants of time, the trajectory of more than one positron in the apparatus.) Alternatively, if it were possible to measure continuously the positron's trajectory (note the use of a counterfactual!), we could claim that, at least within the boundaries of the experiment, the positron under study would have well-defined features and would seem to behave exactly like an individual would—in particular, seeming to have well-defined identity conditions. But how is it possible that such conditions are suddenly acquired in the context of the experiment (assuming Schrödinger's and Weyl's interpretations of quantum mechanics)? Note that the non-individuals approach in principle could account for the appearance of individuality in the particular context of the experiments: as noted, by combining the reflexivity of the indistinguishability relation with the principle that indistinguishable objects satisfy the same properties, identity is yielded. However, the problem is that one still needs to explain which features of *the particular experimental set-up* are responsible for the appearance of identity (and, perhaps, individuality). After all, non-individuals do not in general acquire identity (as noted earlier, on the non-individuals view, indistinguishability does not generally collapse into identity). Thus, the formal account just described of yielding identity (and, perhaps, individuality) does not answer the metaphysical question of why objects that lack identity suddenly acquire it: nothing in the experimental set-up, or, to the best of my knowledge, nothing given by current science, seems to correspond to the formal story. (It is unclear that decoherence provides an answer here, rather than a promissory note.)

Perhaps it could be said that in the case of the trapped ion, the object has been isolated, and thus it has been individuated, at least in the minimal sense that it has been separated from other things. However, the ion, even in the condition of the experiment, is not an individual for it cannot be uniquely determined, uniquely singled out. After all, if the experiment is reset, there is no fact of the matter as to whether the ion that is trapped in the reset experiment is the same or not as the one that had been trapped before (see Krause 2011). It seems to me, however, that to isolate something and to individuate something are separate notions. If one individuates something, one is able to determine which individual that object is; that is, what makes it the individual it is. On this conception, only individuals can be individuated. Thus, in the trapped ion experiment, if the ion is not an individual (and it would minimally need to have well-defined identity conditions to be an individual), it is difficult to see that the object in question has been individuated. It has only been isolated.

With regard to the use of Hacking's argument for entity realism in the context of the three experiments, one of the difficulties is precisely to specify the content of the resulting realism. After all, as soon as one tries to specify which entities one is realist about, one needs to invoke theories to characterize, or, at least, minimally describe, the relevant objects. But these theories, for all we know, may be false and may falsely describe the properties of the objects in question (a point that Hacking, of course, accepts). The problem, however, is that, as a result, there is no way of fully characterizing what one is realist about. Just stating that one is a realist about something (such as, whatever is going on in the machines used in the implementation of the three experiments) and leaving it at that is not enough to fully specify a robust form of realism. In fact, the very concept of an *entity* becomes highly problematic in the context in which objects may lack well-defined identity conditions. As James Ladyman argued, it is an ersatz form of realism that recommends commitment to objects of such a dubious metaphysical status (Ladyman 1998).

11.3.3 TROUBLES WITH TRACKING

As noted, it seems that we can only track an object if we are able to identify the position of the object in each instant of time and have reason to believe that it is still the same object from one instant to another. Without these conditions in place, we could not tell whether we are tracking the same object or different, but indistinguishable, instances of the same type of object. Quantum mechanics, however, does not settle this issue in the context of the experiments. Tracking is left still open.

On the two conceptions of tracking that we discussed (the spatiotemporal and the counterfactual ones) there are difficulties to claim that we track quantum objects. After all, on the spatiotemporal conception, we may think that we are tracking a certain object in space and time, when we are actually tracking a different object altogether. The case of studying DNA using an STM illustrates this possibility very clearly. Researchers thought they were tracking the geometrical configuration of the DNA surface, when they were actually tracking the structure of the underlying carbon. Similar problems emerge in the context of spatiotemporally tracking subatomic particles.

We do not face the same trouble when we consider counterfactual tracking. After all, in that case, using the relevant instruments, (a) if the object were not there, we would not believe it is; (b) if the object were there, we would. Given these conditions, we will not be in a position of thinking that we are tracking some object when we are in fact tracking something else. In particular, condition (a) rules out that possibility. However, it is not clear that we can actually know that these two conditions are met in the context of the experiments discussed. After all, our grip on whether we are counterfactually tracking the objects invoked in the experiments ultimately relies on spatiotemporal tracking (that is, the form of access we do have to the relevant objects), and, as noted earlier, this form of tracking is not always reliable in the context of the experiments in question. As a result, we still face problems with tracking.

Note that the proposal of this chapter is meant to support a form of empiricism that insists that there is a distinction between the observable and the unobservable, but that distinction, in contrast with van Fraassen's approach (1980, 2008), is not drawn in terms of what can be observed with the naked eye. After all, this conceptualization of unobservable seems unable to make proper sense of actual scientific practice in which, for instance, biologists, for good reason, are realist about bacteria, viruses, and ribosomes rather than being agnostic about them. A broader and independently motivated account of the observable is needed for that (see Bueno 2011). On this empiricist account, the counterfactual conditions that are known to hold provide as good a ground as observation to support the relevant conclusions. Electron microscopes are known to satisfy the counterfactual conditions, but this is not the case for the instruments involved in the trapping experiments. Hence, agnosticism about the objects in question is still well motivated.

11.4 Conclusion

Probably the strongest form of realism about quantum mechanics emerges from the experimental results obtained with a variety of sophisticated instruments, from STMs to ion traps. Several powerful arguments for realism emerge from this context. However, as I tried to indicate in this chapter, despite the strength of these arguments, they still fail to establish realism. There are still empiricist responses to them that are reasonable and independently motivated. So far, even experimental physics should not require realism.

Acknowledgments

My thanks go to Jonas Arenhart, Jody Azzouni, Thomas Benda, Newton da Costa, Steven French, Alexandre Guay, Décio Krause, Yi-Wei Liu, Harvey Siegel, and Kenneth Waters for extremely helpful and illuminating discussions. My thanks are also due to two anonymous reviewers for detailed comments on an earlier version of this work, which led to significant changes.

References

Arenhart, Jonas. 2017*a*. The received view on quantum non-individuality: Formal and metaphysical analysis. *Synthese* 194: 1323–1347.
Arenhart, Jonas. 2017*b*. Does weak discernibility determine metaphysics? *Foundations of Science* 32: 109–125.
Azzouni, Jody. 2004. *Deflating existential consequence*. New York: Oxford University Press.
Bueno, Otávio. 2011. When physics and biology meet: The nanoscale case. *Studies in History and Philosophy of Biological and Biomedical Sciences* 42: 180–189.
Bueno, Otávio. 2014*a*. Computer simulations: An inferential conception. *Monist* 97: 378–398.
Bueno, Otávio. 2014*b*. Why identity is fundamental. *American Philosophical Quarterly* 51: 325–332.
Bueno, Otávio. 2016. Epistemology and philosophy of science. In *Oxford handbook in the philosophy of science*, ed. Paul Humphreys, 233–251. Oxford: Oxford University Press.
Bueno, Otávio, and Scott Shalkowski. 2015. Modalism and theoretical virtues: Toward an epistemology of modality. *Philosophical Studies* 172: 671–689.
Chakravartty, Ankan. 2007. *A metaphysics for scientific realism: Knowing the unobservable*. Cambridge: Cambridge University Press.
Cushing, James. 1994. *Quantum mechanics, historical contingency, and the Copenhagen hegemony*. Chicago: University of Chicago Press.
Dehmelt, Hans. 1990. Experiments on the structure of an individual elementary particle. *Science* 247: 539–545.
Eigler, Don, and E. K. Schweizer. 1990. Positioning single atoms with a scanning tunneling microscope. *Nature* 344: 524–526.
French, Steven. 2014. *The structure of the world: Metaphysics and representation*. Oxford: Oxford University Press.
French, Steven, and Décio Krause. 2006. *Identity in physics: A historical, philosophical, and formal analysis*. Oxford: Clarendon Press.
Hacking, Ian. 1983. *Representing and intervening*. Cambridge: Cambridge University Press.
Hume, D. 1738/2000. *A treatise of human nature*, eds. David Norton and Mary Norton. Oxford: Oxford University Press.
Jia, Jin-Feng, Wei-Sheng Yang, and Qi-Kun Xue. 2005. Scanning tunneling microscopy. In *Handbook of microscopy for nanotechnology*, eds. Nan Yao and Zhong Lin Wang, 55–112. Dordrecht: Kluwer Academic Publishers.
Krause, Décio. 2005. Structures and structural realism. *Logic Journal of the IGPL* 13: 113–126.

Krause, Décio. 2010. Logical aspects of quantum (non-)individuality. *Foundations of Science* 15: 79–94.
Krause, Décio. 2011. Is Priscilla, the trapped positron, an individual? Quantum physics, the use of names, and individuation. *Arbor* 187: 61–66.
Ladyman, James. 1998. What is structural realism? *Studies in History and Philosophy of Science* 29: 409–424.
Muller, Fred, and Simon Saunders. 2008. Discerning fermions. *British Journal for the Philosophy of Science* 59: 499–548.
Nozick, Robert. 1981. *Philosophical explanations*. Cambridge, MA: Harvard University Press.
Schenkel, T., A. V. Hamza, A. V. Barnes, and D. H. Schneider. 1999. Interaction of slow, highly charged ions with surfaces. *Progress in Surface Science* 61: 23–84.
Schrödinger, Erwin. 1952. *Science and humanism*. Cambridge: Cambridge University Press.
van Fraassen, Bas. 1980. *The scientific image*. Oxford: Clarendon Press.
van Fraassen, Bas. 1989. *Laws and symmetry*. Oxford: Clarendon Press.
van Fraassen, Bas. 1991. *Quantum mechanics: An empiricist view*. Oxford: Clarendon Press.
van Fraassen, Bas. 2008. *Scientific representation: Paradoxes of perspective*. Oxford: Clarendon Press.
Wallace, David. 2012. *The emergent multiverse: Quantum theory according to the Everett interpretation*. Oxford: Oxford University Press.
Weyl, Hermann. 1931. *The theory of groups and quantum mechanics*. New York: Dover.
Woodward, James. 2003. *Making things happen*. Oxford: Oxford University Press.

{ 12 }

Retail Realism, the Individuation of Theoretical Entities, and the Case of the Muriatic Radical

Jonathon Hricko

12.1 Introduction

The scientific realism debate concerns, among other things, the existence of theoretical entities. Examples of such entities include the atom, phlogiston, caloric, and the electron. When it comes to the existence of such entities, Magnus and Callender (2004) have advocated a position known as *retail realism*. According to that position, we shouldn't attempt to resolve this issue by appealing to wholesale arguments, which are arguments about the existence of theoretical entities in general. Instead, we ought to appeal to retail arguments, which are arguments about the existence of particular kinds of theoretical entities. In that case, retail realists can be realists about some kinds of entities and anti-realists about others. But since realism about a kind of entity can take different forms depending on how one individuates kinds of entities, it's important for retail realists to have something to say about how to do so. I take retail realism to be a promising position, and since it is underdeveloped in this regard, my goal in this chapter is to develop it further by providing some ways in which retail realists can individuate kinds of theoretical entities.

In order to meet that goal, I proceed as follows. In Section 12.2, I introduce retail realism in more detail. In Section 12.3, I introduce the notion of the inclusiveness of individuation. In short, the more inclusively we individuate a kind of entity, the more theories there are in which it appears. In order to demonstrate that this notion bears on actual science, I apply it to a case from the history of chemistry. In Section 12.4, I discuss the details of that case, which involves the hypothetical component of hydrochloric acid known as the muriatic radical. In Section 12.5, I use the case of the muriatic radical to distinguish two forms of realism: a more selective kind, which requires individuating kinds of entities less inclusively, and a more encompassing kind, which does not. In Section 12.6, I argue that the muriatic radical should be individuated in a less inclusive way because doing so allows us to adopt a more selective form of realism. Finally, in Section 12.7, I use the case of the muriatic radical to show how individuating kinds of entities less inclusively also allows for a more encompassing form of realism. The upshot of my arguments in this chapter is that, when retail realists claim to be realists about a kind of entity, they ought to be clear about how inclusively they individuate that kind of entity.

12.2 Retail Realism

In the recent history of the scientific realism debate, the two most influential arguments have been the *no-miracles argument* (NMA), which purports to support realism, and the *pessimistic meta-induction* (PMI), which purports to support anti-realism. According to NMA, the explanatory and predictive success of our best theories in the mature sciences would be miraculous if those theories weren't at least approximately true. In contrast, PMI draws upon the successful-but-false theories of the past as a basis for inferring inductively that our current theories are false as well. The realism debate has been going on for some time now, and the participants in the debate continue to disagree about how it should be resolved.

A number of philosophers have argued that it would be unwise to wait for a resolution and that the realism debate actually requires a dissolution. Among them are Magnus and Callender (2004), who consider what realism might consist in if we were to dissolve the realism debate.[1] Their efforts to do so involve what they call "the wholesale/retail distinction" (2004: 320), which they apply to arguments, positions, and debates.

To begin with, there is the wholesale side of the distinction. NMA and PMI are paradigm cases of what Magnus and Callender call "wholesale arguments," which are "arguments about all or most of the entities posited in our best scientific theories" (2004: 321). Positions like realism and anti-realism can be wholesale as well. "Wholesale realism," for Magnus and Callender, "seeks to explain the success of science in general" (2004: 321), by invoking the existence of the entities posited in our best scientific theories. On the other hand, "wholesale anti-realism seeks to explain the history of science in general" (2004: 321) by showing how we needn't invoke the existence of theoretical entities. Finally, the debate that is conducted in terms of wholesale arguments, and that seeks to determine whether wholesale realism or wholesale anti-realism is correct, is what they call "the wholesale realism debate" (2004: 322). It is this debate that Magnus and Callender want to dissolve.

---

[1] Magnus and Callender argue in favor of dissolution on the grounds that the two most influential arguments in the debate—namely, NMA and PMI—embody the base rate fallacy. Since I'm concerned with their positive proposal—namely, retail realism—in what follows, I'll focus on that, as opposed to their reasons for dissolving the realism debate.

We can now turn to the retail side of the distinction. Magnus and Callender apply the term "retail" primarily to arguments, and "retail arguments," for them, are "arguments about specific kinds of things such as neutrinos, for instance" (2004: 321). The arguments that actual scientists put forward regarding the reality of particular kinds of theoretical entities are examples of retail arguments. One of Magnus and Callender's examples of a retail argument consists in Einstein and Smoluchowski's theory of Brownian motion and Perrin's subsequent experiments, which persuaded scientists to accept the reality of atoms (2004: 321, 334).

At this point, we can get clear on what kind of realism survives the dissolution of the wholesale realism debate. Magnus and Callender argue that dissolving the wholesale debate leaves the retail arguments untouched. As they put it, "there may be good reasons to be a realist about neutrinos, an anti-realist about top quarks, and so on" (2004: 333). Moreover, they see the wholesale–retail distinction as providing a "way of distinguishing profitable realism debates from unprofitable ones" (2004: 334). In short, debates conducted in terms of wholesale arguments are unprofitable, while those conducted in terms of retail arguments are profitable. While Magnus and Callender do not use the term "retail realism" in their paper, other philosophers (e.g., Dicken 2013) have since adopted the term to refer to Magnus and Callender's position. I'll follow their use, and use "retail realism" to refer to the position that the wholesale debate ought to be dissolved and exchanged for a number of retail debates concerning the reality of particular kinds of theoretical entities. Retail realism, then, is the realism with which we are left once we dissolve the wholesale debate, and it allows for the possibility of being a realist about some kinds of theoretical entities and an anti-realist about others.

While Magnus and Callender's paper has not caused the wholesale debate to disappear, it's arguably the case that retail arguments now occupy a more central place in the literature. Even among those philosophers concerned with assessing wholesale positions, many of the arguments they discuss are retail arguments concerning the existence of particular kinds of theoretical entities like the ether and caloric.[2] Moreover, as Dicken (2013) argues, a number of philosophers have defended positions resembling Magnus and Callender's retail realism—according to Dicken, one can find retail realism in the work of Psillos (2009), Saatsi (2010), and Stanford (2006).

I take it that retail realism is a promising position, but that, as it stands, it is underdeveloped. One of my central goals in the remainder of this chapter is to argue that realism about a particular kind of entity can take different forms depending on how one individuates that kind of entity. In that case, when retail realists claim to be realists about a particular kind of entity, they must say something about how they individuate that kind of entity.

---

[2] See, e.g., Psillos (1999: chapt. 6).

12.3 The Inclusiveness of Individuation

My aim in this section is to introduce the notion of the inclusiveness of individuation. Before introducing this notion, I'll first have to be clearer about the kinds of theoretical entities that I'm concerned with individuating. In order to do so, I'll utilize a distinction that Brading and Landry (2006) draw between the presentation and representation of theoretical entities:

> At the semantic level, . . . theoretical objects, as kinds of physical objects, may be *presented* via the shared structure holding between the theoretical models. However, at the ontological level, a physical theory, insofar as it is successful, must also *represent* particular physical objects and/or phenomena and not merely present kinds of physical objects. (2006: 573)

Their reason for introducing this distinction is "to maintain a level of description in which a physical theory can *talk about* electrons, as theoretical objects, without its having to *be about* electrons, as objects that are physically realized in the world" (2006: 573). In what follows, I'll apply this distinction not just to theoretical entities drawn from mathematically formalized theories in physics, as Brading and Landry do, but also to theoretical entities in general, with the caveat that I'll remain noncommittal about how, exactly, theories present kinds of theoretical entities. More qualitative theories may not present kinds of entities in terms of shared structure, and the important point, for my purposes, is not how they present kinds of entities, but just that they do so.

The notion of the inclusiveness of individuation concerns the presentation of kinds of theoretical entities. Suppose we're examining some collection of theories. If we conclude that all of the theories present the same kind of entity, we've individuated that kind of entity in a very inclusive way. In contrast, if we conclude that only one of those theories presents a particular kind of entity, we haven't individuated that kind of entity in a very inclusive way. In between these two extremes, we can talk about degrees of inclusiveness and make comparative judgments regarding how inclusively we individuate particular kinds of entities. For example, if we conclude that five theories present the same kind of entity *A*, but only three theories present the same kind of entity *B*, then we've individuated *A* more inclusively than we've individuated *B*.

I intend this notion to be applicable to kinds of theoretical entities that are not themselves individuals. The substances named by mass terms like "water" are often taken to be paradigm cases of non-individuals (Chauvier 2016: 27–28; Lowe 2016: 51). The central examples of my case study in Section 12.4 are chemical substances, and they all belong to this group. The individuality of other theoretical entities is more controversial. For example, in the philosophy of physics, it's a matter of some debate whether quantum particles like electrons are individuals.[3] It

[3] See, e.g., Ladyman and Ross (2007: 132–140).

may seem problematic to talk about individuating theoretical entities that are not themselves individuals, especially if one takes individuation to be a sufficient condition for individuality. But the problem here is only apparent. In the lab, scientists can individuate samples of chemical substances, as when chemists distinguish between a sample of water and a sample of hydrochloric acid. Analogously, at the level of presentation, we can individuate kinds of such entities. And even if electrons themselves are not individuals, it's still possible to individuate them qua kinds of entities presented by our theories. More generally, we can individuate the kinds of entities that our theories present even when it comes to entities that are not themselves individuals.

In Section 12.5, I'll argue that realism about a particular kind of entity can take different forms depending on how inclusively one individuates that kind of entity. That argument will make use of a case from the history of chemistry, and I'll now turn to that case.

12.4 The Case of the Muriatic Radical

Chemists in the late eighteenth and early nineteenth centuries hypothesized that muriatic acid (now known as hydrochloric acid) contains a substance that they called, among other things, the muriatic radical. I'll begin this section by discussing Scheele's work on muriatic acid in the days before chemists hypothesized the radical. After that, I'll examine Lavoisier's hypothesis of the radical. I'll trace its development through Davy's work on muriatic acid. And finally, I'll discuss Berzelius's views on the radical within the context of his theory of hydracids.

Before doing so, though, it will be necessary to make some remarks about terminology. With the exception of direct quotations, I'll generally use the term "muriatic acid" to refer to the substance that was variously referred to as hydrochloric acid, acid of salt, and marine acid, and I'll generally use the term "oxymuriatic acid" to refer to the substance that was variously referred to as chlorine, dephlogisticated acid of salt, and oxygenated muriatic acid. Moreover, I'll often write of the radical of muriatic acid (or "the radical" for short), even though it's not clear that the chemists I'll discuss were employing the same kind of entity. To be sure, these terminological choices may seem to imply a particular view about how to individuate these kinds of entities. In Sections 12.5, 12.6, and 12.7, I'll turn my attention to this issue. For now, though, I'll adopt this terminology simply for the sake of clarity.

12.4.1 SCHEELE AND THE PREHISTORY OF THE MURIATIC RADICAL

By the late eighteenth century, chemists had been working with muriatic acid for some time. But oxymuriatic acid was not isolated until Scheele did so in 1774. He isolated it by dissolving impure pyrolusite (manganese dioxide), which he called "manganese," in muriatic acid (1774/1931: 20).[4] Scheele called the substance he isolated "dephlogisticated acid of salt" (1774/1931: 31). His choice of terminology follows from his phlogiston theory. According to Scheele, only acids rich in phlogiston can dissolve manganese, and they do so by losing their phlogiston, which, in turn, combines with the manganese to effect its dissolution (1774/1931: 22). Since muriatic acid can dissolve manganese, Scheele concluded that it must contain phlogiston (1774/1931: 28–29). And so, Scheele inferred that the substance that he obtained at the end of the reaction must be a dephlogisticated acid—hence his choice of terminology. For Scheele, then, oxymuriatic acid is a simpler substance than muriatic acid, and one can obtain oxymuriatic acid by extracting the phlogiston from muriatic acid.

Even those, like Berthollet (1788: 276–277), who rejected Scheele's theory acknowledged him as the discoverer of oxymuriatic acid. More recent attributions of the discovery of chlorine to Scheele can be found in the work of early twentieth-century chemists like Lowry (1915: 211). These attributions make some sense in light of the fact that Scheele recognized that he had isolated a hitherto undiscovered substance.[5] However, given the way in which Scheele conceptualized that substance, it's problematic to simply state that he discovered chlorine. Without downplaying the significance of his discovery, we can say that what Scheele really discovered was a way to isolate the substance that later came to be known as chlorine.

12.4.2 Lavoisier's Oxygen Theory of Acidity

Chemists working in the aftermath of Scheele's discovery went on to hypothesize the muriatic radical as a component of both muriatic acid and oxymuriatic acid. And Lavoisier's oxygen theory of acidity provided the motivation for doing so.

Lavoisier's central claim regarding acids is "that oxygen is an element common to them all, which constitutes their acidity" and that "we must therefore, in every acid, carefully distinguish between the acidifiable base, which Mr de Morveau calls the radical, and the acidifying principle or oxygen" (1789/1965: 65). Lavoisier claims that radicals can be either simple substances or compounds (1789/1965: 115). And as Le Grand (1972: 11–12) observes, although Lavoisier's critics often took his theory to involve the claim that oxygen is a sufficient condition for acidity, for Lavoisier himself, oxygen is merely a necessary condition.

We can illustrate Lavoisier's theory by means of some of his examples. While acids like phosphoric acid, sulfuric acid, and carbonic acid all contain oxygen, they

---

[4] See Ashbee (2007: 18), Chang (2010: 51), and Lowry (1915: 210) for the claim that Scheele's "manganese" was pyrolusite. I'll follow Scheele's usage and use the term "manganese" to refer to pyrolusite.

[5] That said, Scheele does entertain the possibility that Stahl may have obtained it previously (1774/1931: 32).

differ from one another insofar as they contain different radicals, namely, phosphorus, sulfur, and carbon, respectively (1789/1965: 60–65). And while sulfurous acid and sulfuric acid contain the same radical, namely, sulfur, they differ in light of the fact that sulfurous acid contains less oxygen than sulfuric acid does (1789/1965: 67–68).

Lavoisier was able to show that a number of acids contain oxygen, but he was unable to show that every acid does. Boracic acid, fluoric acid, and muriatic acid proved to be problematic, but Lavoisier is clear that his theory extends to these substances as well. He claims that these acids are each composed of oxygen combined with a hypothetical radical, and, indeed, one finds the boracic radical, the fluoric radical, and the muriatic radical in Lavoisier's table of simple substances (1789/1965: 175).

Lavoisier hypothesizes the muriatic radical in a particularly striking passage:

> Although we have not yet been able, either to compose or to decompound this acid of sea-salt, we cannot have the smallest doubt that it, like all other acids, is composed by the union of oxygen with an acidifiable base. We have therefore called this unknown substance the *muriatic base*, or *muriatic radical*. (1789/1965:71–72)

While Scheele thought he had decomposed muriatic acid by removing its phlogiston, Lavoisier held that no chemist had succeeded in decomposing it since no one had been able to extract its oxygen. Hence, Lavoisier hypothesized the muriatic radical as that substance which combines with oxygen to form muriatic acid.

Lavoisier and his followers also rejected Scheele's view of oxymuriatic acid. Lavoisier held that this acid is a compound of oxygen and muriatic acid, which he called "oxygenated muriatic acid" (1789/1965: 73). But as Le Grand (1974: 214) and Ashbee (2007: 16) emphasize, it was really Berthollet (1788) who took up the challenge of reinterpreting Scheele's results in terms of the oxygen theory and who thereby did the most to establish Lavoisier's view of the composition of oxymuriatic acid. Importantly, the view of the oxygen theorists is a complete reversal of Scheele's view. While Scheele held that oxymuriatic acid is the simpler substance, for Lavoisier and Berthollet, muriatic acid is the simpler substance. Moreover, the muriatic radical is a component of both acids.

12.4.3 DAVY'S WORK ON THE ACIDS

One of the outstanding problems in chemistry at the time was to decompose muriatic acid and oxymuriatic acid and isolate the muriatic radical. One chemist who attempted to decompose these substances was Davy, who, for some time, held that muriatic acid is composed of water and "a substance, which as yet has never been procured in an uncombined state," and that "oxymuriatic acid is composed of the same substance, . . . united to oxygene [*sic*]" (1809: 468).

But in 1810, the year in which his work on the acids culminated, Davy rejected this view. In a paper read in July of that year, he writes:

> in the usual cases where oxygene [*sic*] is procured from oxymuriatic acid, water is always present, and muriatic acid gas is formed; now, as it is shewn that oxymuriatic acid gas is converted into muriatic acid gas, by combining with hydrogene [*sic*], it is scarcely possible to avoid the conclusion, that the oxygene [*sic*] is derived from the decomposition of water, and, consequently, that the idea of the existence of water in muriatic acid gas, is hypothetical, depending upon an assumption which has not yet been proved—the existence of oxygene [*sic*] in oxymuriatic acid gas. (1810: 236)

As Ashbee (2007: 23) emphasizes, chemists working during this period often misinterpreted their experimental results because they neglected the decomposition of water in their experiments. Davy's achievement in this case was to show that previous conclusions regarding the nature of muriatic acid and oxymuriatic acid did not properly take into account the role of water in various experiments. According to Davy, what happens in these experiments is that oxymuriatic acid combines with the hydrogen of the water to form muriatic acid, leaving oxygen as a byproduct. Since oxymuriatic acid is actually a component of muriatic acid, and since the production of oxygen in various experiments can be traced back to the water, the existence of oxygen in oxymuriatic acid has not been proved. And because of this fact, combined with the fact that the components of muriatic acid are oxymuriatic acid and hydrogen, the existence of water, and of oxygen, in muriatic acid is similarly hypothetical.

In the Bakerian Lecture that he gave in November 1810, Davy goes on to argue for the view that oxymuriatic acid ought to be considered an element.[6] Davy's argument for the elementary nature of this substance depends on his views regarding the nature of elements. As Siegfried (1964: 119) has observed, Davy's views on this matter owe much to Lavoisier's (1789/1965: xxiv) definition of an element as an as-yet-undecompounded body. Davy argues that oxymuriatic acid "has not as yet been decompounded" and is "elementary as far as our knowledge extends" (1811: 1). Golinski (1992: 223) has emphasized that, in making this argument, "Davy brought forward no dramatically new experimental evidence." Davy's point was just that, because existing methods had failed to decompose oxymuriatic acid, it ought to be considered an element. He goes on to push for a change in terminology, on the grounds that "oxymuriatic acid" is a poor name for "a body which is not known to contain oxygene [*sic*], and which cannot contain muriatic acid" (1811: 32). He suggests the names "chlorine" and "chloric gas" be used instead. Davy is thus credited with showing that muriatic acid is composed of hydrogen and

---

[6] Gay-Lussac and Thénard (1809: 358) had entertained this view about a year earlier. But Berthollet convinced them to reject it (Crosland 1980: 102; Gray, Coates, and Åkesson 2007: 45).

chlorine and that the latter is an element. And although he saw no reason to reject the term "muriatic acid" (1811: 33), his work shows why we now refer to that acid as hydrochloric acid.

Interestingly, Davy uses his conclusions in order to argue "that SCHEELE's view . . . may be considered as an expression of facts" (1810: 237). As Davy characterizes Scheele's view, "the illustrious discoverer of the oxymuriatic acid considered it as muriatic acid freed from hydrogene [*sic*]; and the common muriatic acid as a compound of hydrogene [*sic*] and oxymuriatic acid" (1810: 231). Here, Davy is taking advantage of the fact that it was common for phlogiston theorists working in the late eighteenth century to identify phlogiston with inflammable air, which we now call hydrogen.[7] At times, Davy himself engages in some phlogistic speculations and identifies phlogiston with hydrogen.[8] If we identify Scheele's phlogiston with hydrogen, we can see why Davy considers Scheele's view to be correct. However, a number of commentators have noted problems with identifying Scheele's phlogiston with hydrogen (Ashbee 2007: 22, 35; Ladenburg 1900: 19) or with inflammable air (Boantza and Gal 2011: 332–333). Indeed, in later work, Scheele claims that "inflammable Air is composed of heat and phlogiston" (1777/1780: 180), which rules out a direct identification of phlogiston with inflammable air. Hence, Scheele's view is only an expression of facts if one mistakenly identifies his phlogiston with hydrogen.

While Davy praises Scheele's view, he also argues that the view of the oxygen theorists "rests . . . upon hypothetical grounds" (1810: 237). If the acidifying principle—namely, oxygen—isn't present in these two acids, then that theory would seem to be in serious trouble. Indeed, a number of commentators have claimed that Davy's work on the composition of muriatic acid and oxymuriatic acid effectively killed off the oxygen theory of acidity (Brooke 1980: 121–123; Chang 2012: 33; Le Grand 1974: 224).[9] Davy also speculates that oxymuriatic acid may be another principle of acidity and that "on this idea muriatic acid may be considered as having hydrogene [*sic*] for its basis" (i.e., its radical; 1810: 243). Davy, of course, recognized that both oxygen and oxymuriatic acid are attracted to the positive surface of a Voltaic battery, while hydrogen is attracted to the negative surface (1810: 241–242). It's probable that this analogy between oxygen and oxymuriatic acid is among his reasons for speculating that oxymuriatic acid is the acidifying principle, leaving hydrogen as the radical. In any case, the point of Davy's speculations was just that there were now good reasons to doubt the central claim of the oxygen theory of acidity: namely, that oxygen is the sole principle of acidity.

---

[7] Kirwan (1789: 4–5) was a major proponent of this view, and he lists eleven chemists who, at one point or another, identified phlogiston with hydrogen. Cavendish (1766: 145) is another example, though he later gave up this view. Scheele is notably absent from Kirwan's list.

[8] See Siegfried (1964).

[9] But see Gray et al. (2007: 46–52) for the view that, by the time Davy presented his arguments, most chemists no longer strictly adhered to the oxygen theory of acidity.

12.4.4 BERZELIUS ON HYDRACIDS

While Davy is now credited with determining the composition of both muriatic acid and oxymuriatic acid, his views did not go unchallenged. One of his most prominent critics was Berzelius, who, for some time, opposed Davy's views regarding these two substances and defended a form of Lavoisier's oxygen theory of acidity.[10] In accordance with that theory, Berzelius held that the components of muriatic acid are oxygen and the "muriatic radicle" (1813: 254). Berzelius presents his own views on muriatic acid and oxymuriatic acid in terms of both atoms (1813: 254) and volumes (1816: 276) and claims that muriatic acid contains one atom/volume of muriatic radical and two atoms/volumes of oxygen, while oxymuriatic acid contains one atom/volume of muriatic radical and three atoms/volumes of oxygen. Regarding the muriatic radical in particular, Berzelius claims that its nature is still unknown and that chemists have not succeeded in isolating it (1813: 254; 1816: 263).

Berzelius's view of the composition of acids changed as time went on. In 1825, a translation of an appendix from his multivolume *Lärobok i Kemien* appeared as an article in the *Annals of Philosophy*. In that article, Berzelius advocates the view that there are, in general, two kinds of acids: oxygenous acids and hydracids. The oxygenous acids, for Berzelius, are composed of oxygen and an acidifiable base or radical. Berzelius was therefore in agreement with Lavoisier regarding the nature of these acids. In contrast, a hydracid, for Berzelius, is "a combination of a simple or compound body with hydrogen, which, although destitute of oxygen, possesses all the essential characters of the oxygenous acids" (1825: 180). Berzelius claims that "hydracids must be regarded, therefore, as constituted of hydrogen and a peculiar radical, which, as is the case with the oxygen acids, may be either simple or compound" (1825: 180). Hare (1835) quotes some similar passages from what he calls Berzelius's "treatise of Chemistry" (1835: 61), which was most likely a translation of Berzelius's *Lärobok i Kemien*. In one passage, Berzelius defines hydracids as "those acids which contain an electronegative body, combined with hydrogen," and he claims that "hydracids are divided into those which have a simple radical, and those which have a compound radical" (quoted in Hare 1835: 69). Hare also provides a quotation of Berzelius's definition of a radical, according to which "the combustible body contained in an acid, or in a salifiable base, is called the radical of the acid, or of the base" (quoted in Hare 1835: 69). Importantly, Berzelius does not define the radical of an acid as the non-oxygen part of the acid, and, for him, both oxygenous acids and hydracids contain radicals.

One of Berzelius's examples of a hydracid in his 1825 article is muriatic acid, which, by this point, he also calls hydrochloric acid. After distinguishing between "two classes [of hydracids], viz. acids with a *simple*, and acids with a *compound radical*," Berzelius writes: "To the *first* of these classes belong[s] . . . *Hydrochloric*

---

[10] For a good summary of Berzelius's arguments against Davy's views, see Gray et al. (2007: 61–64).

acid; ... *chlorine* ... being regarded as [a] simple substances [*sic*]" (1825: 181). This passage is significant for at least two reasons. First of all, Berzelius is claiming that hydrochloric acid is a compound of hydrogen and elementary chlorine, and this claim represents a significant departure from the position he advocated in the 1813 and 1816 papers discussed earlier. Gray, Coates, and Åkesson (2007: 66) observe that Berzelius avoided the topic of oxymuriatic acid between 1817 and 1820 and that, by the 1820s, his references to chlorine all suggest that it is a simple substance. However, they claim that Berzelius never explained why he made this change nor why he abandoned the terms "muriatic acid" and "oxymuriatic acid" in favor of "hydrochloric acid" and "chlorine," respectively. Moreover, they doubt that Berzelius ever truly converted to the view that chlorine is an element. But it's worth noting that, immediately after the quoted passage, Berzelius goes on to claim that his "principal arguments against the new theory of the constitution of muriatic acid," that is, Davy's theory, "can no longer be considered as valid," in light of "the analogy between muriatic acid and the hydracids" (1825: 181). Berzelius's appeal to this analogy may go some length toward explaining his use of the new theory and the new terminology. The second reason that this passage is significant is that it shows that, for Berzelius, the radical of hydrochloric acid is chlorine. After all, that substance is the nonhydrogen component of hydrochloric acid, and so, based on Berzelius's theory of hydracids, chlorine is the radical of hydrochloric acid.

12.5 Two Forms of Retail Realism

I'll now show how realism about a particular kind of entity can take different forms depending on how inclusively one individuates that kind of entity. It follows that retail realists have to say something about how they individuate the kinds of entities about which they are realists. I'll distinguish between a more selective form and a more encompassing form of realism about a particular kind of entity. And I'll illustrate these different forms of realism by considering two ways of individuating the muriatic radical.

I'll focus my attention on three theories: Lavoisier's oxygen theory of acidity, Davy's new theory regarding the composition of muriatic acid, and Berzelius's theory of hydracids. Lavoisier's theory presents the muriatic radical, or muriatic base, as a hypothetical component of muriatic acid. Davy argued that the components of muriatic acid are in fact hydrogen and chlorine, and he speculated that hydrogen is the basis (or base, in Lavoisier's sense) of muriatic acid. Hence, his theory, combined with his speculations, presents hydrogen as the muriatic radical. Finally, Berzelius's theory presents chlorine as the radical of hydrochloric acid. Each of these theories presents a kind of entity: the radical or basis of what these chemists referred to as muriatic or hydrochloric acid. In short, they present the muriatic radical.

The question at this point is how inclusively we should individuate the muriatic radical. By presenting the muriatic radical, do these three theories all present the same kind of entity, or do they present different kinds of entities? Since inclusiveness is a matter of degree, there are a number of options here. However, in order to illustrate my point, it will suffice to consider the two extreme options: namely, that each theory presents the same kind of entity and that each theory presents a different kind of entity. I'll label these two options *inclusive individuation* and *exclusive individuation*, respectively.

In light of the fact that Lavoisier, Davy, and Berzelius used similar (though not identical) terminology, we might opt for inclusive individuation. If we do, then the three theories I'm considering all present the same kind of entity: the muriatic radical. In that case, realism about that kind of entity amounts to the claim that these theories also represent the same kind of entity, whatever that might be. Opting for inclusive individuation thus results in an encompassing form of realism about the muriatic radical as it is represented by distinct theories.

Alternatively, one might opt for exclusive individuation. In that case, these three theories present three distinct kinds of entities: Lavoisier's muriatic radical, Davy's basis of muriatic acid, and Berzelius's radical of hydrochloric acid. Exclusive individuation allows for a more selective form of realism about these kinds of entities at the level of representation. There are two ways in which this form of realism can be more selective. First of all, it can amount to a kind of pluralist realism, according to which we distinguish kinds of entities both at the level of presentation and at the level of representation. For example, we might be realists about Davy's basis of muriatic acid on the grounds that he identifies it with hydrogen. Similarly, we might be realists about Berzelius's radical of hydrochloric acid on the grounds that he identifies it with chlorine. If so, we have a case in which two theories present two different kinds of entities (Davy's basis and Berzelius's radical) and also represent two different kinds of entities (hydrogen and chlorine). Second, after distinguishing kinds of entities at the level of presentation, we can selectively apply both realist and anti-realist attitudes at the level of representation. For example, we might be anti-realists about Lavoisier's radical on the grounds that his hypothesis regarding the composition of muriatic acid turned out to be false.[11] If so, then, unlike Davy's and Berzelius's theories, Lavoisier's theory presents a kind of entity (his radical) without representing it. By allowing for both of these two possibilities, exclusive individuation allows for a much more selective form of realism than the more encompassing form of realism that results from opting for inclusive individuation.

However, as I'll make clear in Section 12.7, exclusive individuation also allows for the more encompassing form of realism since distinct theories may present

[11] See Chang (2011: 417) for a similar conclusion regarding the existence of Lavoisier's radical. I'll say more about anti-realism regarding Lavoisier's radical in Section 12.6.

different kinds of entities and yet end up representing the same kind of entity. So while inclusive individuation results in the more encompassing form of realism, exclusive individuation doesn't automatically result in the more selective form of realism, but allows for both forms.

I've illustrated these two forms of realism by considering two options for individuation: inclusive individuation and exclusive individuation. But since inclusiveness is really a matter of degree, the two forms of realism differ from one another in degree, not kind. Realism about a kind of entity is more encompassing when there are more theories that represent that kind of entity. And when we individuate a kind of entity more inclusively, the result is a form of realism that is more encompassing. Realism about a kind of entity is more selective when there are fewer theories that represent that kind of entity. And when we individuate a kind of entity less inclusively, we allow for a form of realism that is more selective.

Since realism about a particular kind of entity can take different forms depending on how inclusively one individuates that kind of entity, retail realists must say something about how they individuate the kinds of entities about which they are realists. In short, my claim is that, just as retail realists have the option to be realists about the neutrino and anti-realists about the top quark, they should have the option to be realists about Berzelius's radical and anti-realists about Lavoisier's radical. But, importantly, this latter option requires retail realists to say something about individuation.

While these concerns about individuation are important for retail realists to consider, they may not be all that significant for wholesale realists and anti-realists. Since the arguments they offer focus on theoretical entities in general, the conclusions of those arguments aren't affected much by alternative ways of individuating kinds of entities at the level of presentation. Hence, individuation is a much more pressing issue for retail realists than it is for wholesale realists and anti-realists.

Up until this point, I've argued that individuating kinds of entities more inclusively results in a more encompassing form of realism, while individuating them less inclusively allows for a more selective form of realism. But how should retail realists determine how inclusively to individuate various kinds of entities? In Sections 12.6 and 12.7, I'll answer this question by considering what retail realists should say about two examples: the muriatic radical and oxymuriatic acid.

12.6 Example 1: The Muriatic Radical

What should retail realists say about Lavoisier's muriatic radical and Berzelius's radical of hydrochloric acid? Since we're dealing with two theories, we can consider two options. The first is an inclusive individuation of the radical and a more encompassing form of realism about it. The second is an exclusive individuation of the two radicals and a more selective form of realism. I'll argue that the more selective form of realism is preferable to the more encompassing form. And I'll argue that this constitutes a good reason to opt for exclusive individuation and reject inclusive individuation in this case.

Before doing so, I want to clarify my argumentative strategy by responding to an objection. If opting for the more selective form of realism can count as a reason for choosing exclusive individuation, then perhaps a retail realist can adopt the more selective form of realism before taking a stand on the issue of individuation or even without considering that issue at all. This is contrary to what I argued in Section 12.5. In order to respond to this objection, I'll begin by distinguishing two packages that one must choose between in this case. One consists of an inclusive individuation and the more encompassing form of realism. The other consists of an exclusive individuation and the more selective form of realism. We should consider each of these packages as a whole. If there are good reasons to prefer the more selective form of realism to the more encompassing form, those reasons are *ipso facto* good reasons to prefer exclusive individuation to inclusive individuation. Hence, I'm not committed to the consequence that retail realists can opt for the more selective form of realism without considering the issue of individuation. On the contrary, they ought to consider the issues of realism and of individuation simultaneously.

At this point, we can consider the combination of exclusive individuation and the more selective form of realism. On that package, Lavoisier's theory and Berzelius's theory present distinct kinds of entities: Lavoisier's muriatic radical and Berzelius's radical of hydrochloric acid, respectively. Distinguishing these kinds of entities at the level of presentation allows for the more selective form of realism at the level of representation. If there are good retail arguments for anti-realism about Lavoisier's radical and for realism about Berzelius's radical, respectively, then we have a good reason to opt for this more selective form of realism, and, *ipso facto*, a good reason to opt for exclusive individuation. I'll now argue that there are good retail arguments that support these conclusions.

First of all, Davy's work on muriatic acid supplies us with a good retail argument for anti-realism about Lavoisier's muriatic radical. As we saw in Subsection 12.4.3, Davy showed that the components of muriatic acid are hydrogen and chlorine. He thereby showed that Lavoisier and the oxygen theorists were wrong to claim that muriatic acid is composed of oxygen and a hypothetical radical. And if there is no oxygen in muriatic acid, Lavoisier's oxygen theory of acidity does not provide any reason to identify one component as the radical over the other. Moreover, both muriatic acid and oxymuriatic acid were supposed to contain Lavoisier's radical as a component substance, and Davy showed that there is no reason to think that these two substances share a component. Hence, the only viable option is to adopt an anti-realist attitude toward Lavoisier's muriatic radical. While Lavoisier's theory presents Lavoisier's radical, it doesn't represent it.

Second, Scheele's work on chlorine supplies us with a good retail argument for realism about Berzelius's radical of hydrochloric acid. As we saw in Subsection

12.4.1, Scheele discovered a way to isolate the substance that later came to be known as chlorine by dissolving what he called "manganese" (really, manganese dioxide) in muriatic acid. He conceptualized this substance within the framework of his phlogiston theory and referred to it as "dephlogisticated acid of salt." But even if chemists like Lavoisier took issue with Scheele's theory and choice of terminology, they admitted the reality of the substance that he had isolated, and they advanced a number of theories about it. The retail argument for realism about Berzelius's radical, then, is straightforward. Berzelius's radical is simply the substance that Scheele isolated in 1774. Since realism about this substance was appropriate in 1774, it was surely appropriate in 1825 when Berzelius identified chlorine as the radical of hydrochloric acid. Hence, we ought to be realists about Berzelius's radical. Berzelius's theory of hydracids both presents and represents Berzelius's radical.

The upshot of these retail arguments is a selective form of realism about Berzelius's radical but not about Lavoisier's radical. This selective form of realism requires exclusive individuation (they are part of the same package). This form of realism is inconsistent with inclusive individuation, which would result in a choice between an encompassing realism about the muriatic radical as it is presented by these two theories and a similarly encompassing anti-realism about that kind of entity. The two retail arguments that I've just discussed show that the selective form of realism is preferable to an encompassing realism (or anti-realism) regarding the muriatic radical. Therefore, we have a good reason to prefer exclusive individuation to inclusive individuation in this case.

12.7 Example 2: Oxymuriatic Acid

What should retail realists say about Scheele's dephlogisticated acid of salt, Lavoisier's oxygenated muriatic acid, the early Berzelius's oxymuriatic acid, and Davy's chlorine? Do the theories of these chemists present distinct kinds of entities or the same kind of entity? And what, if anything, do their theories represent? I'll argue that we ought to individuate the kinds of entities less inclusively in this case but that we ought to opt for the more encompassing form of realism. By doing so, I'll show how individuating kinds of entities less inclusively also allows for the more encompassing form of realism.

However, I'll start by arguing for a more inclusive individuation of what was known as oxygenated muriatic acid and oxymuriatic acid. More specifically, Lavoisier's oxygen theory of acidity and the theory that Berzelius defended in his 1813 and 1816 papers both present the same kind of entity, which we can call oxymuriatic acid. Both chemists held that oxymuriatic acid is a compound of oxygen and the muriatic radical. Moreover, Berzelius sees himself as defending a more developed version of Lavoisier's oxygen theory. As Berzelius puts it, the contrast is between "the old theory and that of Sir H. Davy," where the old theory is that of Lavoisier and the oxygen theorists, and he goes on to "give [his] reasons for considering the old theory as the most accurate" (1813: 254). The fact that Berzelius sees himself as defending the old theory of Lavoisier gives us a good reason to individuate oxymuriatic acid more inclusively.

The case of Davy's new theory and Scheele's phlogiston theory presents a useful contrast to the Lavoisier–Berzelius case. Davy's theory and Scheele's theory are best thought of as presenting different kinds of entities: chlorine and dephlogisticated acid of salt, respectively. While Davy claims that Scheele's view "may be considered as an expression of facts," he also emphasizes that it is "obscured by terms derived from a vague and unfounded general theory" (1810: 237). Davy does not see himself as presenting a further development of Scheele's phlogiston theory, and he rejects the view that elementary chlorine is a dephlogisticated acid. In that case, Davy's theory and Scheele's theory present different kinds of entities.

When we consider all four theories together, we arrive at the conclusion that, at the level of presentation, there are three distinct kinds of entities here: oxymuriatic acid, chlorine, and dephlogisticated acid of salt. This is because oxymuriatic acid differs from both chlorine and dephlogisticated acid of salt. An oxygenated acid differs both from a phlogisticated acid and an elementary substance. Moreover, it's significant that Lavoisier and Berzelius defended their oxygen theories in opposition to phlogiston theories like Scheele's and Davy's new theory, respectively. Hence, in this case, we have neither an inclusive individuation nor an exclusive one, but instead one that is, on the whole, not very inclusive.

At the level of representation, these theories all represent the same kind of entity; namely, the substance that we know as chlorine. The retail argument in this case is basically the same as the Scheele-inspired retail argument from Section 12.6. Scheele's real contribution was the discovery of a method for producing the substance that came to be known as chlorine. Chemists working in the aftermath of Scheele's discovery used his method and developed other methods as well. Since Scheele, Lavoisier, Berzelius, and Davy shared various methods and produced the same substance, they also theorized about the same substance. More specifically, Scheele's theory presents dephlogisticated acid of salt, Davy's theory presents chlorine, Lavoisier's and Berzelius's theories present oxymuriatic acid, and all of these theories represent chlorine. Hence, although we haven't individuated the kinds of entities very inclusively at the level of presentation, at the level of representation, we can adopt a more encompassing form of realism about the substance that was variously known as dephlogisticated acid of salt, oxymuriatic acid, and chlorine.

However, my argument faces a potential objection. For some time, philosophers of science have puzzled over the question of whether a theory that is not strictly true of some kind of theoretical entity can nevertheless represent it. Since Scheele's, Lavoisier's, and Berzelius's theories are not even approximately true when it comes to chlorine, perhaps those theories cannot represent chlorine. In order to respond to this objection, it's necessary to first emphasize that Scheele, Lavoisier, and Berzelius were all able to produce relatively pure samples of chlorine, which they used in their experiments. I take it that they couldn't fail to theorize about a substance that they could exercise such control over in their experiments. And since they theorized about chlorine, their theories represent chlorine.

This response to the objection shows that what really does the work in the retail argument for chlorine is Scheele's experiments, not his theory. This kind of view is, of course, not new. Hacking (1983) famously argued that our best guide to the reality of a theoretical entity consists in what scientists can do with it. More recently, Chen (2016: 365) has suggested that, "if a scientist can realize the individuality of an object in a particular experiment, then she has provided the strongest evidence . . . to warrant the reality of the object." And Chang (2011) has argued that chemists' operations for producing various substances (e.g., oxygen gas) justify them in retaining certain theoretical entities (e.g., oxygen) even when the theories in question have been rejected. If retail realists were to go so far as to defend some kind of experimental criterion for the reality of theoretical entities in general, they would have to show how, in doing so, they could avoid relying on the wholesale arguments that they want to reject. Instead of doing that, I'll simply note that experiments are doing the work in this retail argument, while I remain noncommittal about what does the work in other retail arguments.

Two consequences of my arguments in this chapter are (1) the more selective form of realism that we take toward kinds of entities at the level of representation requires that we individuate those kinds of entities less inclusively at the level of presentation, and (2) the more encompassing form of realism that we take toward kinds of entities at the level of representation is possible regardless of how inclusively we individuate kinds of entities at the level of presentation. This second consequence does not entail that the issue of individuation is unimportant when it comes to cases that warrant the more encompassing form of realism. On the contrary, being clear about individuation in such cases allows for a more nuanced realism. In the case of oxymuriatic acid, focusing on the issue of individuation allows us to simultaneously respect the fact that there were important theoretical differences among these four chemists and the fact that they were nonetheless working with the same chemical substance. It's not clear how one can develop such a nuanced realism while ignoring the issue of individuation. By shifting from realism about our best theories to realism about particular kinds of theoretical entities, retail realism allows for a more nuanced realism than wholesale realism does. What I'm suggesting is that, by focusing on the issue of individuation, retail realists can have an even more nuanced view.

12.8 Conclusion

By this point, I hope to have shown that it's important for retail realists to address questions regarding the individuation of kinds of theoretical entities. When they claim to be realists about a particular kind of entity, they ought to be clear about how they individuate that kind of entity and about why they opt for that way of individuating it. By employing the notion of the inclusiveness of individuation, it's possible for them to adopt a more selective form of realism in some cases and a more encompassing form of realism in others.

Acknowledgments

Thanks to Hasok Chang, Ruey-Lin Chen, In-Rae Cho, Melinda Bonnie Fagan, Derek Leben, Min Ouyang, Nicholas Tebben, Karen Yan, and two reviewers for comments on earlier drafts and presentations of the material in this chapter. Thanks to audiences at Kyoto University, National Chung Cheng University, and National Taiwan University for helpful questions and comments. Finally, much of the research for this chapter was completed during my time as a postdoctoral fellow at the Institute of European and American Studies at Academia Sinica, and so I thank the institute, and especially my sponsor, Jih-Ching Ho.

References

Ashbee, Ruth. 2007. The discovery of chlorine: A window on the chemical revolution. In *An element of controversy: The life of chlorine in science, medicine, technology and war*, eds. Hasok Chang and Catherine Jackson, 15–40. London: British Society for the History of Science.

Berthollet, Claude Louis. 1788. Mémoire sur l'Acide Marin Déphlogistiqué. *Histoire de l'Académie Royale des Sciences, Anée 1785*: 276–295.

Berzelius, Jöns Jacob. 1813. On the nature of muriatic acid. *Annals of Philosophy* 2(10): 254–259.

Berzelius, Jöns Jacob. 1816. Comparison of the old and new theories respecting the nature of oxymuriatic acid, to enable us to judge which of the two deserves the preference. *Annals of Philosophy* 7–8(40, 42; 45, 46): 272–280, 429–441, 200–209, 256–264.

Berzelius, Jöns Jacob. 1825. On hydracids, and their combinations with saline bases. *Annals of Philosophy* 10(3): 180–186.

Boantza, Victor, and Ofer Gal. 2011. The "absolute existence" of phlogiston: The losing party's point of view. *The British Journal for the History of Science* 44(3): 317–342.

Brading, Katherine, and Elaine M. Landry. 2006. Scientific structuralism: Presentation and representation. *Philosophy of Science: Proceedings of the 2004 Biennial Meeting of the Philosophy of Science Association, Part II: Symposia Papers*, ed. Miriam Solomon, 73(5): 571–581.

Brooke, John Hedley. 1980. Davy's chemical outlook: The acid test. In *Science and the sons of genius: Studies on Humphry Davy*, ed. Sophie Forgan, 121–175. London: Science Reviews.

Cavendish, Henry. 1766. Three papers, containing experiments on factitious air. *Philosophical Transactions* 56: 141–184.

Chang, Hasok. 2010. The hidden history of phlogiston: How philosophical failure can generate historiographical refinement. *HYLE—International Journal for Philosophy of Chemistry* 16(2): 47–79.

Chang, Hasok. 2011. The persistence of epistemic objects through scientific change. *Erkenntnis* 75(3): 413–429.

Chang, Hasok. 2012. *Is water H_2O? Evidence, realism and pluralism*. Dordrecht: Springer.

Chauvier, Stéphane. 2016. Why individuality matters. In *Individuals across the sciences*, eds. Alexandre Guay and Thomas Pradeu , 25–45. New York: Oxford University Press.

Chen, Ruey-Lin. 2016. Experimental realization of individuality. In *Individuals across the sciences*, eds. Alexandre Guay and Thomas Pradeu, 348–370. New York: Oxford University Press.

Crosland, Maurice P. 1980. Davy and Gay-Lussac: Competition and contrast. In *Science and the sons of genius: Studies on Humphry Davy*, ed. Sophie Forgan, 95–120. London: Science Reviews.

Davy, Humphry. 1809. New analytical researches on the nature of certain bodies, being an appendix to the Bakerian lecture for 1808. *Philosophical Transactions of the Royal Society of London* 99: 450–470.

Davy, Humphry. 1810. Researches on the oxymuriatic acid, its nature and combinations; and on the elements of the muriatic acid. With some experiments on sulphur and phosphorus, made in the laboratory of the Royal Institution. *Philosophical Transactions of the Royal Society of London* 100: 231–257.

Davy, Humphry. 1811. The Bakerian lecture [for 1810]: On some of the combinations of oxymuriatic gas and oxygene, and on the chemical relations of these principles, to inflammable bodies. *Philosophical Transactions of the Royal Society of London* 101: 1–35.

Dicken, Paul. 2013. Normativity, the base-rate fallacy, and some problems for retail realism. *Studies in History and Philosophy of Science* 44(4): 563–570.

Gay-Lussac, Joseph Louis, and Louis Jacques Thénard. 1809. Extrait Des Mémoires lus à l'Institut national, depuis le 7 mars 1808 jusqu'au 27 février 1809. *Mémoires de Physique et de Chimie de la Société d'Arcueil* 2: 295–358.

Golinski, Jan. 1992. *Science as public culture: Chemistry and the enlightenment in Britain, 1760–1820*. Cambridge: Cambridge University Press.

Gray, Tamsin, Rosemary Coates, and Mårten Åkesson. 2007. The elementary nature of chlorine. In *An element of controversy: The life of chlorine in science, medicine, technology and war*, eds. Hasok Chang and Catherine Jackson, 41–72. London: British Society for the History of Science.

Hacking, Ian. 1983. *Representing and intervening: Introductory topics in the philosophy of natural science*. Cambridge: Cambridge University Press.

Hare, Robert. 1835. Some encomiums upon the excellent treatise of chemistry, by Berzelius; Also objections to his nomenclature, and suggestions respecting a substitute, deemed preferable, in a letter to Professor Silliman. *The American Journal of Science and Arts* 27(1): 61–74.

Kirwan, Richard. 1789. *An essay on phlogiston and the constitution of acids*, 2nd ed. London: J. Johnson.

Ladenburg, Albert. 1900. *Lectures on the history of the development of chemistry since the time of Lavoisier*. Edinburgh: The Alembic Club.

Ladyman, James, and Don Ross. 2007. *Every thing must go: Metaphysics naturalized*. With John Collier and David Spurrett. New York: Oxford University Press.

Lavoisier, Antoine Laurent. 1789/1965. *Elements of chemistry*. New York: Dover.

Le Grand, H. E. 1972. Lavoisier's oxygen theory of acidity. *Annals of Science* 29(1): 1–18.

Le Grand, H. E. 1974. Ideas on the composition of muriatic acid and their relevance to the oxygen theory of acidity. *Annals of Science* 31(3): 213–225.

Lowe, E. J. 2016. Non-individuals. In *Individuals across the sciences*, eds. Alexandre Guay and Thomas Pradeu, 49–60. New York: Oxford University Press.

Lowry, Thomas Martin. 1915. *Historical introduction to chemistry*. London: Macmillan.

Magnus, P. D., and Craig Callender. 2004. Realist ennui and the base rate fallacy. *Philosophy of Science* 71(3): 320–338.

Psillos, Stathis. 1999. *Scientific realism: How science tracks truth*. London: Routledge.

Psillos, Stathis. 2009. *Knowing the structure of nature: Essays on realism and explanation*. Basingstoke: Palgrave Macmillan.

Saatsi, Juha. 2010. Form-driven vs. content-driven arguments for realism. In *New waves in philosophy of science*, eds. P. D. Magnus and Jacob Busch, 8–28. Basingstoke: Palgrave Macmillan.

Scheele, Carl Wilhelm. 1777/1780. *Chemical observations and experiments on air and fire*. London: J. Johnson.

Scheele, Carl Wilhelm. 1774/1931. On manganese or magnesia; and its properties. In *The collected papers of Charles Wilhelm Scheele, translated from the Swedish and German originals by Leonard Dobbin*, 17–49. London: G. Bell and Sons.

Siegfried, Robert. 1964. The phlogistic conjectures of Humphry Davy. *Chymia* 9: 117–124.

Stanford, P. Kyle. 2006. *Exceeding our grasp: Science, history, and the problem of unconceived alternatives*. New York: Oxford University Press.

{13}

Is Aldo Leopold's "Land Community" an Individual?

Roberta L. Millstein

13.1 Introduction

In his 1949 essay "The Land Ethic," Aldo Leopold—a twentieth-century forester, hunter, game/wildlife manager, and professor who has been very influential in environmental ethics and conservation biology—famously stated:

> A thing is right when it tends to preserve the integrity, stability, and beauty of the *biotic community*. It is wrong when it tends otherwise. (Leopold 1949: 224–225; emphasis added)

This passage, and Leopold's land ethic more generally, implies that the biotic community is a locus of direct moral obligation, or even, some argue, an entity with intrinsic value (Callicott 1987, 2013). But what did Leopold mean by "biotic community?" Interestingly, Leopold considered the biotic community to include not only biotic components but also abiotic components: "soils, waters, plants, and animals, or collectively: the land" (Leopold 1949: 204).[1] So, "biotic community" is a somewhat misleading term; *land community*, another term that Leopold employs in "The Land Ethic," seems more appropriate (and thus I will use it for the remainder of this chapter).[2]

---

[1] Indeed, in an earlier work, Leopold clarifies that he uses the term "biota" to include "not only plants and animals, but soils and waters as well" (1939: 727).

[2] Leopold's reference to "the" biotic community or "the" land community might lead one to think that there was only one such community, consisting of the entire earth. However, I do not think that was Leopold's meaning; I think it was just a poetic manner of speaking, one which I echo in this essay. Evidence for my claim can be found in "The Land Ethic" itself, where Leopold refers to marshes, bogs, dunes, and deserts as examples of "entire" biotic communities that are seen as lacking economic value (1949: 212).

Some authors have raised concerns about making the land community the ethical locus of the land ethic, regardless of what we call it. Kristin Shrader-Frechette sums up these worries well:

> Nor is it obvious how to define the system at issue. The ecological problem of defining the system at issue is analogous to the economic problem of defining a theory of social choice and choosing some "whole" that aggregates or represents numerous individual choices. Defining an ecological "whole" to which Callicott and Leopold can refer is especially problematic, both because the biologists (e.g., Clements, Elton, Forbes) cited by Callicott to explicate his [Leopold's] views are no longer accepted by most contemporary scientists as having correct views about ecological communities, and because the contemporary variant of Clements's position, the GAIA hypothesis, has been rejected by most ecologists as an unproved metaphor or mere speculation. At best it is an hypothesis. They admit the scientific facts of interconnectedness and coevolution on a small scale, but they point out that particular ecosystems and communities do not *persist* through time. Hence, there is no clear referent for the alleged "dynamic stability" of an ecosystem or community. (Shrader-Frechette 1996: 60)

From this passage, I glean the following concerns:

1. It is not clear how to define a land community, which Shrader-Frechette seems to think of in terms of an ecological community or an ecosystem.
2. The concept of "land community," at least as explicated by J. Baird Callicott in his earlier work,[3] is outdated and rejected by contemporary science, and the closest contemporary view has also been rejected by most contemporary scientists.
3. Particular land communities are not things that persist through long periods of time.
4. Thus, there is no clear referent for the land ethic and the stability that it seeks to promote.

Indeed, even Callicott, who has been called the "leading philosophical exponent of Aldo Leopold's land ethic" (Norton 2002: 127), thinks that particular land communities cannot be clearly identified, adding to Shrader-Frechette's list of concerns the following:

5. The boundaries of communities and ecosystems are not fixed by nature, but rather determined by the scientific questions that ecologists pose (see, e.g., Callicott 2013; Eliot 2013).

---

[3] See Callicott (2013) for a revised and updated account.

Some of these concerns can be understood as concerns that the land community is not an *individual*. Yet Leopold did seem to think that the land community was an individual, even if he didn't use the word "individual" specifically; for example, in some essays (e.g., Leopold 1923/1991, 1934/1991, 1944/1991) he explored the idea that the community is an organism. In "The Land Ethic" (1949), he seemed to de-emphasize (although not eliminate) his characterization of communities as organisms, but even there he still maintained that the members of communities are interdependent; referred to soils, waters, plants, and animals *collectively*; and spoke of the organization (or, in degraded situations, the *dis*organization) of land. Relatedly, in an unpublished essay written in 1944, Leopold "sketch[ed] the concept of land-as-a-whole" (Leopold 1944/1991: 310), described land-health (or land-illness) as an attribute of the community as a whole, and suggested that "the components of land have a collective as well as separate welfare" (Leopold 1944/1991: 316). These all seem to be ways of getting at what we would today term "individuality." Can these views of Leopold's be defended, or is the land ethic undercut because the concept of "land community" cannot be characterized, as Shrader-Frechette suggests?

In what follows, I will examine the concept of a "land community" focusing on the following two questions in particular:

1. Is the concept of "land community"—which, as we shall see, blends ideas that would more typically belong to either community ecology or ecosystem ecology—hopelessly misguided or outdated? Does any past or contemporary work support such an entity?
2. If the concept of a "land community" can be defended, is it coherent enough to be a locus of direct moral obligation or an entity with intrinsic value?[4] In other words, is it an *individual*?[5]

My ultimate goal is to see if there is a *defensible* concept of "land community" as an individual that is close to Leopold's stated views; I will argue that there is. This will not establish that the land community *is* a locus of direct moral obligation or an entity with intrinsic value—it will only establish that it is a *candidate* for being a morally considerable entity (i.e., I seek to establish only a necessary condition, not

[4] Here I set aside the question of whether there is such a thing as intrinsic value (see, e.g., O'Neill [1992], McShane [2007], and Zimmerman [2015] for general discussion). The question I am asking is rather, assuming that intrinsic value exists, does a land community have the necessary characteristics for intrinsic value? Or, if there is no such thing as intrinsic value, does a land community have the necessary characteristics for moral consideration?

[5] See Odenbaugh (2007, 2010) and Eliot (2011, 2013) for discussion of ecological communities and ecosystems as individuals. The view I defend here has some similarities to both accounts; see also Levins and Lewontin (1980, 1985). Lean (2015) argues against individuality for communities but defends a related concept, which he calls an "indexical community." However, these accounts do not address the question of whether communities and ecosystems can or should be combined and whether the combination would be an individual.

a sufficient one). But rather than solely looking to Clements, Elton, and others to understand Leopold—influences are important, but influence is never complete—let us begin with Leopold himself.

13.2 Land Communities as Blended Community-Ecosystems

Land, Leopold tells us, "is not merely soil; it is a fountain of energy flowing through a circuit of soils, plants, and animals. Food chains are the living channels which conduct energy upward; death and decay return it to the soil" (1949: 216). So, Leopold's concept of a land community not only included abiotic components, but was also at least partially characterized in terms of *matter and energy flow*. This was represented by a biotic pyramid (visually depicted in a diagram in his 1939 essay, "A Biotic View of Land") that he described as follows:

> Plants absorb energy from the sun. This energy flows through a circuit called the biota, which may be represented by a pyramid consisting of layers. The bottom layer is the soil. A plant layer rests on the soil, an insect layer on the plants, a bird and rodent layer on the insects, and so on up through various animal groups to the apex layer, which consists of the larger carnivores. (Leopold 1949: 215)

However, Leopold also emphasized that a land community is composed of *interdependent parts*. Drawing on the work of community ecologist Charles Elton (see Newton [2006] for discussion), Leopold described a complex tangle of lines of dependency for food and other "services" such as shade (see Figure 13.1). Food chains are sometimes thought of as just an energy conduit, but for Leopold they also represented trophic (feeding) relationships and other types of relationships between members of different species. Leopold thus stressed the interactions between organisms and the way in which changes in some species affect other species, and he did so throughout "The Land Ethic" and elsewhere. For example, he analyzed the consequences of deer overbrowsing in the absence of predators (Leopold 1943).

Leopold's land community concept thus emphasizes matter and energy flow through organisms and abiotic components and also emphasizes the interdependence among organisms and abiotic components. This is notable because some authors (e.g., Callicott and Mumford 1997; Odenbaugh 2007) have differentiated between two types of entities and two types of approaches in ecology: ecological communities and a community ecology approach on the one hand, ecosystems and an ecosystem ecology approach on the other. It is the latter approach, the ecosystem approach, that includes abiotic components and invokes matter and energy flow, de-emphasizing (or even disregarding) organisms and populations. This

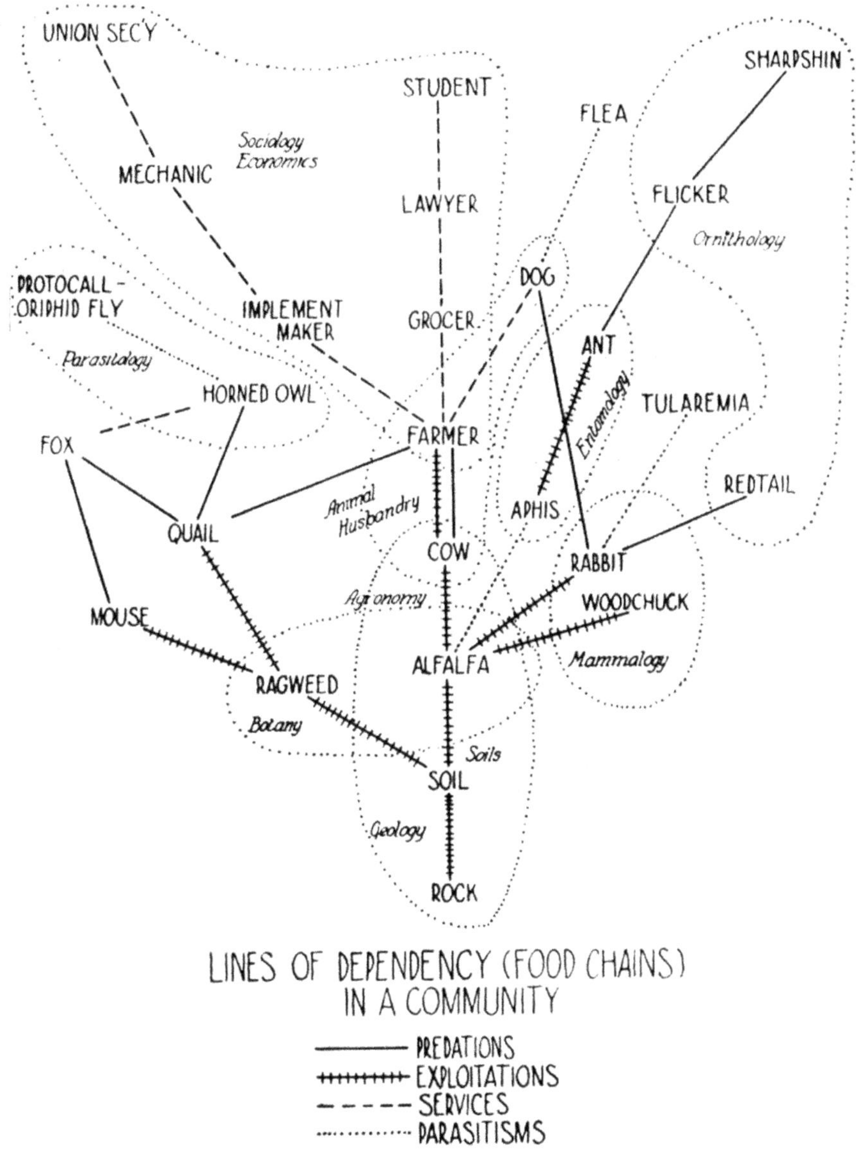

FIGURE 13.1 *How Leopold visualized the tangled "lines of dependency" in a land community.*

Reprinted from Leopold, Aldo.1942. The role of wildlife in a liberal education. *Transactions of the Seventh North American Wildlife Conference* April 8–10, 1942: 485–489. Copyright 1942 by the Wildlife Management Institute. Reprinted with permission.

approach contrasts with the former approach, the community ecology approach, which emphasizes interactions between organisms.[6]

Therefore, Leopold's "land community" concept combines aspects of the concept of "ecological community" as it is typically conceived with aspects of the concept of "ecosystem" as it is typically conceived.[7] But is this combination tenable? Does it complicate the case for the land community as an individual? In order to answer these questions, we need to have a better understanding of how Leopold's land community concept is situated in the history of these terms. Was he bucking a cogent scientific consensus (was he just a lone wolf?), or is his stance understandable given the historical flow of ideas? I seek to address these questions because there seems little point in spending time discussing whether an idiosyncratic scientific concept satisfies a philosophical conception of individuality, especially if the concept turned out to be misguided or outdated (concern (2) of the list in Section 13.1) with respect to our contemporary understandings. However, as we shall see, it is not idiosyncratic, misguided, or even outdated to blend the concept of ecological community with the concept of ecosystem. Moreover, the examination of others who have had similar ideas to Leopold's will help both to flesh out the land community concept and to highlight potential challenges.

13.3 Early Community and Ecosystem Concepts

What follows is a potted timeline of central figures in the development of ecological community and ecosystem concepts, with an eye toward situating the concept of a Leopoldean land community.

Frederic Clements (1916) characterized multispecies groupings as *communities*.[8] He is particularly (and somewhat notoriously) known for thinking of communities as organisms, but Eliot (2011) has given good reason to think that Clements's commitment to communities as organisms has been overstated. Regardless, as mentioned earlier, there are traces of thinking of communities as organisms in Leopold (Callicott 1987).

---

[6] In what follows, I will use the term "land community" when I mean to refer to Leopold's conception or my elaboration of it; when I use just "community" or "ecological community" I am referring instead to its meaning in community ecology.

[7] To be clear (and as will be discussed further later), not *all* ecologists conceive of "ecosystem" and "community" in these strictly divergent ways; some include population interactions in their study of ecosystems and some include abiotic components in their study of communities. Indeed, I hope to debunk the notion that these are always disparate approaches.

[8] Lynn Nyhart (2009) traces the idea of a biological community even earlier, to Karl August Möbius in the late nineteenth century, who characterized these living communities (*Lebensgemeinschaft*) in terms of the dependence of their members on one another and on their physical conditions of existence.

Charles Elton (1927) believed that plants and animals are not mere assemblages of species living together, but rather that they form closely knit communities[9] with interdependent members comparable to our own human communities; he also believed that these other plant/animal communities include humans. He saw relations between animals as largely food relations, giving rise to food chains, the food-cycle, and the "pyramid of numbers" (Elton 1927). Leopold met Elton in 1931 (Meine 2010) and was very influenced by him (Newton 2006), not only in terms of Elton's concept of community but also by his ideas concerning food chains and the pyramid of numbers (what Leopold called the "land pyramid"). The parallels between the two sets of ideas are obvious and striking.

Arthur Tansley (1935) is generally credited with the ecosystem concept. He argued that plants and animals are too different to be considered part of the same community and so rejected the community concept altogether. He did believe that biomes (*sensu* Clements), "the whole webs of life adjusted to particular complexes of environmental factors," are real "wholes," often integrated wholes—but he did not believe that they are organisms. Rather, he maintained that biomes, together with all of the physical factors involved, are *systems* (ecosystems); this, he suggested, is the more fundamental conception. According to Betty Jean Craige (2002), Leopold's concept of "land" as a system with biotic and abiotic components predated Tansley in Leopold's 1933 essay, "The Conservation Ethic" (1933/1991); according to Callicott (2013), Leopold anticipated many of Tansley's ideas of an ecosystem in his 1939 essay, "Biotic View of Land." In other words, it appears that Leopold included abiotic components and energy flow independently of (and perhaps prior to) Tansley. Thus, Leopold may be less in a Tansleyan tradition and more in an Eltonian one. Callicott (2013) speculates that Leopold may have influenced Lindeman (perhaps through Evelyn Hutchinson) although Lindeman doesn't cite Leopold.

Raymond Lindeman (1942) was a key developer of the ecosystem concept—but for him, trophic or "energy-availing" relationships occur *within the community*. He thought that the discrimination between living organisms as parts of the "biotic community" and dead organisms and inorganic nutritives as parts of the "environment" seemed "arbitrary and unnatural."

Eugene Odum (1971), like Lindeman, included a community of interacting organisms leading to a flow of energy in his ecosystem concept:[10]

> the community cannot exist without the cycling of materials and the flow of energy in the ecosystem. An ecosystem is any unit that includes all of the organisms (i.e., the "community") in a given area interacting with the

[9] Odenbaugh (2007, 2010) attributes to Elton an ecosystem concept rather than a community concept, but I don't see evidence for this (see also Hagen 1992).

[10] Odenbaugh (2007) says that Odum characterized ecosystems purely in terms of their energetics. But there still seems to be a strong community component to Odum's concept.

> physical environment so that a flow of energy leads to clearly defined trophic structure, biotic diversity, and material cycles (i.e., exchange of materials between living and nonliving parts) within the system. (1971: 8)

Perhaps not by accident; he was influenced by Leopold (Craige 2002) and cited Leopold explicitly.

So, Odum's and Lindeman's *ecosystem* ideas, by incorporating *community* elements, are actually very much in line with the idea of a Leopoldean land community and could thus potentially be used to flesh out Leopold's concept. Moreover, the division between community ecology and ecosystem ecology is not as clean as, for example, Callicott and Mumford (1997) would have us think.[11] The history of the two is entangled (Hagen 1992), and using a concept drawing from both areas of ecology is not a rogue eccentricity of Leopold's.

But a lot has changed since 1971, flagged by Donald Worster's (1990) critique of Odum's (and Clements's) notion that nature moves toward order and harmony:

> Ecology is not the same as it was. A rather drastic change has been going on in this science of late—a radical shifting away from the thinking of Eugene Odum's generation, away from its assumptions of order and predictability, a shifting toward what we might call a new *ecology of chaos*. (Worster 1990: 162)[12]

And even if Worster's critique is off the mark, we might still reasonably wonder whether there are *contemporary* candidates for a land community that combine community and ecosystem elements, with or without assumptions of stability.

13.4 Contemporary Community–Ecosystem Concepts

The short answer is "yes": O'Neill (2001), Chapin et al. (2011), Schultze, Beck, and Müller-Hohenstein (2005), and Hastings and Gross (2012) all combine ecological community and ecosystem elements (again, interactions between organisms and matter/energy flow, respectively) in describing the entities that they study, as I describe in what follows.[13]

O'Neill (2001) argues that we must recognize the simple empirical fact that ecosystems are collections of interacting populations, with component populations shaped by natural selection (O'Neill 2001: 3278). He suggests that the resulting biotic potential determines ecosystem dynamics just as much as chemical and

---

[11] A caveat: the *practice* of community ecology and ecosystem ecology may be more divergent than ecologists' stated concepts of community and ecosystem are. My focus for now is on the concepts.

[12] I don't fully accept Worster's characterization of past or contemporary ecology, but I will grant it for the purposes of this chapter.

[13] Biodiversity and ecosystem functioning (BEF) research also seems to embrace a blended community–ecosystem approach; see discussion and references in Dussault and Bouchard (2017).

physical constraints. He points out that while various problems can be solved by viewing an ecosystem purely in terms of functional groups that recover to the same rate processes, feedbacks, and complex organization (e.g., accounting for situations where species are moving in and out of an area over time), such an approach creates its own problems (for example, a matter/energy flow focus overlooks ecotones, i.e., zones where one vegetation type suddenly changes into another, and also minimizes the role of natural selection). In other words, we *need* to study both community and ecosystem elements. He further maintains that the critical property of an ecosystem is not stability but rather the ability to change state in response to a continuous spectrum of change and variability (sustainability).

Chapin et al. (2011) maintain the view that ecosystem ecology addresses the *interactions* between organisms and their environment as an integrated system—that it addresses the interactions that link biotic systems, of which people are an integral part, with the physical systems on which they depend. On their view, an ecosystem consists of all the organisms and the abiotic pools with which they interact; ecosystem processes are the transfers of energy and materials from one pool to another. They state that they are taking a "nonequilibrium perspective," recognizing that most ecosystems exhibit unbalanced inputs and losses, their dynamics are influenced by varying external and internal factors, they exhibit no single stable equilibrium, disturbance is a natural component of their dynamics, and human activities exert a pervasive influence. So, although the definition of ecosystem that Chapin et al. provide is a bit unfocused, they do allow for interaction together with a nonequilibrium perspective and also usefully distinguish the *processes* of an ecosystem from what an ecosystem *is*.

Schulze et al. (2005) characterize ecosystems as *networks of interrelations* between organisms and their environment in a defined space. They assert that the limits of an ecosystem must extend so far that the essential parts of material turnover per ground area (e.g., carbon assimilation, nitrogen mineralization, formation of ground water, etc.) are taken into account quantitatively. This suggests that there can be mistaken ways of characterizing the boundaries of an ecosystem because to leave out areas of significant material turnover would cause a misdescription of the characteristics and limits of the partial system. More on this point later.

Hastings and Gross (2012) characterize an *ecosystem* as a system composed of both the organisms (animal, plant, microbe) and the abiotic environment and all interactions among and between these components. On the other hand, they characterize an ecosystem *model* as a model designed to capture the pools and fluxes of mass (and sometimes energy) in an ecosystem. This nicely separates what an ecosystem *is* from the particular aspects of an ecosystem (flow of materials and energy) that an ecologist might study.

In summary, here are some important and useful insights that can be drawn from these contemporary ecologists who combine community and ecosystem approaches. First, both the "matter/energy flow" alone approach creates problems as does the "population interaction" alone approach (O'Neill 2001). Second, sustainability, rather than stability, may be the relevant property (O'Neill 2001), so that the worries that Worster raises become moot; Chapin et al. (2011) also explicitly advocate for a nonequilibrium approach. Third, what an ecosystem *is* may be different from its *processes* (Chapin et al. 2011) and its *models* (Hastings and Gross 2012). Fourth, some purported boundaries may exclude relevant processes and thus be inappropriate (Schulze et al. 2005). Fifth, these ecologists collectively challenge the picture of community ecology and ecosystem ecology as distinct approaches.

13.5 Potential Problems with a Combined Community–Ecosystem Concept

But there are some potential problems with a combined community–ecosystem approach, flagged even by those who endorse such an approach. Post et al. (2007) maintain that boundaries are set by discontinuities or steep gradients in the flux and flow of material and energy and/or by discontinuities or steep gradients in interactions between populations of different species. However, they point out that whereas some systems are *well-bounded* (which is not to say "closed"), others are *open* (Post et al. 2007). In well-bounded systems (e.g., lakes, islands) these two criteria coincide—and coincide with physical boundaries as well—making delineating ecosystem boundaries relatively straightforward. In such systems, "interactions among organisms are typically stronger and cycling of material and energy is typically tighter within than across the physical boundaries of these ecosystems" (Post et al. 2007: 115). On the other hand, in open systems (e.g., most terrestrial habitats, estuaries, and streams), the two approaches do not coincide, as when resources come from areas where species are not interacting (e.g., upstream). The problem, then, is how to understand the boundaries of open systems. As indicated in Chapter 1 of this volume, one important aspect of epistemic individuation is demarcating an individual from other things and environments. In biology, abiotic elements are often considered "the environment," but with the land community, the abiotic components are part of the entity, so the question is even more challenging than usual.

Post et al. (2007) describe various challenging scenarios for understanding the boundaries of open systems. Suppose, for example, large "inputs" are coming from the "outside" at short temporal scales, as is the case when, for example, highly mobile organisms such as geese and migrating fish move large amounts of nutrients around the landscape, with lakes, wetlands, and streams receiving these nutrients.[14] Then we should recognize that the system is larger than we had initially thought.

[14] O'Neill et al. (1986) imagine a similar situation, but where the community is more extensive than the matter/energy flows.

Alternatively, suppose that "internal" cycling of matter/energy is stronger than "external" inputs, as is the case when, for example, a watershed is visited by few mobile organisms or when it has a very high productivity. Then we should consider "internal" cycling to dictate the boundary.

Post et al. conclude: "In open ecosystems where there is little or no congruence among physical and functional boundaries . . . each different question may dictate very different definitions of ecosystem boundaries" (2007: 122). Although this is an interesting position worth considering, it doesn't seem to me that Post and colleagues have made the case for it. That is, they don't seem to have described situations where different questions would indicate different boundaries; on the contrary, as I describe in the preceding paragraph, they seem to have offered solutions to such cases where one might *think* that such a problem arose. Perhaps they mean simply to suggest, as they suggest at various points in the paper, that different temporal scales might affect whether we consider a system to be open or well-bounded, with all systems being open given a long enough time scale. While this is an important consideration, and while relevant time scale needs to be factored into questions concerning boundaries, it does not seem to follow that different questions dictate different system boundaries. To put the point another way, it seems that they have offered a scheme for understanding how, *given* a time scale, system boundaries ought to be characterized.

But even taking Post et al.'s stated conclusion at face value, a question remains: are open systems where different questions dictate different ecosystem definitions and different ecosystem boundaries coherent enough to be entities that we owe direct obligations to or to be entities that have intrinsic value? (This is concern (5) of the list in Section 13.1.) This question has an ontological component and an ethical component, which I will discuss in turn.

In order address the ontological component—namely, the status of interest-relative entities—it will be instructive to consider similar views. Callicott likewise holds that ecosystems "are in effect defined, both spatially and temporally, by the ecological question posed" (2013: 3), yet he maintains that they are "real, existing entities" (2013: 94)—at least in part. That is, he maintains that "when we come to isolate them, to bound them, for purposes of ecological study, we partly create them" similar to the way, he says, that "electrons emerge fully into existence when quantum physicists measure them" (2013: 41). Setting aside questions about his interpretation of physics and the strangeness of this analogy, it is difficult to understand what Callicott means. Does he mean that, in the absence of investigators, there really are no ecosystems or communities, at least not in a full sense? Perhaps not, but then it seems as though one cannot continue to defend ecosystems and communities as individuals or as "real, existing entities."

Eliot (2013) offers a more sophisticated version of the argument that our interests partially determine whether something is a community. According to Eliot, boundaries are determined by the set of causal relations relevant to some interest; furthermore, a community is "a real object, in so far as its component populations are connected by a particular kind of causal connection" (2013: 8).[15] Odenbaugh (2010) similarly claims that different causal relations may pick out different ecosystems, although he does not tie this claim to our human interests. However, we don't seem to have such loose causal relation requirements for other putative individuals, such as organisms.[16] For example, in the human body, "circulates blood" does not fully coincide in physical space with "circulates oxygen," yet we think of those causal relations as picking out the same individual (the same organism), not two different individuals. This is presumably because the system that circulates blood and the system that circulates oxygen are tightly interconnected with other systems from which they are not fully separable. So we should at least consider whether the same is true for putative communities/ecosystems.

Turning to the ethical component of the question, if (contra to what I argued earlier) Callicott's understanding of the ontology is correct—if ecosystems are real yet interest-relative—then it seems like the moral considerability of an ecosystem is dependent on an ecologist studying it. If the ecosystem loses moral considerability (indeed, ceases to fully *be* an ecosystem) when it is not being studied, then that is a weak notion of moral considerability indeed. So, Callicott's understanding of "ecosystem" is insufficient for his (and our) ethical purposes, making a continued search for a possible workable alternative desirable.

On the other hand, Eliot's and Odenbaugh's alternative approaches toward characterizing communities/ecosystems, which are ultimately grounded in causal relations, also run into problems concerning moral considerability. Here, the problem arises because of the multiplicity of possible boundaries. To be clear, the worry here is *not* that boundaries may be a bit fuzzy, since surely many real, existing entities have fuzzy boundaries; for example, to be an organism is to constantly lose and gain cells, yet human organisms are surely "real, existing entities." (We thus ought not be surprised when land communities turn out to have fuzzy boundaries as well, as they do.) Rather, the problem is one (as highlighted by Russow 1981 for the case of species) of how many entities there are. Considering a given geographical area, do we have one land community, two land communities, or more? Perhaps a lack of a definitive answer to that question is not problematic on its own, but what if different ways of drawing boundaries for a given geographical area cross-cut each other, so that, in some cases, we have one land community as a subset of another, whereas with others, one land community overlaps with another? It makes it unclear as to what we have moral obligations to, especially if we have to make choices between different purported land communities. Can we eliminate or add moral obligations simply by asking different sorts of

[15] See also McShane (2004) on this point.

[16] This is not to deny that characterizing organisms is without challenges (see, e.g., Dupré [2010] and [Clarke 2011] for some of the complications), just to suggest that organisms are often taken to be prototypical individuals, thus making it reasonable to argue using them as analogy.

scientific questions? Are all possible ways of drawing boundaries equally legitimate? It would seem that there are an infinite number of possible ways of drawing boundaries for a given geographic area;[17] do our moral obligations shift with each possible drawing? That seems untenable and unworkable.

Moreover, would it be wise to try to treat purported land communities well while failing to consider some of the population interactions or energy flows relevant to their sustainability? It seems as though one would run into practical problems if one did so. To use the analogy of the human body again, it would be akin to trying to benefit one's arm muscles without consideration of one's heart and lungs; eventually, the arms will fail when the body fails from ill health. So again, we ought to consider more closely the question of whether there are ways that boundaries of open systems can be delineated that are more systematic and defensible.[18]

13.6 Response to Problems

Indeed, there seem to be (at least) three ways of handling open systems, systems where the spatial area of the densely interacting populations is larger than that of the dense matter/energy flow—or vice versa:

One possibility is that *the land community exists within the* larger *of the two areas.* In other words, we always "go big"; if the spatial area of the densely interacting populations is larger than that of the dense matter/energy flow, the land community consists of the area covered by the densely interacting populations, whereas if the area of the dense matter/energy flow is larger than that of the densely interacting populations, the land community consists of the area covered by the dense matter/energy flow. However, a possible problem with this approach is that we would lose the concept of the ecosystem as a focal level, going beyond locales that lend themselves to concrete study in the field—perhaps to biomes (Currie 2011). This doesn't strike me as a devastating objection—we could just acknowledge that what we study is always a subset of the entity itself—but the objection is worth taking seriously, especially if there are better alternatives.

A second possibility is that *the land community exists within the* smaller *of the two areas.* In other words, we always "go small"; if the spatial area of the densely interacting populations is smaller than that of the dense matter/energy flow, the land community consists of the area covered by the densely interacting populations, whereas if the area of the dense matter/energy flow is smaller than

[17] Perhaps Post et al., Callicott, Eliot, and/or Odenbaugh have some way of limiting the questions asked and the boundaries drawn, but I am not sure what those would be, given what they have said.

[18] To my knowledge, Leopold did not address the question of how to determine the boundaries of land communities. Thus, my hope here is to develop an account that is Leopoldean in spirit, with land communities as individuals, emphasizing interdependencies between organisms.

that of the densely interacting populations, the land community consists of the area covered by the dense matter/energy flow. The problem with this proposal is that it might exclude *causally relevant factors* for the future states of populations and abiotic components and thus give a misleading picture that would be subject to error, making this proposal completely untenable, in my view. Note that this problem is similar to one faced by drawing multiple boundaries, discussed at the end of the last section.

A third possibility is that *the land community includes interactions or matter/energy flows from the larger area if and only if those interactions or matter/energy flows are stronger or larger than those of the smaller area.* This, in essence, seems to be how Post et al. handled cases such as those where mobile organisms brought large amounts of nutrients into lakes, wetlands, and streams. This is the solution that I am inclined to accept; it promises to preserve land communities as objects of study while taking into account most of the important causal processes that affect the land community's future ("most of" because, since no biological system is closed, there is always the potential for a rare but strong causal influence from the outside).

Here it might be objected that we do not even have sufficient congruence to constitute ecological communities and ecosystems, much less congruence between ecological communities and ecosystems. Kim Sterelny, for example, raises concerns about the lack of congruence within purported ecological communities, using Black Mountain (a bush reserve near Australian National University) as an example:

> "Black Mountain" names a quite heterogeneous region of about 10 square kilometers with gentle variation from patch to patch. As a consequence of these gradual changes in character, the different populations might not be correlated. A local brushtail possum population may overlap with a local ringtail possum population, a local boobook owl population, a greater glider population, and a number of eucalyptus populations. For on Black Mountain, there are no sharp changes that matter to all of these species, keeping local populations congruent with one another. (Sterelny 2006: 225)

However, this lack of congruence does not invalidate the third solution to the problem of boundaries. It is not necessary that all the populations of a community be located in the same place; what matters is that there is there is continuity of causal interaction across the populations, even if, for example, the local ringtail population is not interacting with the local boobook owl population. As long as the interactions among the listed populations are stronger than other, "external" interactions, they are all part of the community. This is analogous to the case of a continuous population (Millstein 2010), where the endpoints of a population spread over space do not interact with each other even though there is interaction among the organisms across the entire space, forming one population.

Moreover, a quick comparison to organisms shows that the parts of an individual need not be congruent; just as a heart muscle and a leg muscle are not congruent, the populations (the "parts") of a community need not be congruent.[19] The objection is puzzling.

Sterelny raises a further sort of worry about drawing boundaries for an interactionist account of ecological communities specifically:[20]

> the interaction patterns of different components of putative communities may well not coincide. Even if communities are networks of interacting populations, they are typically demographically open. Migrants move in and out of most habitat patches. As we saw in Section 2, such movements are likely to have stabilizing effects. We have two populations rather than one if organisms of the same type are related by metapopulation dynamics rather than competition. The echidnas in Black Mountain are part of a different population, and hence a different community from the echidnas on the O'Connor Ridge (about a kilometer to the north) if they are a source population for the O'Connor Ridge echidnas. They then buffer that group against population collapses rather than competing with them for scarce resources. Prima facie, though, there is not much reason to expect the dynamics of echidna populations to match those of larger and more mobile organisms, or those of smaller and less mobile ones. (Sterelny 2006: 217)

Again, however, I do not see that this is an insurmountable objection.[21] If we have correctly identified the echidnas as forming a metapopulation, then the interactions between those two populations are rare (see Millstein [2010] for a discussion of the metapopulation concept). So, even though these rare interactions may sometimes turn out to be significant (as in the case that Sterelny describes, where one population recolonizes a location where another population has gone extinct or nearly extinct), there is no difficulty in saying that the echidnas in Black Mountain are part of one community and the echidnas on the O'Connor Ridge are part of another. They still represent a situation where there are continuous interactions among Black Mountain populations and among O'Connor Ridge populations with discontinuities between—discontinuities do not imply that there are *no* interactions, only that they are fewer and weaker. And if it were to turn out, contra to supposition, that there were significant migrations and interactions between the echidnas on Black Mountain and the echidnas on the O'Connor Ridge, then we have misidentified the echidnas as a metapopulation; they would instead be a patchy population (see Millstein 2010), and we would then have a case for considering all of the populations (consisting of different species) of O'Connor Ridge and Black Mountain to be one community (since, again, it is not required, as

[19] Thanks to Gregory Mikkleson for this point.

[20] Sterelny is responding to Levins and Lewontin (1985).

[21] In fairness, Sterelny does admit that "It is hard to tell just how serious this problem is" (2006: 225).

discussed in my response to the first of Sterelny's objections, that every population interact with every other population or that they be in the same place).

I have focused on the objection that the populations within ecological communities lack sufficient congruence for us to identify their boundaries, but the same sort of responses can be made to those who might claim that the matter/energy flows of ecosystems lack congruence. Individuality does not require location in the same space; the University of California, Davis, is an individual even though part of it is in Davis, part is in Sacramento, and part is in Bodega Bay. So there can be, say, flow of nutrient *X* in one area and flow of nutrient *Y* in another, but those flows would be part of the same ecosystem so long as there is continuity of flow between them.

Recall, however, that the goal of this chapter is not to defend an ecological community concept or an ecosystem concept per se, but rather to defend a combined ecological community–ecosystem concept (i.e., a land community). And I have already described how to address lack of congruence between an ecological community and an ecosystem: the land community includes interactions or matter/energy flows from the larger area if and only if those interactions or matter/energy flows are stronger or larger than those of the smaller area.

13.7 Toward a Leopoldean Land Community Concept

Insights from Sections 13.2–13.6 lead me to propose:

> A *Leopoldean land community* consists of populations[22] of different species interacting with each other and with their abiotic environment; these *survival-relevant interactions* often produce a *flow of energy and materials* between biotic components and between biotic components and abiotic components (and vice versa).[23]

Let me elaborate on this proposal a bit more.

Survival-relevant interactions between the populations include competition for scarce resources, predator/prey, parasite/host, pollinator/pollinated, and provision of shade or shelter. Note that this list is reminiscent of Leopold's own typology of dependencies, as shown in Figure 13.1: predations, exploitations, services, and

---

[22] Too often (e.g., O'Neill 2001; Odenbaugh 2007; Post et al. 2007) communities are spoken of as being composed of species and species interactions, but this is loose language (i.e., I don't think any of these authors actually believe that the entire species need be present or interacting). A species may be composed of multiple populations, not all of which are part of the same community. This is true even if the populations form a metapopulation, that is, if they are weakly connected (Millstein 2010). Damuth's (1985) term "avatar," referring to the population of a species found in a particular community, would be appropriate here.

[23] These survival-relevant interactions give rise to the interdependencies that Leopold spoke of. I will provide a more complete account of Leopold's concept of "interdependency" in a future work.

parasitisms. *Relevant flows of materials and energy* include primary production (photosynthesis, chemosynthesis), secondary production, evapotranspiration, decomposition, and nutrient cycling. These are not meant to be controversial, or even original, parts of my proposal; they are simply the typical interactions and matter/energy flows identified by ecologists. Survival-relevant interactions between the populations can produce flows of materials and energy, but flows of material and energy can also produce or affect survival-relevant interactions between populations. *Food webs* are of particular importance to a combined community–ecosystem approach because they can represent species interactions within a community *and* energy flow through those species (Post et al. 2007); they are thus of particular importance to a land community.

Land community boundaries[24] for *well-bounded systems* are where discontinuities or steep gradients in the flow of material and energy coincide with discontinuities or steep gradients in species interactions. Land community boundaries for *open systems* are at a minimum delineated by the smaller of the two types of discontinuities or steep gradients, including the more extensive interactions or matter/energy flows *if and only if* those interactions or matter/energy flows are stronger or larger than those of the smaller area. This approach has the advantage of including all significant causally relevant factors for the future states of populations and abiotic components (interdependencies). It may mean that there are fewer land communities than one might have thought; however, I am not sure that this is a problem. Ecologists may reasonably choose to study subsets (including particular types of interactions or particular matter or energy flows) of these for various pragmatic reasons, but such choices do not affect the ontology of land communities.

Although here I have drawn on Post et al. (2007) in using discontinuities or steep gradients to identify boundaries, this approach is similar to Simon's (2002) account of "nearly complete decomposability," which I have used elsewhere in characterizing the concept of population (Millstein 2009, 2010). It is different from Simon's approach in focusing on the strength of interactions and flows—where "strength" can be understood as the size of the effect that changes in one population produce in another population or the size of the effect on abiotic components—rather than their rate. This difference should not be seen as a crucial one; *differences in the rate of interactions are likewise relevant for boundaries of land communities.*

This approach has the added benefit of addressing Eliot's (2011) concern that interactions alone do not offer a basis for differentiating particular communities (in context, he is criticizing Odenbaugh's interactionist account of communities) from the global community of all organisms if all or most organisms are connected by interactions. On the view defended here, weak, small, or infrequent interactions

---

[24] Note that land community boundaries are fuzzy rather than sharp.

would not suffice to add an "(n + 1)th" population to a land community. Moreover, even if a strongly negative interacting population (say, a population of super-predators like the snakehead fish in the Potomac River System; see Odenkirk and Owens 2007) were added to a land community or a strongly positive interacting keystone population (say, wolves in Yellowstone; see Ripple and Beschta 2007) were eliminated from a land community, as long as other interacting populations persist, the land community persists. It might not persist in a healthy state (which is where conservationist concerns would enter in), but it persists; thus, the account I have proposed here is not subject to the "(n + 1)th" or "(n − 1)th" problems that Eliot (2011) describes.

The composition of species in a land community may change over time; moreover, the populations that it contains may evolve over time. The land community is the same entity if, and only if, there is continuity of interaction and matter/energy flow within the entity through time.[25] Thus, since their members may change, land communities may or may not be stable in the sense of "stasis" or "equilibrium." Sustainability (similar to what *Leopold* meant by stability or land health; see Newton 2006) may be a more pertinent trait or feature (O'Neill 2001). For example, Leopold (1944) traced four epochs within Southwestern Wisconsin,[26] but his concern was land health in the face of different practices, not change of species (although he was concerned that the latter often negatively affected the former, while acknowledging many cases in which it did not do so).

Land communities also have beginnings and endings in time. Although I lack the space to fully develop the ideas here, I would use a similar approach to the one I developed elsewhere for thinking about populations through time (Millstein 2015) for thinking about land communities through time. However, I suspect that one difference will be that whereas populations come into existence and go out of existence fairly frequently, that land communities do so only relatively rarely. There would have to be a complete loss of continuity (a complete absence of interaction and matter/energy flow) for a land community to go "extinct." On the other hand, given the current state of the environment and current societal practices, new land communities might most commonly be formed by human-caused splitting of an existing land community.

A land community so described would be an *individual* in the Ghiselin-Hull sense (see, e.g., Ghiselin 1974, 1997; Hull 1976, 1978)[27]; it would:

- Be a *particular* thing, not a class; it would be a *spatiotemporally restricted* entity (i.e., located in a particular place and time, even if not continuous in space).

---

[25] See Odenbaugh (2007) on this point with respect to the ecological community concept.

[26] Fur trade epoch, 1680–1832; Fire epoch, 1750–1850; Wheat epoch, 1832–1878; Dairy epoch, 1872–(his) present.

[27] See also Hamilton, Smith, and Haber (2009) and Millstein (2009).

- Not merely be an assemblage; it would be an *integrated*, cohesive entity because of the *causal interactions* among the parts, giving the parts (to some extent) a shared fate. (This is the most important criterion, in my view; recall Leopold's claim that the components of the land community have a collective as well as separate welfare.)
- Have *beginnings* and *endings* in time.
- Be *continuous* through time, allowing for change over time.

This characterization challenges the so-called Gleasonian picture of communities as mere coincidental assemblages of whatever organisms happen to be located in a particular place at a particular time. Some philosophers (e.g., Regan 1983) cast doubt on the idea that mere "collections" can have moral rights, so this is a salient point. While it is an empirical question as to whether there are causal interactions among populations of different species that affect their survival (and so, an empirical question as to whether there are land communities in the sense I have described) or whether they are mere assemblages, there is certainly plenty of evidence (more than I could reasonably capture here) that such causal interactions exist. Indeed, Eliot (2011) argues persuasively that Gleason (e.g., Gleason 1917) has been interpreted too radically, noting that "Every ecologist, including Gleason, recognizes interactions among organisms, including that some require others, to survive" (Eliot 2011: 102). Once one considers, for example, that trophic interactions (which affect both the eaten and the eater) are sufficient, it becomes virtually impossible to deny the existence of survival-relevant interactions. And recent work suggests that these and other interactions are important enough that order and timing of species immigration during community assembly can affect species composition and abundances (Fukami 2015). On other hand, if Gleason turns out to have been right that the compositions of species in an area change frequently, nothing I have argued for is challenged here since it is the causal interactions and matter/energy flows that are essential to the land community's continuity and individuality and not any particular composition of species (addressing concern (3) of the list in Section 13.1).

On this view, *individuality comes in degrees*, especially with respect to integration, thus eliminating the need for a separate term for "wholes" (see Odenbaugh 2007). Here I set aside the question of whether the land community is an organism or whether it is a unit of selection,[28] both of which bring additional complications. I also set aside questions that Sterelny (2006) raises over whether communities are

[28] That is, I am not considering whether the land community satisfies a criterion of *evolutionary* individuality, as do, e.g., Clarke (2011) and Ereshefsky and Pedroso (2013). Nothing I have said in this chapter presupposes that land communities are units of selection; for example, interdependence between organisms could evolve via co-evolution of populations. Note that there are still other concepts of individuality in the literature; however, my claims in this chapter should be understood only to apply to the concept of individuality characterized here.

"internally regulated" or the extent to which they have emergent properties; even Sterelny acknowledges:

> Of course local populations do not live completely independently of one another. Life at Black Mountain is diffusely interdependent; the organisms that live there are not autonomous islands of life. Plants often depend on animals for pollination and seed dispersal, on symbiotic partners for crucial nutrients, and on detrivores for nutrient cycling. Consumer guilds—herbivores and carnivores—are obviously dependent on other organisms. (Sterelny 2006: 216)

This interdependence, stressed by Leopold in "The Land Ethic" and elsewhere, is the essential piece needed to make the case for the individuality of land communities. Organismality, internal regulation, being a unit of selection, and/or emergent properties might make for a more "robust" individuality, but they are not necessary for it. That is, they are not necessary for differentiating an individual from an abstract type or a mere set or a mere assemblage. Moreover, the concept of individuality relied on here is sufficient to pick out an entity that can be distinguished from other entities of its type and which persists in time, allowing for it to be a candidate for direct moral obligation and intrinsic value.

13.8 Conclusion

There are, admittedly, further issues to be worked out. For example, does it matter whether the interactions that compose the land community are between populations, or are interactions between the populations' organisms sufficient? If it does matter, are interactions such as predator/prey interactions genuinely population-level rather than organism-level? Also, is there a *general* way to specify these interactions? Eliot (2011) suggests that Odenbaugh's (2007) specification that the interactions be "ecological" or "biotic" is vague. In Section 13.6, I specified that the interactions should be population-level survival interactions. Is that sufficient to address Eliot's concern about vagueness? Another sort of worry is whether I have left open the possibility of land communities at different scales; one might think, for example, that there could be local land communities, regional land communities, and the like. If my account cannot accommodate this possibility, is that a problem? Perhaps not; ecologists speak of meta-communities (Leibold et al. 2004); there might be meta–land communities or meta–meta–land-communities.

Nonetheless, I hope to have shown that there is some reason to think that a Leopoldean concept of a land community is consistent with some contemporary ecology, to have given a more precise characterization of it, and to have demonstrated that if any such entity exists in the world, it would be an individual. If this is correct, then a Leopoldean land community is at least a candidate for direct moral obligation and intrinsic value; the next step would be to show not only that it has the necessary characteristics for direct moral obligation and intrinsic value, but also that it has sufficient ones. I leave that next step for a future work.

Acknowledgments

Thanks to Antoine Dussault and two anonymous reviewers for their extremely thoughtful comments on an earlier draft. Thanks also to members of the audience at venues where earlier versions of this chapter were presented: the 2015 meeting of the International Society for the History, Philosophy, and Social Studies of Biology, a colloquium hosted by the Department of Philosophy at Simon Fraser University; the International Society for Environmental Ethics session at the 2016 Pacific Division Meeting of the American Philosophical Association; and the 2016 Texas Tech Philosophy Graduate Student Conference.

References

Callicott, J. Baird. 1987. The conceptual foundations of the land ethic. In *Companion to* A Sand County almanac, ed. J. Baird Callicott, 186–217. Madison: University of Wisconsin Press.

Callicott, J. Baird. 2013. *Thinking like a planet: The land ethic and the Earth ethic.* Oxford: Oxford University Press.

Callicott, J. Baird, and Karen Mumford. 1997. Ecological sustainability as a conservation concept. *Conservation Biology* 11(1): 32–40.

Chapin III, F. Stuart, Pamela A. Matson, and Peter Vitousek. 2011. *Principles of terrestrial ecosystem ecology*. New York: Springer Science & Business Media.

Clarke, Ellen. 2011. The problem of biological individuality. *Biological Theory* 5(4): 312–325.

Clements, Frederic E. 1916. *Plant succession: An analysis of the development of vegetation.* Washington, DC: Carnegie Institution of Washington.

Craige, Betty Jean. 2002. *Eugene Odum: Ecosystem ecologist and environmentalist.* Athens: University of Georgia Press.

Currie, William S. 2011. Units of nature or processes across scales? The ecosystem concept at age 75. *New Phytologist* (190): 21–34.

Damuth, John. 1985. Selection among "species": A formulation in terms of natural functional units. *Evolution* 39(5): 1132–1146.

Dupré, John. 2010. The polygenomic organism. *The Sociological Review* 58: 19–31.

Dussault, Antoine C., and Frédéric Bouchard. 2017. A persistence enhancing propensity account of ecological function to explain ecosystem evolution. *Synthese* 194: 1115–1145.

Eliot, Christopher H. 2011. The legend of order and chaos: Communities and early community ecology. In *Handbook of the philosophy of science: Philosophy of ecology*, eds. Kevin deLaplante, Bryson Brown, and Kent A. Peacock, 49–108. North Holland: Elsevier.

Eliot, Christopher H. 2013. Ecological objects for environmental ethics. In *Linking ecology and ethics for a changing world*, eds. Ricardo Rozzi, S. T. A. Pickett, Clare Palmer, Juan J. Armesto, and J. Baird Callicott, 219–229. Dordrecht: Springer.

Elton, Charles. 1927. *Animal ecology*. New York: The Macmillan Company.

Ereshefsky, Marc, and Makmiller Pedroso. 2013. Biological individuality: The case of biofilms. *Biology and Philosophy* 28(2): 331–349.

Fukami, Tadashi. 2015. Historical contingency in community assembly: Integrating niches, species pools, and priority effects. *Annual Review of Ecology, Evolution, and Systematics* 46: 1–23.

Ghiselin, Michael T. 1974. A radical solution to the species problem. *Systematic Zoology* 23(4): 536–544.

Ghiselin, Michael T. 1997. *Metaphysics and the origin of species*. Albany: State University of New York Press.

Gleason, Henry Allan. 1917. The structure and development of the plant association. *Bulletin of the Torrey Botanical Club* 44(10): 463–481.

Hagen, Joel Bartholemew. 1992. *An entangled bank: The origins of ecosystem ecology*. New Brunswick: Rutgers University Press.

Hamilton, Andrew, Nathan R. Smith, and Matthew H. Haber. 2009. Social insects and the individuality thesis: Cohesion and the colony as a selectable individual. In *Organization of insect societies: From genome to sociocomplexity*, eds. J. Gadau and J. Fewell, 572–589. Cambridge, MA: Harvard University Press.

Hastings, Alan, and Louis Gross. 2012. *Encyclopedia of theoretical ecology*. Berkeley: University of California Press.

Hull, David L. 1976. Are species really individuals? *Systematic Zoology* 25: 174–191.

Hull, David L. 1978. A matter of individuality. *Philosophy of Science* 45: 335–360.

Lean, Christopher Hunter. 2015. Ecological communities are robust population networks. In *Biennial meeting of the International Society for the History, Philosophy, and Social Studies of Biology*. Montreal, Canada.

Leibold, Mathew A., M. Holyoak, N. Mouquet, P. Amarasekare, J. M. Chase, M. F. Hoopes, et al. 2004. The metacommunity concept: A framework for multi-scale community ecology. *Ecology Letters* 7(7): 601–613.

Leopold, Aldo. 1923/1991. Some fundamentals of conservation in the southwest. In *The river of the mother of god: And other essays by Aldo Leopold*, eds. Susan L. Flader and J. Baird Callicott, 86–97. Madison: University of Wisconsin Press.

Leopold, Aldo. 1933/1991. The conservation ethic. In *The river of the mother of god: And other essays by Aldo Leopold*, eds. Susan L. Flader and J. Baird Callicott, 181–192. Madison: University of Wisconsin Press.

Leopold, Aldo. 1934/1991. The arboretum and the university. In *The river of the mother of god: And other essays by Aldo Leopold*, eds. Susan L. Flader and J. Baird Callicott, 209–211. Madison: University of Wisconsin Press.

Leopold, Aldo. 1939. A biotic view of land. *Journal of Forestry* 37(9): 727–730.

Leopold, Aldo. 1939/1991. A biotic view of land. In *The river of the mother of god: And other essays by Aldo Leopold*, eds. Susan L. Flader and J. Baird Callicott, 266–273. Madison: University of Wisconsin Press.

Leopold, Aldo. 1942. The role of wildlife in a liberal education. *Transactions of the Seventh North American Wildlife Conference* April 8–10, 1942: 485–489.

Leopold, Aldo. 1943. Deer irruptions. *Natural Resources Committee, Wisconsin Academy of Sciences, Arts, and Letters*, 351–366.

Leopold, Aldo. 1944. Land-health in S. W. Wisconsin. *The Aldo Leopold Archives*, http://digital.library.wisc.edu/1711.dl/AldoLeopold Lectures 1933–1947: 868–870.

Leopold, Aldo. 1944/1991. Conservation: In whole or in part? In *The river of the mother of god: And other essays by Aldo Leopold*, eds. Susan L. Flader and J. Baird Callicott, 310–319. Madison: University of Wisconsin Press.

Leopold, Aldo. 1949. The land ethic. In *A Sand County almanac and sketches here and there*, 201–226. New York: Oxford University Press.

Levins, Richard, and Richard C. Lewontin. 1980. Dialectics and reductionism in ecology. *Synthese* 43: 47–78.

Levins, Richard, and Richard C. Lewontin. 1985. Dialectics and reduction in ecology. In *The dialectical biologist*, eds. Richard Levins and Richard C. Lewontin, 132–160. Cambridge, MA: Harvard University Press.

Lindeman, Raymond L. 1942. The trophic-dynamic aspect of ecology. *Ecology* 23(4): 399–417.

McShane, Katie. 2004. Ecosystem health. *Environmental Ethics* 26: 227–245.

McShane, Katie. 2007. Why environmental ethics shouldn't give up on intrinsic value. *Environmental Ethics* 29(1): 43–61.

Meine, Curt. 2010. *Aldo Leopold: His life and work*. Madison: University of Wisconsin Press.

Millstein, Roberta L. 2009. Populations as individuals. *Biological Theory* 4(3): 267–273.

Millstein, Roberta L. 2010. The concepts of population and metapopulation in evolutionary biology and ecology. In *Evolution since Darwin: The first 150 years*, eds. M. A. Bell, D. J. Futuyma, W. F. Eanes, and J. S. Levinton, 61–86. Sunderland: Sinauer.

Millstein, Roberta L. 2015. Thinking about populations and races in time. *Studies in History and Philosophy of Biological and Biomedical Sciences: Special Issue on Genomics and Philosophy of Race* 52: 5–11, guest edited by R. G. Winther, R. L. Millstein, and R. Nielsen.

Newton, Julanne Lutz. 2006. *Aldo Leopold's odyssey: Rediscovering the author of a Sand County almanac*. Washington, DC: Island Press.

Norton, Bryan. 2002. Epistemology and environmental values. In *Land, value, community: Callicott and environmental philosophy*, eds. Wayne Ouderkirk and Jim Hill, 123–132. Albany: State University of New York Press.

Nyhart, Lynn K. 2009. *Modern nature: The rise of the biological perspective in Germany*. Chicago: University of Chicago Press.

O'Neill, John. 1992. The varieties of intrinsic value. *The Monist* 75(2): 119–137.

O'Neill, Robert V. 2001. Is it time to bury the ecosystem concept (with full military honors, of course!). *Ecology* 82(12): 3275–3284.

O'Neill, Robert V., Donald Lee Deangelis, J. B. Waide, and Garland E. Allen. 1986. *A hierarchical concept of ecosystems*. Princeton: Princeton University Press.

Odenbaugh, Jay. 2007. Seeing the forest and the trees: Realism about communities and ecosystems. *Philosophy of Science* 74(5): 628–641.

Odenbaugh, Jay. 2010. On the very idea of an ecosystem. *New Waves in Metaphysics*: 240–258.

Odenkirk, John, and Steve Owens. 2007. Expansion of a Northern Snakehead population in the Potomac River system. *Transactions of the American Fisheries Society* 136(6): 1633–1639.

Odum, Eugene. P. 1971. *Fundamentals of ecology*, 3rd ed. Philadelphia: W. B. Saunders.

Post, David M., Martin W. Doyle, John L. Sabo, and Jacques C. Finlay. 2007. The problem of boundaries in defining ecosystems: A potential landmine for uniting geomorphology and ecology. *Geomorphology* 89(1): 111–126.

Regan, Tom. 1983. *The case for animal rights*. Berkeley: University of California Press.

Ripple, William J., and Robert L. Beschta. 2007. Restoring Yellowstone's Aspen with wolves. *Biological Conservation* 138(3): 514–519.

Russow, Lily. M. 1981. Why do species matter? *Environmental Ethics* 3(2): 101–112.

Schulze, Ernst-Detlef, Erwin Beck, and Klaus Müller-Hohenstein. 2005. *Plant ecology*. Berlin: Springer.

Shrader-Frechette, Kristin. 1996. Individualism, holism, and environmental ethics. *Ethics and the Environment* 1: 55–69.

Simon, Herbert A. 2002. Near decomposability and the speed of evolution. *Industrial and Corporate Change* 11: 587–599.

Sterelny, Kim. 2006. Local ecological communities. *Philosophy of Science* 73: 215–231.

Tansley, Arthur G. 1935. The use and abuse of vegetational concepts and terms. *Ecology* 16(3): 284–307.

Worster, Donald. 1990. The ecology of order and chaos. *Environmental History Review* 14: 1–18.

Zimmerman, Michael J. 2015. Intrinsic vs. extrinsic value. In *The Stanford encyclopedia of philosophy* (spring 2015 edition), ed. Edward N. Zalta. http://plato.stanford.edu/archives/spr2015/entries/value-intrinsic-extrinsic/.

{ INDEX }

Milton Keynes UK
Ingram Content Group UK Ltd.
UKHW020949290224
438411UK00003BA/26

9 780190 636814